LDPC CODE DESIGNS, CONSTRUCTIONS, AND UNIFICATION

Written by leading experts, this self-contained text provides systematic coverage of low-density parity-check (LDPC) codes and their construction techniques, unifying both algebraic- and graph-based approaches into a single theoretical framework (the superposition construction). An algebraic method for constructing protograph LDPC codes is described, and entirely new codes and techniques are presented. These include a new class of LDPC codes with doubly quasi-cyclic structure, as well as algebraic methods for constructing spatially and globally coupled LDPC codes.

Authoritative, yet written using accessible language, this is essential reading for electrical engineers, computer scientists, and mathematicians working in communications and information theory.

JUANE LI is a postdoctoral researcher in the Department of Electrical and Computer Engineering at the University of California, Davis. Her current research interests are channel coding for communications and storage systems, and hardware implementation of encoders and decoders for LDPC codes.

SHU LIN is an Adjunct Professor in the Department of Electrical and Computer Engineering at the University of California, Davis, and an IEEE Life Fellow. He is the co-author of *Channel Codes: Classical and Modern* (Cambridge, 2009) and *Error Control Coding: Fundamentals and Applications* (2nd ed., Prentice Hall, 2004).

KHALED ABDEL-GHAFFAR is a Professor of Electrical and Computer Engineering at the University of California, Davis. He has held research positions at the IBM Almaden Research Center, Delft University of Technology, and the University of Bergen.

WILLIAM E. RYAN is an IEEE Fellow and currently works for Zeta Associates, USA. He was previously a Professor at New Mexico State University and the University of Arizona. He is the co-author of the book *Channel Codes: Classical and Modern* (Cambridge, 2009).

DANIEL J. COSTELLO, JR. is the Leonard Bettex Professor Emeritus of Electrical Engineering at the University of Notre Dame, Indiana, and an IEEE Fellow. He is the co-author of the textbook *Error Control Coding: Fundamentals and Applications* (2nd ed., Prentice Hall, 2004).

"This book provides an in-depth survey of recently developed quasi-cyclic LDPC codes. It is a treasured reference on practical channel coding methods for both theorists and practitioners working in communications and information theory."

Lara Dolecek, *University of California, Los Angeles*

"The importance of LDPC codes in numerous applications and their capacity-approaching performance has led to an explosion in research into their construction and analysis over the past decade. The numerous effective constructions of them can be broadly classified as algebraic and graphical, including the important superposition, protograph, and spatial coupling techniques.
This timely volume explains, unifies, and greatly clarifies these diverse approaches and lays a solid foundation that will be invaluable to researchers, practitioners, and students alike."

Ian F. Blake, *University of British Columbia*

"A book from the leaders in the field of error-correcting codes. Superposition – a unified framework for low-density parity check code construction – makes a description of codes of various classes rather simple."

Bane Vasic, *University of Arizona, Tucson*

LDPC CODE DESIGNS, CONSTRUCTIONS, AND UNIFICATION

JUANE LI
University of California, Davis

SHU LIN
University of California, Davis

KHALED ABDEL-GHAFFAR
University of California, Davis

WILLIAM E. RYAN
Zeta Associates, Colorado

DANIEL J. COSTELLO, JR.
University of Notre Dame, Indiana

CAMBRIDGE
UNIVERSITY PRESS

CAMBRIDGE
UNIVERSITY PRESS

Shaftesbury Road, Cambridge CB2 8EA, United Kingdom

One Liberty Plaza, 20th Floor, New York, NY 10006, USA

477 Williamstown Road, Port Melbourne, VIC 3207, Australia

314–321, 3rd Floor, Plot 3, Splendor Forum, Jasola District Centre, New Delhi – 110025, India

103 Penang Road, #05–06/07, Visioncrest Commercial, Singapore 238467

Cambridge University Press is part of Cambridge University Press & Assessment, a department of the University of Cambridge.

We share the University's mission to contribute to society through the pursuit of education, learning and research at the highest international levels of excellence.

www.cambridge.org
Information on this title: www.cambridge.org/9781107175686

10.1017/9781316780152

First published 2017

A catalogue record for this publication is available from the British Library

Library of Congress Cataloging-in-Publication data
Names: Li, Juane, author.
Title: LDPC code designs, constructions, and unification / Juane Li, University of California, Davis [and 4 others].
Other titles: Low-density parity-check code designs, constructions, and unification
Description: New York : Cambridge University Press, 2017. | Includes bibliographical references and index.
Identifiers: LCCN 2016032809 | ISBN 9781107175686 (hardback)
Subjects: LCSH: Error-correcting codes (Information theory) | Coding theory.
Classification: LCC QA268 .L53 2017 | DDC 005.7/17–dc23
LC record available at https://lccn.loc.gov/2016032809

ISBN 978-1-107-17568-6 Hardback

Contents

Preface

Error control codes protect the accuracy of data in modern information systems, including computing, communication, and storage systems. Low-density parity-check (LDPC) codes and their relatives represent the state of the art in error control coding and are renowned for their ability to perform close to the theoretical limits. This book presents recent results on various LDPC code designs, making strong connections between two prominent design approaches, the *algebraic-based* and the *graph-theoretic-based* constructions. New codes and code construction techniques are presented.

Most methods for constructing LDPC codes can be classified into two general categories, the algebraic-based and the graph-theoretic-based constructions. The two best-known graph-theoretic-based construction methods are the *progressive edge-growth* (PEG) and the *protograph-based* (PTG-based) methods, devised in 2001 and 2003, respectively. Both of these techniques involve computer-aided design. One of the earliest algebraic-based methods for constructing LDPC codes is the *superposition* (SP) construction, proposed in 2002. In this book, the algebraic-based construction method is re-interpreted from both the algebraic and the graph-theoretic perspectives. From the algebraic point of view, it is shown that the SP-construction of LDPC codes includes, as special cases, most of the major algebraic construction methods developed since 2002. From the graph-theoretic point of view, it is shown that the SP-construction also includes the PTG-based construction as a special case. Based on this PTG/SP connection, an algebraic method is developed here to construct PTG-based LDPC codes.

There are advantages to putting the algebraic-based and the PTG-based constructions into a *single framework*, the SP framework. One advantage is that SP descriptions of codes tend to be relatively compact, enabling simple code specifications in standards and textbooks. Another advantage to studying LDPC codes under the SP framework is that students and practitioners need only learn a single code design approach rather than the myriad approaches that exist in the published literature.

Both binary and nonbinary code constructions will be presented under the SP framework. The SP-construction also leads to a new class of LDPC codes with a *doubly quasi-cyclic* (QC) structure as well as algebraic methods for constructing *spatially* and *globally coupled* LDPC codes. The globally coupled codes will be shown to possess a highly effective burst-erasure correction capability.

A good number of new LDPC codes are constructed and simulated over the binary-input additive white Gaussian noise channel and the binary erasure channel.

This book will open the door for readers to understand many topics in modern LDPC codes that are scattered in the literature. It is intended as a self-study guide for students, researchers, and engineers interested in LDPC codes and their variations. The book explains the different design methodologies in detail and provides an ample number of code constructions along with simulations. The book shows that code design and construction are more of an art rather than science. Hopefully, after reading this book, the reader may gain enough artistic experience to produce codes that not only meet required specifications but also improve upon those reported here. To make the material widely accessible, the authors have kept the presentation as clear as possible and assumed only basic knowledge of terminology and results that are commonly covered in textbooks on coding theory.

Acknowledgment

This book is the climax of more than fifteen years of work on LDPC codes by graduate students of the last four authors. In particular, the second and third authors would like to thank Drs. Lei Chen, Qiuju Diao, Ivana Djurdjevic, Qin Huang, Jingyu Kang, Yu Kou, Lan Lan, Keke Liu, Shumei Song, Ying Yu Tai, Heng Tang, Jun Xu, Lingqi Zeng, Li Zhang, and Bo Zhou. Each of them spent thousands of hours investigating choices of code parameters in order to optimize performance. We are also grateful to the federal agencies NASA and NSF and to the industrial support from Cadence, Denali Software, Intel, LSI, Northrop Grumman Space Technology, and San Disk. Many thanks to Dr. Julie Lancashire at Cambridge University Press for her constant encouragement to write this book. We also acknowledge Ms. Isabella Yeh, whose painting on the front cover was created specifically for this book. Finally, the first author would like to thank her father and mother for their love, and the last four authors would like to thank their wives for their patience and understanding. To the six of them, this work is dedicated.

1

Introduction

The ever-growing need for cheaper, faster, and more reliable communication and storage systems has forced many researchers to seek means to attain the ultimate limits on reliable information transmission and storage. Low-density parity-check (LDPC) codes are currently the most promising coding technique to achieve the channel capacities (or Shannon limits) for a wide range of channels. Discovered by Gallager in 1962 [40], these codes were rediscovered in the late 1990s [83, 81]. Ever since their rediscovery, a great deal of research effort has been expended in design, construction, encoding, decoding algorithms, structural analysis, performance analysis, generalizations, and applications of these remarkable codes. Many LDPC codes have been adopted as standard codes for various current and next-generation communication systems, such as wireless, optical, satellite, space, digital video broadcast (DVB), multi-media broadcast (MMB) systems, and others. Applications to high-density data storage systems, such as flash memories, are now being seriously investigated. In fact, they appear in some recent data storage products. This rapid dominance of LDPC codes in applications is due to their capacity-approaching performance, which can be achieved with *practically implementable iterative decoding algorithms*. More applications of these codes are expected to come and their future is promising. However, further research is needed to better understand the structural properties and performance characteristics of these codes.

Most methods to construct LDPC codes can be classified into two general categories: *graph-theoretic-based* and *algebraic-based* (or *matrix-theoretic-based*) methods. The best-known graph-theoretic-based construction methods are the *progressive edge-growth* (PEG) [43, 44] and the *protograph* (PTG) based methods [105]. The algebraic-based methods for constructing LDPC codes were first introduced in 2000 [56, 55, 38, 57]. Since then, various algebraic methods for constructing LDPC codes, binary and nonbinary, have been developed using mathematical tools such as finite geometries, finite fields, and combinatorial designs [58, 76, 110, 109, 35, 107, 3, 19, 39, 74, 111, 102, 101, 112, 64, 65, 116, 97, 100, 50, 99, 113, 114, 25, 46, 26, 70, 68, 77, 78]. Most of the algebraic constructions have several important ingredients including *base matrices*, *matrix dispersion* (or *matrix expansion*), and *masking*. By a proper choice and combination of these ingredients, algebraic LDPC codes with excellent overall error performance can be constructed. Algebraic LDPC codes have, in general, much lower error-floors than randomly constructed LDPC codes. For example, algebraic LDPC codes have been recently reported that can achieve a decoded bit error rate as low as 10^{-15} without visible error-floors over the additive white Gaussian noise channel (AWGNC) [70, 68, 77, 78].

One of the earliest algebraic-based methods proposed for constructing LDPC codes is the *SP-construction* (also called the *hybrid construction*). This specific and very flexible algebraic-based method for constructing LDPC codes was devised in 2002 [76]. Since then, several other powerful algebraic-based methods have been proposed. These methods basically evolved from the SP-construction method, although the connection went unnoticed.

The SP-construction method presented in [76] starts with a small base matrix $\mathbf{B}$ and a set $\boldsymbol{R}$ of *sparse matrices* of the same size, not necessarily square. Then, each nonzero entry in $\mathbf{B}$ is replaced by a member matrix in $\boldsymbol{R}$ following certain *replacement rules* (or *constraints*) [76, 110, 109, 111, 97] and each zero entry is replaced by a zero matrix (ZM) of the same size as that of a member matrix in $\boldsymbol{R}$. The replacement operation expands the base matrix $\mathbf{B}$ into an array $\mathbf{H}$ of sparse matrices in $\boldsymbol{R}$ and/or ZMs. The null space of $\mathbf{H}$ gives an LDPC code, called an *SP-LDPC code*. We refer to this matrix replacement of a nonzero entry in $\mathbf{B}$ as *superposition*. The matrices in $\boldsymbol{R}$ are called *member* (or *constituent*) matrices and the set $\boldsymbol{R}$ is called the *replacement set*. In [110], the structural properties of the Tanner graph [103] associated with the array $\mathbf{H}$ was investigated. It was shown that the Tanner graph associated with the array $\mathbf{H}$ is an *expansion* of the Tanner graph associated with the base matrix $\mathbf{B}$.

The PTG-based method for constructing LDPC codes was introduced by Thorpe in 2003 [105]. This method was devised to construct the Tanner graph of an LDPC code of large size using a relatively small well-designed bipartite graph, called a *protograph*, as the base graph. In the construction, the first step is to choose a protograph with a *near-capacity iterative decoding threshold* [80, 34] as a building block (or as a base). The second step is to create copies of the chosen protograph. The third step is to permute the edges of the copies according to certain rules to connect them into a Tanner graph of larger size. The null space of the adjacency matrix of the resultant Tanner graph gives an LDPC code, called a *PTG-based LDPC code* (or *PTG-LDPC code*). The second and third steps of the construction together form a *graph expansion* process. Since the introduction of this graphical method for constructing LDPC codes, it has been extensively investigated over the last 12 years [105, 80, 34, 30, 31, 32, 33, 1, 11, 2, 88, 85] (see also their references). Impressive theoretical results (in terms of code ensemble properties) have been developed [34, 31, 1, 2, 85] and many PTG-LDPC codes with good error performance have been constructed. Most recently, a new class of LDPC codes has emerged, called *spatially coupled LDPC codes*, that has attracted a great amount of research enthusiasm and may be regarded as an evolution of PTG-LDPC codes from the graph-theoretical point of view.

From the above brief descriptions of the PTG-based and the SP-based methods of LDPC code construction, we can see that both methods have *two key ingredients*, *a base* and *an expansion of the base*. If we interpret these two construction methods from the same point of view, either from the graph-theoretic point of view or from the algebraic point of view, we see that they are closely related, i.e., there is a *strong connection* between the SP-construction and the PTG-based construction.

In this book, we view the SP-based and PTG-based constructions from a *broader perspective* and interpret them from both the algebraic (or matrix-theoretic) and the graph-theoretic points of view. From the algebraic point of view, we show that the SP-construction of LDPC codes includes, as *special cases*, many of the algebraic construction methods developed since 2002. From the graph-theoretic point of view, we show that the PTG-based construction of LDPC codes is a special case of the SP-construction. Furthermore, we develop an algebraic method for constructing PTG-LDPC codes.

We note that, while SP-LDPC codes contain PTG-LDPC codes as special cases from both the graph-theoretic and the matrix-theoretic points of view, their design approaches have historically been very different. SP-LDPC codes (and their relatives) have been designed using algebraic approaches with an eye toward very low decoding error-floors and highly structured (lower-complexity) decoders. By contrast, PTG-LDPC codes have a long history of being designed by searching "good" ensembles and choosing codes from the ensembles. An ensemble is considered to be good in the sense of decoding threshold [80, 34] or minimum distance [31, 32, 1]. Whether or not these ensemble approaches can be extended to SP-LDPC codes which are not PTG-LDPC codes will require further research.

Also in this book, we unify all the major algebraic methods for constructing LDPC codes based on *matrix dispersions* of base matrices under a *single framework* in terms of the SP-construction. We also introduce a new class of LDPC code with a *doubly* QC structure as well as algebraic methods for constructing *spatially coupled* (SC) and *globally coupled* (GC) LDPC codes. The constructions of these codes are also special cases of the SP-construction.

Although the focus of this book is on binary LDPC codes, all the developments, interpretations, and constructions presented for binary LDPC codes can be generalized to their nonbinary (NB) counterparts.

The rest of this book is organized as follows. Chapter 2 gives some definitions and basic concepts of matrices and introduces some fundamental structural properties and performance characteristics of LDPC codes which will be used in the later chapters. In Chapter 3, we give a brief review of the PTG-based method for constructing binary LDPC codes from the conventional graph-theoretic point of view. In Chapter 4, an algebraic method for constructing binary PTG-LDPC codes is presented. In Chapter 5, we first present the SP-construction of LDPC codes from a broader perspective than that given in [76, 109, 111]. Then, we give a graph-theoretic interpretation of the SP-construction of LDPC codes and show that the PTG-based construction is actually a special case of the SP-construction. Chapter 6 presents various constructions of base matrices and replacement sets for the SP-construction of LDPC codes. Chapter 7 presents a special type of the SP-method for constructing QC-LDPC codes. This construction method, called *matrix dispersion*, is based on dispersing (or expanding) the nonzero entries of an algebraically constructed base matrix over a finite field into circulants of the same size. We also give the necessary and sufficient conditions on a base matrix to ensure that the dispersion results in a Tanner graph with girth at least 6 or at least 8. This chapter basically puts all the major algebraic methods for constructing LDPC codes under the framework of the SP-construction. In Chapter 8, a class of LDPC codes with a doubly QC-structure is given. Chapter 9 presents algebraic methods for constructing *spatially coupled* (SC) QC-LDPC codes, both terminated and not terminated.

In Chapter 10, a new type of LDPC code, called the GC-LDPC code, is presented. Two specific methods for constructing this new type of code are presented. The first method is devised based on cyclic base arrays of matrices over NB fields. The second method is based on the *direct product* of two LDPC codes. We show that GC-LDPC codes in product form are effective at correcting erasures clustered in bursts. Also presented in this chapter is a *reduced-complexity local/global two-phase iterative decoding scheme* for GC-LDPC codes, which allows correction of *local* as well as *global* random errors and erasures. Chapter 11 generalizes the SP-construction of binary LDPC codes to construct NB LDPC codes. Several effective methods for constructing NB LDPC codes are presented. Chapter 12 concludes the book with some remarks on possible future research.

Throughout the book, examples of constructing LDPC codes and their error performance over the AWGNC and the binary erasure channel (BEC) are presented. Also, at the end of each chapter, there is a discussion on some related or unsolved issues and possible research directions.

Furthermore, three appendices are included at the end of the book. In Appendix A, two classes of arrays of circulant permutation matrices constructed based on two types of finite geometries are given. These arrays can be used to construct base matrices and replacement sets of matrices for the SP-construction of LDPC codes. They can also be used to construct well performing LDPC codes directly. Appendix B gives an algorithm for searching compatible masking and base matrices for the construction of QC-LDPC codes. Appendix C presents an iterative algorithm for decoding NB LDPC codes.

2

Definitions, Concepts, and Fundamental Characteristics of LDPC Codes

In this chapter, we first present some definitions and concepts of matrices which are relevant for later developments of various algebraic-based and graph-theoretic-based constructions of LDPC codes. Then, we give a brief presentation of LDPC codes and characterize their fundamental structural properties that affect their error performance over the AWGNC and the BEC. Good coverage of the fundamentals of LDPC codes and the major iterative algorithms for decoding these codes can be found in [74, 97, 95].

2.1 Matrices and Matrix Dispersions of Finite Field Elements

Let GF(q) be a finite field with q elements, where q is a power of a prime [73]. GF(q) is called a NB field if $q \neq 2$ and a binary field if $q = 2$. A matrix $\mathbf{A}$ is said to be *regular* if every row has the same weight, called the row weight, every column has the same weight, called the column weight, and the row weight equals the column weight. If the row weight (or column weight) of a regular matrix $\mathbf{A}$ is w, we say that $\mathbf{A}$ has weight w and call it a weight-w regular matrix. For $w \neq 0$, a weight-w regular matrix must be a square matrix. A binary weight-1 regular matrix is a *permutation matrix* (PM). It is clear that a weight-w regular matrix is the sum of w disjoint PMs. More generally, a rectangular matrix with constant column weight w_c and constant row weight w_r is said to be (w_c, w_r)-regular.

A square matrix over GF(q) is called a *circulant* if every row of the matrix is the cyclic-shift of the row above it one place to the right, and the top row is the cyclic-shift of the last row one place to the right. It is clear that a circulant is a regular matrix whose weight equals the weight of its top row. The top row of a circulant is called the *generator* of the circulant. A binary circulant of weight 1 is a binary *circulant permutation matrix* (CPM). It is clear that a weight-w circulant is the sum of w disjoint CPMs.

Two different matrices of the same size are said to be *mutually disjoint* if there is no common location where they both have nonzero entries, i.e., the set of locations with nonzero entries in one matrix and the set of locations with nonzero entries in the other matrix are disjoint. It is clear that the sum of two mutually disjoint $k \times k$ regular matrices of weights w_1 and w_2 is a $k \times k$ regular matrix of weight $w_1 + w_2$.

A matrix over GF(q) for which no two columns (or two rows) have more than one location where they both have nonzero entries is said to satisfy the *row-column (RC) constraint*. Such a matrix is called an *RC-constrained matrix*. For an RC-constrained matrix, every 2×2 submatrix contains at least one zero entry. The Tanner graph [103] of an RC-constrained matrix has a girth of at least 6 [58, 76, 109, 74, 111, 64, 26]. Two RC-constrained matrices $\mathbf{A}$ and $\mathbf{B}$ of the same size are said to satisfy the *pair-wise (PW) RC-constraint* if the matrix formed by $\mathbf{A}$ and $\mathbf{B}$ arranged either in a row-block $[\mathbf{A}\ \mathbf{B}]$ or in a column-block $[\mathbf{A}^T\ \mathbf{B}^T]^T$ satisfies the RC-constraint. Clearly, all the different binary CPMs of the same size satisfy the PW-RC-constraint. Let $\mathbf{A}$, $\mathbf{B}$, and $\mathbf{C}$ be three PW-RC-constrained matrices of the same size. Suppose we form a 2×2 array with $\mathbf{A}$, $\mathbf{B}$, $\mathbf{C}$, and a ZM of the same size as the constituent matrices. Then the array, as a matrix, satisfies the RC-constraint.

Let $\mathbf{H}$ be an array of matrices of the same size. Suppose all the constituent matrices of $\mathbf{H}$ satisfy the RC-constraint and any two constituent matrices in a row-block (or a column-block) of $\mathbf{H}$ satisfy the PW-RC-constraint. If any 2×2 subarray of $\mathbf{H}$ satisfies the RC-constraint or contains at least one ZM, then $\mathbf{H}$ satisfies the RC-constraint. This simply follows from the RC- and the PW-RC-constrained structures of the constituent matrices of $\mathbf{H}$. An array $\mathbf{H}$ is referred to as a 2×2 *RC-constrained array* if any 2×2 subarray of $\mathbf{H}$ satisfies the RC-constraint. RC-constrained matrices or arrays play an important role in the construction of LDPC codes and their error performance. In designing an array $\mathbf{H}$, the constraint that requires any 2×2 subarray of $\mathbf{H}$ to satisfy the RC-constraint or to contain at least one ZM is referred to as the 2×2 *array RC-constraint*.

Hereafter, for a $k \times t$ matrix $\mathbf{A}$ over GF(q), we label its rows and columns from 0 to $k-1$ and 0 to $t-1$, respectively, and express the matrix in the form of $\mathbf{A} = [a_{i,j}]_{0 \le i < k, 0 \le j < t}$, where $a_{i,j}$ is the entry at location (i, j), the intersection of the i-th row and the j-th column of the matrix.

In many constructions of binary QC LDPC codes, a nonzero element in GF(q) is represented by a binary CPM. Let α be a primitive element of GF(q). Then, the powers of α ($\alpha^0 = 1, \alpha, \alpha^2, \ldots, \alpha^{q-2}$) give all the nonzero elements of GF(q). Commonly, the 0-element of GF(q) is represented by $\alpha^{-\infty}$, i.e., $0 = \alpha^{-\infty}$. For $0 \le j < q-1$, we represent the element α^j by a binary CPM, denoted by $\mathbf{A}(\alpha^j)$, of size $(q-1) \times (q-1)$, whose generator has the unit-element

"1" of GF(q) as its single nonzero component at position j. It is clear that the representation of the element α^j by the binary CPM $\mathbf{A}(\alpha^j)$ of size $(q-1) \times (q-1)$ is unique and the mapping between α^j and $\mathbf{A}(\alpha^j)$ is *one-to-one*. This matrix representation of α^j is referred to as the *binary CPM-dispersion* of α^j. Note that the size of the binary CPM $\mathbf{A}(\alpha^j)$ is the order of α. The 0-element of GF(q) is represented by a $(q-1) \times (q-1)$ ZM.

Let β be a nonzero element in GF(q) of order k, where k is a proper factor of $q-1$. The k elements $\beta^0 = 1, \beta, \beta^2, \ldots, \beta^{k-1}$ form a cyclic subgroup of GF(q) [73]. For $0 \leq j < k$, we can represent the element β^j by a binary CPM $\mathbf{A}(\beta^j)$ of size $k \times k$ whose generator has its single nonzero component at location j. Clearly, the mapping between β^j and $\mathbf{A}(\beta^j)$ for $0 \leq j < k$ is one-to-one. In this case, we disperse an element of a cyclic subgroup of GF(q) into a binary CPM of smaller size.

Suppose GF(q)$\backslash\{0\}$ is a cyclic subgroup of a larger field GF(q'), i.e., GF(q) is a subfield of GF(q'). Let ζ be a primitive element GF(q'). Then, an element ω in GF(q)$\backslash\{0\}$ can be expressed as a power of ζ, say $\omega = \zeta^\ell$, with $0 \leq \ell < q'-1$. With respect to the larger field GF(q'), we can disperse ω into a binary CPM $\mathbf{A}(\omega)$ of size $(q'-1) \times (q'-1)$ whose generator has its single nonzero component at location ℓ. Again, the mapping between ω and $\mathbf{A}(\omega)$ is one-to-one. In this case, an element of GF(q) is dispersed into a binary CPM of larger size.

The above simply shows that an element of a field GF(q) can be one-to-one dispersed into a CPM of size equal to, smaller than, or larger than $q-1$. This one-to-one CPM-dispersion of a field element will be used in the QC-LDPC code constructions presented in later chapters.

2.2 Fundamental Structural Properties and Performance Characteristics of LDPC Codes

An LDPC code over a finite field GF(q) is a q-ary linear block code given by the null space of a *sparse* parity-check matrix $\mathbf{H}$ over GF(q). An LDPC code is said to be *regular* if its parity-check matrix $\mathbf{H}$ has constant column weight, say γ, and constant row weight, say ρ. Such an LDPC code is said to be (γ, ρ)-regular. If the columns and/or rows of the parity-check matrix $\mathbf{H}$ have multiple weights, then the null space of the parity-check matrix $\mathbf{H}$ gives an *irregular* LDPC code.

If $\mathbf{H}$ is an array of sparse circulants of the same size over GF(q), then the null space over GF(q) of the parity-check array $\mathbf{H}$ gives a q-ary *QC-LDPC code*. If $\mathbf{H}$ consists of a single sparse circulant or a column of sparse circulants, then its null space gives a *cyclic* LDPC code. For $q = 2$, the null space of $\mathbf{H}$ over GF(2) gives a binary LDPC code.

An LDPC code C is commonly represented by a *special graph* for decoding, design of decoding algorithms, code performance analysis, and code design. Suppose the parity-check matrix $\mathbf{H}$ of the LDPC code C is an $m \times n$ matrix over GF(q):

$$\mathbf{H} = \begin{bmatrix} h_{0,0} & h_{0,1} & \cdots & h_{0,n-1} \\ h_{1,0} & h_{1,1} & \cdots & h_{1,n-1} \\ \vdots & \vdots & \ddots & \vdots \\ h_{m-1,0} & h_{m-1,1} & \cdots & h_{m-1,n-1} \end{bmatrix}. \tag{2.1}$$

Consider a graph $\mathcal{G} = (\mathbf{V}, \mathbf{C}, \mathbf{E})$ that consists of two sets of nodes, $\mathbf{V} = \{v_0, v_1, \ldots, v_{n-1}\}$ and $\mathbf{C} = \{c_0, c_1, \ldots, c_{m-1}\}$, and a set of edges $\mathbf{E} = \{(j,i)\}$. The n nodes in $\mathbf{V}$ represent the columns of the parity-check matrix $\mathbf{H}$, called *variable nodes* (VNs). The m nodes in $\mathbf{C}$ represent the rows of $\mathbf{H}$, called *check nodes* (CNs). VNs can only connect to CNs and vice versa. Therefore, $\mathcal{G}$ is a *bipartite graph*. The VN v_j is connected to the CN c_i by an edge (j,i) if and only if the entry $h_{i,j}$ at the intersection of the i-th row and the j-th column of $\mathbf{H}$ is a nonzero element in GF(q), say β. Such an edge is labeled with β. For the binary field GF(2), the edges in $\mathcal{G}$ are not labeled; an edge connecting VN v_j and CN c_i simply implies $h_{i,j} = 1$. Note that a VN and a CN in $\mathcal{G}$ are either not connected or connected by one and only one edge, i.e., no parallel edges in $\mathcal{G}$.

The bipartite graph $\mathcal{G}$ constructed above is a graphical representation of the parity-check matrix $\mathbf{H}$ of the LDPC code C. Conversely, the parity-check matrix $\mathbf{H}$ of the LDPC code is the *adjacency* matrix of the graph $\mathcal{G}$. Since the LDPC code C is the null space of $\mathbf{H}$, the n VNs actually represent the n code symbols of a codeword in C and the m CNs represent the m *parity-check constraints* that the code symbols must satisfy. Hence, the bipartite graph $\mathcal{G}$ is a graphical representation of the LDPC code C and is called the *Tanner graph* [103] of the code (or the Tanner graph associated with the parity-check matrix $\mathbf{H}$).

The performance of an LDPC code decoded using iterative decoding algorithms based on belief propagation (BP), such as the *sum-product algorithm* (SPA) [81, 74, 97], the *min-sum algorithm* (MSA) [18, 74, 97], or their simplified variations [74, 97], very much depends on several important structural properties of its Tanner graph *collectively*. In the following, we describe these structural properties.

The number of edges that are incident with a VN v_j (or CN c_i) in the Tanner graph $\mathcal{G}$ of an LDPC code is called the *degree* of v_j (or c_i). Following from the construction of $\mathcal{G}$, it is clear that the degree of a VN v_j equals the weight of the j-th column of $\mathbf{H}$ and the degree of c_i equals the weight of the i-th row of $\mathbf{H}$. A *path* in $\mathcal{G}$ is defined as an alternating sequence of nodes and edges, beginning and ending with nodes, such that each edge is incident with nodes preceding and following it, and no node appears more than once except possibly for the first and the last nodes. The number of edges on a path is called the *length* of the path. A closed path that begins and ends at the same node is called a *cycle*. A cycle in $\mathcal{G}$ contains an even number of edges, i.e., the length of a cycle in $\mathcal{G}$ is even. Since a VN and a CN in $\mathcal{G}$ are connected by at most one edge, $\mathcal{G}$ contains no cycle of length 2. The length of the shortest cycles in $\mathcal{G}$ is called the *girth* of the graph.

The Tanner graph $\mathcal{G}$ of an LDPC code C with girth g may contain many cycles of lengths equal to or larger than g. The numbers of cycles of various lengths $(g, g+2, g+4, \ldots)$ in $\mathcal{G}$ are referred to as the *cycle distribution* of $\mathcal{G}$. If the parity-check matrix $\mathbf{H}$ satisfies the RC-constraint as defined in Section 2.1, then the Tanner graph $\mathcal{G}$ of the code C contains no cycle of length 4 and its girth is at least 6 [58, 76, 109, 74, 111, 64, 26, 97].

Let γ_i denote the fraction of VNs in $\mathcal{G}$ with degree i and d_v denote the maximum VN degree. Then, the polynomial

$$\gamma(X) = \sum_{i=1}^{d_v} \gamma_i X^{i-1}$$

is called the *degree distribution* of VNs in $\mathcal{G}$ [96, 74, 97]. Let ρ_i denote the fraction of CNs in $\mathcal{G}$ with degree i and d_c denote the maximum CN degree. Then, the polynomial

$$\rho(X) = \sum_{i=1}^{d_c} \rho_i X^{i-1}$$

is called the *degree distribution* of CNs in $\mathcal{G}$ [96, 74, 97]. For a regular LDPC code, all the VNs in $\mathcal{G}$ have the same degree and all the CNs in $\mathcal{G}$ have the same degree. It follows from the correspondence between the degree of a VN (or CN) in $\mathcal{G}$ and the weight of its corresponding column (or row) of $\mathbf{H}$ that $\gamma(X)$ and $\rho(X)$ also give the column and row weight distributions of the adjacency matrix $\mathbf{H}$ of the graph $\mathcal{G}$.

Example 2.1. Consider a binary $(10, 5)$ linear LDPC block code which is given by the null space of the following 5×10 parity-check matrix $\mathbf{H}$ over GF(2):

$$\mathbf{H} = \begin{bmatrix} 1 & 1 & 1 & 1 & 0 & 0 & 0 & 0 & 0 & 0 \\ 1 & 0 & 0 & 0 & 1 & 1 & 1 & 0 & 0 & 0 \\ 0 & 1 & 0 & 0 & 1 & 0 & 0 & 1 & 1 & 0 \\ 0 & 0 & 1 & 0 & 0 & 1 & 0 & 1 & 0 & 1 \\ 0 & 0 & 0 & 1 & 0 & 0 & 1 & 0 & 1 & 1 \end{bmatrix}. \tag{2.2}$$

$\mathbf{H}$ has constant column weight 2 and constant row weight 4. By checking, we find that $\mathbf{H}$ satisfies the RC-constraint. The Tanner graph $\mathcal{G}$ associated with $\mathbf{H}$ is shown in Fig. 2.1. Since $\mathbf{H}$ satisfies the RC-constraint, $\mathcal{G}$ contains no cycle of length 4 and hence its girth is at least 6. In fact, its girth is exactly 6. A cycle of

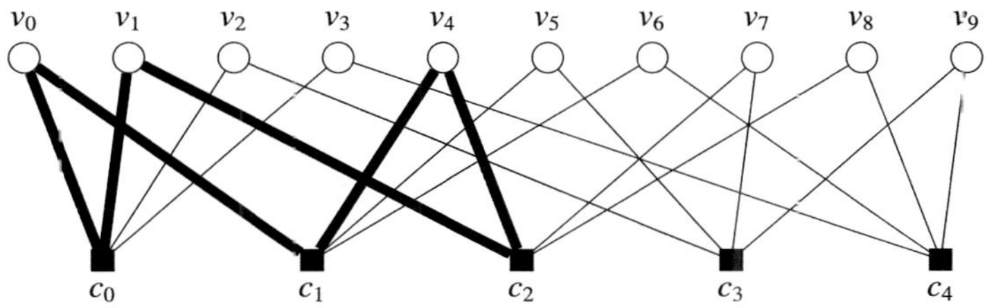

Figure 2.1. The Tanner graph $\mathcal{G}$ of the (10, 5) linear LDPC block code given in Example 2.1.

length 6 is displayed by the bold line in Fig. 2.1. Moreover, this LDPC code has the following degree distribution polynomials: $\gamma(X) = X$, $\rho(X) = X^3$.

In the Tanner graph $\mathcal{G}$ of an LDPC code C, a VN may be connected to other VNs by paths of length 2. Given a VN v_j in a Tanner graph $\mathcal{G}$, the number of VNs that connect to the VN v_j by paths of length 2 is defined as the *connection number* of the VN v_j. The connection numbers of VNs in the Tanner graph $\mathcal{G}$ give a measurement of the *connectivity* of $\mathcal{G}$. If the parity-check matrix $\mathbf{H}$ of the LDPC code C has constant column weight and row weight, all the VNs in the Tanner graph $\mathcal{G}$ of C have the same connection number; otherwise, they have multiple connection numbers. By checking the Tanner graph shown in Fig. 2.1, we find that the connection number of each VN is 6 since the parity-check matrix $\mathbf{H}$ given by (2.1) is RC-constrained and has constant column weight 2 and constant row weight 4. The connection numbers of VNs in a Tanner graph $\mathcal{G}$ give the *VN-connection distribution* of $\mathcal{G}$. The VN-connection distribution of the Tanner graph $\mathcal{G}$ in Fig. 2.1 is (6, 6, 6, 6, 6, 6, 6, 6, 6, 6).

Another important structural property of the Tanner graph $\mathcal{G}$ of an LDPC code given by the null space of an $m \times n$ parity-check matrix $\mathbf{H}$ is the *trapping set* structure [94]. For $1 \leq \kappa \leq n$ and $0 \leq \tau \leq m$, a (κ, τ) trapping set of $\mathcal{G}$ is a subset $\mathbf{T}$ of VNs with $|\mathbf{T}| = \kappa$ such that the subgraph of $\mathcal{G}$ induced by the set $\mathbf{T}$, denoted by $\mathcal{G}[\mathbf{T}]$, has exactly τ odd degree CNs (and an arbitrary number of even degree CNs), where κ is referred to as the size of the trapping set. A trapping set is said to be *elementary* if all the associated CNs in $\mathcal{G}[\mathbf{T}]$ have degree 1 or 2. A trapping set is called small if $\kappa \leq \sqrt{n}$ and $\tau/\kappa \leq 4$ [62].

Let $\mathbf{V}_s$ be a set of VNs in $\mathcal{G}$ and $\mathbf{C}_s$ be a set of CNs in $\mathcal{G}$ that are adjacent to the VNs in $\mathbf{V}_s$, i.e., each CN in $\mathbf{C}_s$ is connected to at least one VN in $\mathbf{V}_s$. The CNs in $\mathbf{C}_s$ are called the *neighbors* of the VNs in $\mathbf{V}_s$. The set $\mathbf{V}_s$ of VNs in $\mathcal{G}$ is called a *stopping set* [24] of $\mathcal{G}$ if each CN in $\mathbf{C}_s$ is connected to at least two VNs in $\mathbf{V}_s$.

In many cases, the parity-check matrix $\mathbf{H}$ of an LDPC code C contains *redundant rows*, i.e., the number of rows of $\mathbf{H}$ is greater than its rank. These redundant rows result in redundant CNs in the Tanner graph $\mathcal{G}$ of the code.

They are either added by design to enhance the performance of the code or they come naturally by the construction of the parity-check matrix **H**. In general, a parity-check matrix constructed based on a finite geometry or a finite field has a large row-redundancy (CN-redundancy) [58, 46, 28], as a consequence of its *highly structured* properties.

Numerous studies of LDPC codes since their rediscovery in the late 1990s show that the error performance of an LDPC code over the AWGNC decoded with an iterative BP-decoding algorithm very much depends on the girth, the cycle distribution, the degree distributions of VNs and CNs, the VN-connectivity, the CN-redundancy, the trapping set structure of the Tanner graph $\mathcal{G}$, and the minimum distance of the code *collectively*. No one of these structural characteristics dominates the performance of all LDPC codes, although one characteristic might be responsible for the performance of a single code. For example, for an LDPC code to perform well over the AWGNC decoded with an iterative BP-decoding algorithm, the girth of its Tanner graph must be at least 6. As an another example, the performance of an LDPC code over the BEC very much depends on the stopping set structure and the minimum distance of the code. Studies of these structural properties of the Tanner graph can be found in [103, 46, 28, 77, 56, 24, 63, 94] (and in their references).

LDPC codes perform amazingly well with an iterative BP-decoding algorithm, such as the SPA or the MSA. However, with iterative decoding, most LDPC codes have a common severe weakness, known as *error-floor* [94]. The error-floor of an LDPC code is characterized by the phenomenon that as the signal-to-noise ratio (SNR) continues to increase, the error probability suddenly decreases at a much slower rate than that in the region of low to moderate SNRs. A high error-floor may preclude LDPC codes from applications where very low error rates are required. For low error rate applications such as high-speed optical communication and data storage, the LDPC code (and decoder) must be designed to achieve the required error probability without an error-floor.

The overall performance of an LDPC code decoded with an iterative BP-decoding algorithm is in general measured by: (1) the decoded error probability (bit, symbol, or block), (2) the error-floor, and (3) the rate of *decoding convergence*. In the code design, it is desirable to achieve a good *balance* among these three performance measurements.

With iterative BP-decoding, the decoded error probability performance of an LDPC code is in general judged by how close the code performs to the *Shannon limit*, the *decoding threshold*, or the *sphere packing bound* (SPB). The error probability performance, especially in the region of low to moderate SNRs, is mainly determined by the girth, the cycle distribution, and the VN and CN degree distributions of the Tanner graph of the code. The girth g needs to be reasonably large (at least 6) and the numbers of cycles of lengths g, $g+2$, $g+4$ (called *short cycles*) should be small. The VN and CN degree distributions should be designed to make the error probability performance of the code close to the decoding threshold or the Shannon limit, especially for an irregular LDPC code.

The rate of decoding convergence is measured by *how fast* the decoding of an LDPC code converges to its performance limit, i.e., how many decoding iterations are needed. Fast decoding convergence reduces computational complexity, power consumption, and decoding delay. Fast decoding convergence is required for very high-speed communication and high-density storage systems. The rate of decoding convergence of an LDPC code very much depends on the VN-connectivity and the CN-redundancy of its Tanner graph. For an LDPC code with high-degree VN-connectivity, each VN is connected to a large number of other VNs with paths of length 2. Hence, in each decoding iteration, each VN receives extrinsic information from many other VNs. As a result, in a few iterations, each VN message processing unit (VN-MPU) collects enough extrinsic information to update its log-likelihood ratio (LLR) to a level sufficient to make a correct hard decision with high probability.

The error-floor of an LDPC code on the AWGNC is a measure of how low an error probability the code can achieve. It is mostly determined by the trapping set structure of the Tanner graph and the minimum distance of the code. A trapping set simply corresponds to an *error pattern* that prevents the iterative LDPC code decoder from converging. The *most harmful* trapping sets are small trapping sets (corresponding to error patterns with small numbers of errors) because they are more probable. If a code has a reasonably large minimum distance and its Tanner graph contains no trapping sets with sizes smaller than its minimum distance, the code will have a low error-floor. The CN-redundancy (or row-redundancy of the parity-check matrix of the code) can also push the error-floor down. Designing codes to avoid harmful trapping sets to mitigate the error-floor is a challenging combinatorial problem. High error-floors most commonly occur for random or pseudo random LDPC codes constructed by computer search. LDPC codes constructed using algebraic methods in general have much lower error-floors. It has been proved that Tanner graphs of several subclasses of algebraic-based LDPC codes do not contain harmful trapping sets with sizes smaller than their minimum distances (which are in general reasonably large) [46, 28, 77]. The error-floor of an LDPC code can be lowered by taking a decoder based strategy to remove or reduce the effect of harmful trapping sets. Several such decoder based strategies have been proposed, such as the prior- and the post-processing decoding strategies [41, 115, 90, 91, 108, 51]. Among them, the most effective decoding strategy is the backtracking iterative decoding algorithm [51].

How to design LDPC codes with good waterfall error performance, very low error-floor, and fast rate of decoding convergence without excessive decoding complexity is a hard and challenging problem. The design techniques in this book represent some very effective solutions to this problem.

2.3 Discussion and Remarks

In this chapter, we merely presented some important structural properties and characteristics of an LDPC code that affect the performance of the code decoded with an iterative BP-decoding algorithm. The major objectives of this book are to bring two major categories of LDPC code constructions, the algebraic-based and the PTG-based, together under a single framework, namely the superposition, and to present new designs and constructions of SP-LDPC codes for both the AWGNC and the BEC. Since the focus is on the design of LDPC codes, we will not discuss various iterative BP-decoding algorithms and schemes for these codes in this book. Presentations of major iterative BP-decoding algorithms and schemes can be found in [74, 97, 95, 81, 18].

In the following, we will give some discussion and remarks about the structural properties presented in Section 2.2 based on our observation, analysis, and experimental results over the last 15 years. First, to the best of our knowledge, VN-connectivity and its impact on the rate of decoding convergence were never discussed before in the public literature on LDPC codes. It is a dominant structural property for fast decoding convergence of long and high-rate LDPC codes, as will be shown in examples given in later chapters. However, high degrees of VN-connectivity have a disadvantage. In decoder hardware implementation, high degrees of VN-connectivity require more wires for message passing between the VN-MPUs and the CN-MPUs and a larger memory for storing messages. In practical applications, a trade-off between the rate of decoding convergence and the decoder complexity is needed.

A large girth in general gives good error performance of an LDPC code in the waterfall region. However, a large girth does not necessarily guarantee a good error-floor performance. In fact, a large girth may result in a high error-floor if a code lacks other good structural properties. Many experimental results show that if an LDPC code has high degrees of VN-connectivity, large CN-redundancy, and no small trapping sets, a girth of 6 is sufficient to give a good overall performance in terms of error probability, error-floor performance, and rate of decoding convergence, as will be shown by examples in later chapters.

To achieve a low error-floor, the parity-check matrix of an LDPC code in general requires a column weight (or an average column weight) of at least 3.

For an RC-constrained LDPC code, the larger the column weight the lower the error-floor. To achieve a bit error rate/probability of 10^{-15} without a visible error-floor, a column weight of 6 is needed. A large column weight also allows an RC-constrained LDPC code to perform hard-decision one-step majority-logic decoding [74, 97] to correct residue errors after a few decoding iterations. This may also bring the iterative decoding process to a quick conclusion. However, a large column weight has a disadvantage. It in general pushes the error performance of a code away from the Shannon limit or the decoding threshold, and also increases the decoder complexity. In the code design, the choice of column weight (column weight distribution) depends on the required error rate (or the location of error-floor).

Finite geometry (FG) codes [58, 74, 97] form an amazing class of structured LDPC codes. They are cyclic and hence can be encoded easily with simple feedback registers. They have large minimum distances, no trapping sets with size smaller than their minimum distances, high degrees of VN-connectivity, and large helpful CN-redundancy [58, 46, 28]. With all these good structural properties, FG-LDPC codes give good overall performance in terms of error probability, error-floor, and rate of decoding convergence. However, the major disadvantage of these codes is the high decoder complexity. Recently, two reduced-complexity iterative schemes for decoding QC-LDPC codes were reported [79, 69, 68, 78]. These two decoding schemes significantly reduce the decoder complexity. Since an FG-LDPC code is cyclic and can be arranged in QC-form [46], these two decoding schemes can be applied to decode an FG-LDPC code to reduce its decoder complexity significantly.

Two special classes of finite geometries and their associated LDPC codes will be discussed in Appendix A.

3

A Review of PTG-Based Construction of LDPC Codes

In this chapter, we give a review of the PTG-based construction of binary LDPC codes from the conventional graph-theoretic point of view. The objective of this review is to provide a background to develop a method for constructing PTG-LDPC codes from an algebraic point of view and to relate the PTG-based construction to the SP-construction of LDPC codes.

3.1 PTG-LDPC Code Construction

The PTG-based construction of LDPC codes was introduced by Thorpe in 2003 [105]. The construction consists of 3 steps. Step 1 of the construction starts with a properly chosen bipartite graph, denoted by $\mathcal{G}_{ptg} = (\mathbf{V}, \mathbf{C}, \mathbf{E})$, of a relatively small size that consists of a set $\mathbf{V} = \{v_0, v_1, \ldots, v_{n-1}\}$ of n VNs, a set $\mathbf{C} = \{c_0, c_1, \ldots, c_{m-1}\}$ of m CNs, and a set $\mathbf{E} = \{(j, i)\}$ of edges, where (j, i) denotes the edge that connects the j-th VN v_j to the i-th CN c_i. (The subscript "*ptg*" of $\mathcal{G}_{ptg}$ stands for "*protograph.*") Let λ denote the number of edges in $\mathbf{E}$. Step 2 of the construction is to create k copies of $\mathcal{G}_{ptg}$. We denote these k copies by $\mathcal{G}_{ptg,0}, \mathcal{G}_{ptg,1}, \ldots, \mathcal{G}_{ptg,k-1}$. For $0 \leq j < n$, we group the copies of the j-th VN v_j in the k copies of $\mathcal{G}_{ptg}$ into a set of k VNs, denoted by $\Phi_j = \{v_{0,j}, v_{1,j} \ldots, v_{k-1,j}\}$. We refer to the k VNs in Φ_j as the type-j VNs. For $0 \leq i < m$, we group the copies of the i-th CN c_i in the k copies of $\mathcal{G}_{ptg}$ into a set of k CNs, denoted by $\Omega_i = \{c_{0,i}, c_{1,i}, \ldots, c_{k-1,i}\}$. We refer to the k CNs in Ω_i as the type-i CNs. The $k\lambda$ edges in the k copies of $\mathcal{G}_{ptg}$ are grouped into λ sets, $\mathbf{E}_0, \mathbf{E}_1, \ldots, \mathbf{E}_{\lambda-1}$, each consisting of k copies of an edge in the set $\mathbf{E}$ of $\mathcal{G}_{ptg}$. For $0 \leq t < \lambda$, if the edges in $\mathbf{E}_t$ are copies of the edge (j, i), then the edges in $\mathbf{E}_t$ are said to be type-(j, i) edges. Step 3 of the construction is to connect the k copies $\mathcal{G}_{ptg,0}, \mathcal{G}_{ptg,1}, \ldots, \mathcal{G}_{ptg,k-1}$ by permuting the edges of the individual copies of $\mathcal{G}_{ptg}$ among themselves. The permutation is carried out under the *restriction*: for $0 \leq j < n$ and $0 \leq i < m$, type-j VNs in Φ_j can only connect to type-i CNs in Ω_i by edges of type-(j,i). The edge permutations among the copies of $\mathcal{G}_{ptg}$ result in a connected bipartite graph, denoted by $\mathcal{G}_{ptg}(k,k)$, with kn VNs, km CNs, and $k\lambda$ edges, which has the *same degree distributions* as the original graph $\mathcal{G}_{ptg}$. The pair (k,k) in $\mathcal{G}_{ptg}(k,k)$ indicates that each VN in $\mathcal{G}_{ptg}$ is expanded into k VNs in $\mathcal{G}_{ptg}(k,k)$ and each CN in $\mathcal{G}_{ptg}$ is expanded into k CNs in $\mathcal{G}_{ptg}(k,k)$. Steps 2 and 3 of the construction are commonly referred to as the *copy-and-permute operation*.

The above copy-and-permute operation expands a small bipartite graph $\mathcal{G}_{ptg}$ into a larger bipartite graph $\mathcal{G}_{ptg}(k,k)$ by a factor of k. The parameter k is called the *expansion factor* (or *lifting degree*). The graph $\mathcal{G}_{ptg}$ is called the *protograph* [105] for the code construction. In choosing (or designing) a protograph $\mathcal{G}_{ptg}$ for the LDPC code construction, *parallel edges* between a VN and a CN are permitted, but the edge permutations must be performed in such a way that the expanded graph $\mathcal{G}_{ptg}(k,k)$ does not contain parallel edges between a VN and a CN. Since each VN, each CN, and each edge of the protograph $\mathcal{G}_{ptg}$ are expanded by the same factor k, we refer to the expansion of $\mathcal{G}_{ptg}$ to $\mathcal{G}_{ptg}(k,k)$ as *a uniform graph expansion*.

Let $\mathbf{B}_{ptg} = [b_{i,j}]_{0 \le i < m, 0 \le j < n}$ denote the adjacency matrix of the protograph $\mathcal{G}_{ptg}$. It is an $m \times n$ matrix over nonnegative integers with rows corresponding to the m CNs of $\mathcal{G}_{ptg}$ and columns corresponding to the n VNs of $\mathcal{G}_{ptg}$. If the entry $b_{i,j}$ at location (i,j) of $\mathbf{B}_{ptg}$ is a positive integer e, then there are e parallel edges connecting the VN v_j and the CN c_i in $\mathcal{G}_{ptg}$. The adjacency matrix $\mathbf{B}_{ptg}$ is called the *base matrix* of the PTG-based code construction. The adjacency matrix of the expanded graph $\mathcal{G}_{ptg}(k,k)$, denoted by $\mathbf{H}_{ptg}(k,k)$, is a binary $mk \times nk$ matrix. If $\mathbf{H}_{ptg}(k,k)$ is sparse, the null space of $\mathbf{H}_{ptg}(k,k)$ gives an LDPC code C_{ptg}, called a *PTG-based LDPC code*. The matrix $\mathbf{H}_{ptg}(k,k)$, obtained by expanding the protograph $\mathcal{G}_{ptg}$ as described above, is the parity-check matrix of the PTG-based LDPC code C_{ptg}. Hereafter, we refer to a PTG-based LDPC code simply as a PTG-LDPC code. The bipartite graph $\mathcal{G}_{ptg}(k,k)$ is the Tanner graph of C_{ptg} (or the Tanner graph associated with the parity-check matrix $\mathbf{H}_{ptg}(k,k)$).

Based on the constraint on the copy-and-permute operation performed on the copies of the protograph $\mathcal{G}_{ptg}$, we see that the connections of the k VNs of type-j in Φ_j to the k CNs of type-i in Ω_i via edges of type-(j,i) can be specified by a $k \times k$ regular matrix $\mathbf{A}_{i,j}$, called a *connection matrix*. Let δ be the number of nonzero entries in the base matrix $\mathbf{B}_{ptg}$. Then, at most δ distinct regular matrices of size $k \times k$ are needed to connect the k copies of the protograph $\mathcal{G}_{ptg}$ to form the expanded graph $\mathcal{G}_{ptg}(k,k)$. Hence, the adjacency matrix $\mathbf{H}_{ptg}(k,k) = [\mathbf{A}_{i,j}]_{0 \le i < m, 0 \le j < n}$ of $\mathcal{G}_{ptg}(k,k)$ is an $m \times n$ array of regular matrices of size $k \times k$. If the VN v_j and the CN c_i in the protograph $\mathcal{G}_{ptg}$ are connected by e

parallel edges, then the connection matrix $\mathbf{A}_{i,j}$ that specifies the connections of the k VNs of type-j in Φ_j to the k CNs of type-i in Ω_i is a weight-e regular matrix. Preferably, the copy-and-permute operation on the copies of a protograph is carried out in such a way that the connection matrices are binary circulants. In this case, the parity-check matrix $\mathbf{H}_{ptg}(k,k)$ is an $m \times n$ array of binary circulants and/or ZMs of size $k \times k$ and the PTG-LDPC code C_{ptg} given by the null space of $\mathbf{H}_{ptg}(k,k)$ is a binary QC-LDPC code, denoted as a QC-PTG-LDPC code. QC-LDPC codes, in general, have advantages over other types of LDPC codes in terms of encoding and decoding implementation [72, 20, 68, 78, 79, 69].

The performance of a PTG-LDPC code depends on the choice of the protograph, its VN and CN degree distributions, and the permutations of the edges among the copies (to avoid short cycles, such as cycles of length 4, in the Tanner graph of the resultant code). Protographs are generally designed with good *decoding thresholds* based on a computer-aided *EXIT chart-based* search procedure [80, 34]. In general, a good decoding threshold gives good waterfall performance, *but not necessarily a low error-floor or a fast rate of decoding convergence*. The PEG algorithm [43, 44] is often used to help design a PTG-LDPC code from a protograph.

Given a protograph $\mathcal{G}_{ptg}$ which is designed for a fixed rate R_c with a good iterative decoding threshold and an expansion factor (or lifting degree) k, there are many ways to permute the edges among the k copies of $\mathcal{G}_{ptg}$ to connect them into a bipartite graph $\mathcal{G}_{ptg}(k,k)$. Different edge permutations result in different connected bipartite graphs and different cycle distributions. In general, a bipartite graph with a reasonably large girth and a small number of short cycles gives an LDPC code with a better error performance when it is decoded with an iterative BP-decoding algorithm. Commonly, the PEG algorithm is used to find a good permutation. A girth of at least 6 is normally needed to ensure a reasonably good error performance.

In the following, we use a small example to illustrate the PTG-based construction process.

Example 3.1. Consider the protograph $\mathcal{G}_{ptg}$ shown in Fig. 3.1(a), that consists of 3 VNs, 2 CNs, and 6 edges. We see that VN v_0 is connected to CN c_0 by two parallel edges. The base (or adjacency) matrix of $\mathcal{G}_{ptg}$ is given by

$$\mathbf{B}_{ptg} = \begin{bmatrix} 2 & 0 & 1 \\ 1 & 1 & 1 \end{bmatrix}.$$

Suppose we take 3 copies of $\mathcal{G}_{ptg}$ (i.e., $k=3$), permute the edges of these copies (under the permutation restriction), and connect the three types of VNs and CNs

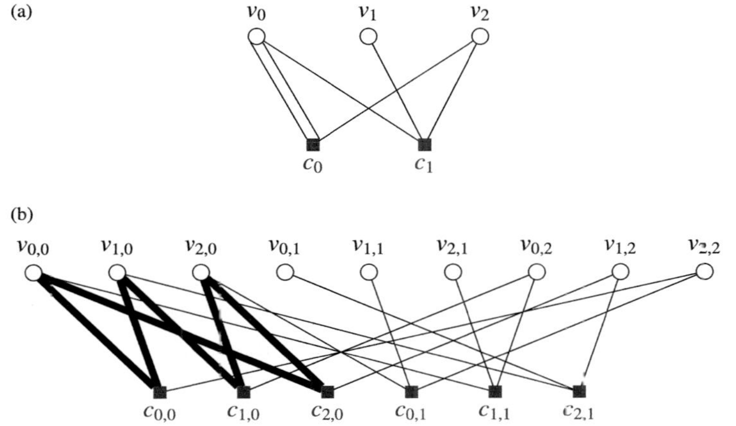

Figure 3.1. (a) The protograph $\mathcal{G}_{ptg}$ given in Example 3.1; (b) the Tanner graph $\mathcal{G}_{ptg}(3,3)$ given in Example 3.1.

using the following 6 connection matrices:

$$\mathbf{A}_{0,0}=\begin{bmatrix}1&1&0\\0&1&1\\1&0&1\end{bmatrix},\ \mathbf{A}_{0,1}=\begin{bmatrix}0&0&0\\0&0&0\\0&0&0\end{bmatrix},\ \mathbf{A}_{0,2}=\begin{bmatrix}0&0&1\\1&0&0\\0&1&0\end{bmatrix},$$

$$\mathbf{A}_{1,0}=\begin{bmatrix}0&0&1\\1&0&0\\0&1&0\end{bmatrix},\ \mathbf{A}_{1,1}=\begin{bmatrix}0&1&0\\0&0&1\\1&0&0\end{bmatrix},\ \mathbf{A}_{1,2}=\begin{bmatrix}0&0&1\\1&0&0\\0&1&0\end{bmatrix}.$$

The copy-and-permute process based on the above connection matrices results in a connected bipartite graph $\mathcal{G}_{ptg}(3,3)$ shown in Fig. 3.1(b). Since the VN v_0 and the CN c_0 are connected by two parallel edges, the connection matrix $\mathbf{A}_{0,0}$, which connects the three type-0 VNs and the three type-0 CNs, is a weight-2 regular matrix. The adjacency matrix $\mathbf{H}_{ptg}(3,3)$ of $\mathcal{G}_{ptg}(3,3)$ is a 2×3 array of regular matrices of size 3×3 given below:

$$\mathbf{H}_{ptg}(3,3)=\begin{bmatrix}1&1&0&0&0&0&0&0&1\\0&1&1&0&0&0&1&0&0\\1&0&1&0&0&0&0&1&0\\0&0&1&0&1&0&0&0&1\\1&0&0&0&0&1&1&0&0\\0&1&0&1&0&0&0&1&0\end{bmatrix}.$$

We can easily check that $\mathbf{H}_{ptg}(3,3)$ satisfies the RC-constraint. Hence, the Tanner graph $\mathcal{G}_{ptg}(3,3)$ associated with $\mathbf{H}_{ptg}(3,3)$ has girth at least 6. From

Fig. 3.1(b), we find a cycle of length 6 and hence the girth of $\mathcal{G}_{ptg}(3,3)$ is 6. Furthermore, since $\mathbf{H}_{ptg}(3,3)$ is a 2×3 array of circulants, the null space of $\mathbf{H}_{ptg}(3,3)$ gives a QC-PTG-LDPC code of length 9 and rate $1/3$, whose Tanner graph has girth 6.

3.2 Conclusion and Remarks

Ensembles of PTG-LDPC codes with *asymptotically* good decoding thresholds and good minimum distance properties have been constructed in [34, 31, 32, 2]. For large enough k, a PTG-LDPC code chosen randomly from a good ensemble in general gives good error performance, especially in the waterfall region (close to the decoding threshold). Note that, in each ensemble, the parity-check matrix of a PTG-LDPC code is an array of regular matrices (not necessarily circulants). Besides random construction, there is no systematic method (or guideline) for constructing a PTG-LDPC code that not only performs well but also has structure to facilitate its decoder implementation. Finding such a method may be a good research problem.

4

An Algebraic Method for Constructing QC-PTG-LDPC Codes and Code Ensembles

In this chapter, we consider QC-PTG-LDPC codes from an algebraic point of view and present a *novel algebraic approach* to design these codes using a technique that we call *base matrix decomposition*.

4.1 Construction of QC-PTG-LDPC Codes by Decomposing Base Matrices

Suppose we choose a protograph $\mathcal{G}_{ptg}$ with n VNs, $v_0, v_1, \ldots, v_{n-1}$, and m CNs, $c_0, c_1, \ldots, c_{m-1}$, and expand it by a factor k. Let $\mathbf{B}_{ptg} = [b_{i,j}]_{0 \leq i < m, 0 \leq j < n}$ be the adjacency matrix of $\mathcal{G}_{ptg}$, an $m \times n$ matrix over nonnegative integers. The first step in constructing a QC-PTG-LDPC code is to decompose the base matrix $\mathbf{B}_{ptg}$ into k matrices $\mathbf{D}_0, \mathbf{D}_1, \ldots, \mathbf{D}_{k-1}$, each of size $m \times n$. In decomposing the base matrix $\mathbf{B}_{ptg}$, we distribute its entries among the k matrices $\mathbf{D}_0, \mathbf{D}_1, \ldots, \mathbf{D}_{k-1}$ in such a way that none of them contains any integer *greater than* one, i.e., the entries of each matrix are binary. If the integer-entry $b_{i,j}$ at location (i, j) in $\mathbf{B}_{ptg}$ is greater than one, then it is split into $b_{i,j}$ one-entries which are distributed in $b_{i,j}$ different matrices among $\mathbf{D}_0, \mathbf{D}_1, \ldots, \mathbf{D}_{k-1}$ at location (i, j)s. The above constraint on decomposing the base matrix $\mathbf{B}_{ptg}$ is called the *decomposition constraint* and the parameter k is called the *decomposition factor*. In the decomposition of $\mathbf{B}_{ptg}$, ZMs are allowed, i.e., there may be ZMs among the matrices $\mathbf{D}_0, \mathbf{D}_1, \ldots, \mathbf{D}_{k-1}$. Therefore, for $0 \leq e < k$, each matrix $\mathbf{D}_e$ is either an $m \times n$ matrix with binary entries or an $m \times n$ ZM. Furthermore, we require that the *sum* (using integer addition) of the matrices $\mathbf{D}_0, \mathbf{D}_1, \ldots, \mathbf{D}_{k-1}$ gives the base matrix $\mathbf{B}_{ptg}$ of the protograph $\mathcal{G}_{ptg}$. The matrices $\mathbf{D}_0, \mathbf{D}_1, \ldots, \mathbf{D}_{k-1}$ are referred to as the *constituent matrices* of $\mathbf{B}_{ptg}$ in the decomposition. The set $\Psi = \{\mathbf{D}_0, \mathbf{D}_1, \ldots, \mathbf{D}_{k-1}\}$ is called the *decomposition set* and its elements (member matrices) need not to be distinct. Note that, if the protograph $\mathcal{G}_{ptg}$ contains no parallel edges, its base matrix $\mathbf{B}_{ptg}$ is a binary matrix. In this case, the constituent matrices $\mathbf{D}_0, \mathbf{D}_1, \ldots, \mathbf{D}_{k-1}$ of $\mathbf{B}_{ptg}$ are mutually disjoint. The matrix $\mathbf{B}_{ptg}$ is called the *decomposition base matrix*.

Next, we arrange the k matrices $\mathbf{D}_0, \mathbf{D}_1, \ldots, \mathbf{D}_{k-1}$ into a *row-block*, then cyclically shift this row-block $k-1$ times, each time by n positions (or one constituent matrix) to the right. This results in a $k \times k$ array $\mathbf{H}_{ptg,cyc}(m,n)$ of $m \times n$ matrices with a *block-cyclic structure* shown below:

$$\mathbf{H}_{ptg,cyc}(m,n) = \begin{bmatrix} \mathbf{D}_0 & \mathbf{D}_1 & \cdots & \mathbf{D}_{k-1} \\ \mathbf{D}_{k-1} & \mathbf{D}_0 & \cdots & \mathbf{D}_{k-2} \\ \vdots & \vdots & \ddots & \vdots \\ \mathbf{D}_1 & \mathbf{D}_2 & \cdots & \mathbf{D}_0 \end{bmatrix}. \tag{4.1}$$

The array $\mathbf{H}_{ptg,cyc}(m,n)$ consists of k row-blocks and k column-blocks of matrices of size $m \times n$. (Note that the pair (m,n) in $\mathbf{H}_{ptg,cyc}(m,n)$ denotes the size of constituent matrices in $\mathbf{H}_{ptg,cyc}(m,n)$.) Each row-block is the *cyclic-shift* of the row-block above it one constituent matrix to the right, and the top row-block is the cyclic-shift of the last row-block one constituent matrix to the right. Each column-block is the cyclic-shift of the column-block on its left one constituent matrix downward, and the first column-block is the cyclic-shift of the last column-block one constituent matrix downward. The subscript "*cyc*" in $\mathbf{H}_{ptg,cyc}(m,n)$ stands for "*block-cyclic structure*." Since the constituent matrices of each row-block (or column-block) are the same, i.e., $\mathbf{D}_0, \mathbf{D}_1, \ldots, \mathbf{D}_{k-1}$, the integer sum of the constituent matrices in each row-block (or each column-block) of $\mathbf{H}_{ptg,cyc}(m,n)$ is a copy of the base matrix $\mathbf{B}_{ptg}$ of the protograph $\mathcal{G}_{ptg}$. Therefore, the block-cyclic array $\mathbf{H}_{ptg,cyc}(m,n)$ is composed of k copies of the decomposition base matrix $\mathbf{B}_{ptg}$ of the protograph $\mathcal{G}_{ptg}$. It is a matrix expansion of $\mathbf{B}_{ptg}$. The top row-block of the cyclic array $\mathbf{H}_{ptg,cyc}(m,n)$ is called the *generator row-block*.

Let $\mathcal{G}_{ptg,cyc}(k,k)$ denote the Tanner graph associated with the $k \times k$ block-cyclic array $\mathbf{H}_{ptg,cyc}(m,n)$. Then, $\mathcal{G}_{ptg}(k,k)$ is composed of k copies of the protograph $\mathcal{G}_{ptg}$ interconnected by the edges of the copies. Suppose we take each column-block of $\mathbf{H}_{ptg,cyc}(m,n)$ as a copy of the base matrix $\mathbf{B}_{ptg}$ of the protograph $\mathcal{G}_{ptg}$ (in the decomposed form). Then, the interconnections of the k copies of the protograph $\mathcal{G}_{ptg}$ are based on the rows of $\mathbf{H}_{ptg,cyc}(m,n)$. If we take each row-block of $\mathbf{H}_{ptg,cyc}(m,n)$ as a copy of the base matrix $\mathbf{B}_{ptg}$, then the interconnections of the k copies of the protograph $\mathcal{G}_{ptg}$ are based on the columns of $\mathbf{H}_{ptg,cyc}(m,n)$. Hence, the Tanner graph $\mathcal{G}_{ptg,cyc}(k,k)$ associated with the $k \times k$ block-cyclic array $\mathbf{H}_{ptg,cyc}(m,n)$ is an expansion of the protograph $\mathcal{G}_{ptg}$ by a factor of k, where the expansion factor of a protograph is given by the decomposition factor of its base matrix.

Using the $k \times k$ block-cyclic array $\mathbf{H}_{ptg,cyc}(m,n)$ as a parity-check matrix, its null space gives a QC-PTG-LDPC code, denoted by $C_{ptg,cyc}$, of length nk in the form for which cyclically shifting each codeword n positions to the right also gives a codeword in $C_{ptg,cyc}$. This simply follows from the block-cyclic structure of $\mathbf{H}_{ptg,cyc}(m,n)$ and represents the definition of QC codes first introduced in the 1960s [74, 106, 17, 52, 89, 53].

Now we convert the array $\mathbf{H}_{ptg,cyc}(m,n)$ to an alternate QC-form, as an array of circulants of the same size. Label the rows and columns of $\mathbf{H}_{ptg,cyc}(m,n)$ from 0 to $mk-1$ and 0 to $nk-1$, respectively. Define the following index sequences

$$\pi_{\text{row}}^{(0)} = [0, m, 2m, \ldots, (k-1)m], \tag{4.2}$$

$$\pi_{\text{row}} = [\pi_{\text{row}}^{(0)}, \pi_{\text{row}}^{(0)}+1, \ldots, \pi_{\text{row}}^{(0)}+m-1], \tag{4.3}$$

$$\pi_{\text{col}}^{(0)} = [0, n, 2n, \ldots, (k-1)n], \tag{4.4}$$

$$\pi_{\text{col}} = [\pi_{\text{col}}^{(0)}, \pi_{\text{col}}^{(0)}+1, \ldots, \pi_{\text{col}}^{(0)}+n-1]. \tag{4.5}$$

Then, π_{row} and π_{col} define a permutation of the rows and a permutation of the columns of $\mathbf{H}_{ptg,cyc}(m,n)$, respectively. If we first permute the rows of $\mathbf{H}_{ptg,cyc}(m,n)$ based on π_{row} and then permute the columns of $\mathbf{H}_{ptg,cyc}(m,n)$ based on π_{col}, we obtain an $m \times n$ array $\mathbf{H}_{ptg,qc}(k,k)$ of $k \times k$ matrices as follows:

$$\mathbf{H}_{ptg,qc}(k,k) = \begin{bmatrix} \mathbf{A}_{0,0} & \mathbf{A}_{0,1} & \cdots & \mathbf{A}_{0,n-1} \\ \mathbf{A}_{1,0} & \mathbf{A}_{1,1} & \cdots & \mathbf{A}_{1,n-1} \\ \vdots & \vdots & \ddots & \vdots \\ \mathbf{A}_{m-1,0} & \mathbf{A}_{m-1,1} & \cdots & \mathbf{A}_{m-1,n-1} \end{bmatrix}. \tag{4.6}$$

Based on the block-cyclic structure of the array $\mathbf{H}_{ptg,cyc}(m,n)$ given by (4.1) and the row and column permutations π_{row} and π_{col} performed on $\mathbf{H}_{ptg,cyc}(m,n)$, we can see that each constituent matrix $\mathbf{A}_{i,j}$ in the array $\mathbf{H}_{ptg,qc}(k,k)$ is either a circulant or a ZM of size $k \times k$. For $0 \le i < m, 0 \le j < n$ and $0 \le e < k$, let $g_{i,j,e}$ be the entry at location (i, j) of the e-th constituent matrix $\mathbf{D}_e$ in the generator (top) row-block of the block-cyclic array $\mathbf{H}_{ptg,cyc}(m,n)$. Then, the k-tuple $\mathbf{g}_{i,j} = (g_{i,j,0}, g_{i,j,1}, \ldots, g_{i,j,k-1})$ is the *generator* of the constituent circulant $\mathbf{A}_{i,j}$ of the array $\mathbf{H}_{ptg,qc}(k,k)$. (If $\mathbf{g}_{i,j}$ is a zero k-tuple, $\mathbf{A}_{i,j}$ is a ZM of size $k \times k$.) Hence, we can construct the array $\mathbf{H}_{ptg,qc}(k,k)$ *directly* from these circulant generators without performing the row and column permutations on $\mathbf{H}_{ptg,cyc}(m,n)$. This is done simply by replacing the nonzero entry $b_{i,j}$ at location (i, j) of the base matrix $\mathbf{B}_{ptg} = [b_{i,j}]_{0 \le i < m, 0 \le j < n}$ of the protograph $\mathcal{G}_{ptg}$ with the $k \times k$ circulant $\mathbf{A}_{i,j}$ generated by $\mathbf{g}_{i,j} = (g_{i,j,0}, g_{i,j,1}, \ldots, g_{i,j,k-1})$. If $b_{i,j} = 0$, $\mathbf{A}_{i,j}$ is a $k \times k$ ZM. Let δ be the number of nonzero entries in the base matrix $\mathbf{B}_{ptg}$. Then, the array

$\mathbf{H}_{ptg,qc}(k,k)$ consists of δ circulants of size $k \times k$ (they may not be all distinct). If the entries of the base matrix $\mathbf{B}_{ptg}$ are binary, then $\mathbf{H}_{ptg,qc}(k,k)$ is an $m \times n$ array of CPMs and/or ZMs of size $k \times k$.

Let $\mathcal{G}_{ptg,qc}(k,k)$ denote the Tanner graph associated with the $m \times n$ array $\mathbf{H}_{ptg,qc}(k,k)$. Then, $\mathcal{G}_{ptg,qc}(k,k)$ and the Tanner graph $\mathcal{G}_{ptg,cyc}(k,k)$ associated with the $k \times k$ block-cyclic array $\mathbf{H}_{ptg,cyc}(m,n)$ are isomorphic (or structurally identical). One can be obtained from the other by permuting the locations of VNs and CNs. With $\mathbf{H}_{ptg,qc}(k,k)$ as the parity-check matrix, we obtain a QC-PTG-LDPC code $C_{ptg,qc}$ which is isomorphic (combinatorially equivalent) to the QC-PTG-LDPC code $C_{ptg,cyc}$ given by the null space of $\mathbf{H}_{ptg,cyc}(m,n)$. A codeword $\mathbf{v}$ in $C_{ptg,qc}$ consists of n sections, each containing k bits. If we cyclically shift all n sections of the codeword $\mathbf{v}$ *simultaneously* one position to the right *within* each section, we obtain another codeword in $C_{ptg,qc}$. This QC-structure is referred to as a *section-wise cyclic structure*. Therefore, a QC-PTG-LDPC code can be put either in a block-cyclic form or in a section-wise cyclic form.

The QC-PTG-LDPC code $C_{ptg,cyc}$ in block-cyclic form can be decoded with the reduced-complexity iterative decoding scheme proposed in [79] using the generator row-block $\mathbf{H}_{ptg,cyc,dec} = [\mathbf{D}_0, \ \mathbf{D}_1, \ \ldots, \mathbf{D}_{k-1}]$ of its parity-check matrix $\mathbf{H}_{ptg,cyc}(n,m)$ given by (4.1) as the *decoding matrix*, where the subscript "*dec*" stands for "*decoding*." This decoding matrix $\mathbf{H}_{ptg,cyc,dec}$ is an $m \times nk$ submatrix of $\mathbf{H}_{ptg,cyc}(n,m)$. Decoding $C_{ptg,cyc}$ based on $\mathbf{H}_{ptg,cyc,dec}$ reduces the complexity of hardware implementation of an iterative BP-decoder for $C_{ptg,cyc}$ by a factor of k. The QC-PTG-LDPC code $C_{ptg,qc}$ in section-wise cyclic form can be decoded using the reduced-complexity iterative decoding scheme proposed in [69] based on an $m \times nk$ submatrix $\mathbf{H}_{ptg,qc,dec}$ that consists of the first (top) rows of the m row-blocks of the parity-check matrix $\mathbf{H}_{ptg,qc}(k,k)$ of $C_{ptg,qc}$. Decoding $C_{ptg,qc}$ based on $\mathbf{H}_{ptg,qc,dec}$ also reduces the complexity of hardware implementation of an iterative BP-decoder for $C_{ptg,qc}$ by a factor of k. The reduction in decoder complexity is at the expense of a reduction in decoding speed. Various trade-offs between the decoder complexity and the decoding speed can be achieved by using decoding matrices of various sizes [79, 69].

If the parity-check matrix $\mathbf{H}_{ptg,cyc}(n,m)$ in block-cyclic form (or $\mathbf{H}_{ptg,qc}(k,k)$ in section-wise cyclic form) also has a section-wise cyclic structure (or a block-cyclic structure), then the QC-PTG-LDPC code $C_{ptg,cyc}$ (or $C_{ptg,qc}$) has a *doubly QC-structure* (i.e., both block-cyclic and section-wise cyclic structures at the same time without performing row and column permutations). We refer to such a QC-LDPC code as a *doubly QC-LDPC code*. More on doubly QC-LDPC codes will be discussed in Chapter 8. An array of matrices of the same size with a doubly QC-structure can be used to construct spatially coupled (SC) QC-LDPC codes, which will be discussed in Chapter 9.

With the above construction of a QC-PTG-LDPC code, the graphical process of edge permutations and connections of the copies of a protograph is replaced by a *matrix decomposition-and-replacement* process. It is clear that there are many possible decompositions of the base matrix $\mathbf{B}_{ptg}$, each resulting in a different $k \times k$ array $\mathbf{H}_{ptg,cyc}(m,n)$ (or an $m \times n$ array $\mathbf{H}_{ptg,qc}(k,k)$) and hence a different QC-PTG-LDPC code $C_{ptg,cyc}$ (or $C_{ptg,qc}$). One extreme decomposition of $\mathbf{B}_{ptg}$ is that each nonzero matrix $\mathbf{D}_e$ in the decomposition set $\Psi = \{\mathbf{D}_0, \mathbf{D}_1, \ldots, \mathbf{D}_{k-1}\}$ contains a single 1-entry. Hence, in any decomposition of the base matrix $\mathbf{B}_{ptg}$, the maximum number of nonzero matrices in the set Ψ equals the integer sum of the nonzero entries in $\mathbf{B}_{ptg}$ (the total number of edges in the protograph $\mathcal{G}_{ptg}$).

In the above algebraic construction of a QC-PTG-LDPC code, there is no upper limit on the expansion factor k. However, k must be greater than or equal to the largest integer in the base matrix $\mathbf{B}_{ptg}$ of the chosen protograph $\mathcal{G}_{ptg}$. This ensures that all the nonzero constituent matrices in the decomposition set Ψ are binary, which is necessary to avoid parallel edges in the Tanner graph associated with $\mathbf{H}_{ptg,cyc}(m,n)$ or $\mathbf{H}_{ptg,qc}(k,k)$.

We can label the edge(s) connecting the j-th VN and the i-th CN of the protograph $\mathcal{G}_{ptg}$ with the generator $\mathbf{g}_{i,j} = (g_{i,j,0}, g_{i,j,1}, \ldots, g_{i,j,k-1})$ of the constituent circulant $\mathbf{A}_{i,j}$ of $\mathbf{H}_{ptg,qc}(k,k)$ at location (i,j). In this case, $\mathbf{g}_{i,j} = (g_{i,j,0}, g_{i,j,1}, \ldots, g_{i,j,k-1})$ is called the *vector label* of the edge (j,i). The number of nonzero components in $\mathbf{g}_{i,j}$ equals the number of parallel edges connecting the j-th VN and the i-th CN of $\mathcal{G}_{ptg}$. From the viewpoint of the graphical construction of a PTG-LDPC code, the vector labels of the edges in $\mathcal{G}_{ptg}$ give precise information of how to permute the edges of the copies of $\mathcal{G}_{ptg}$ and connect them into the Tanner graph of the QC-PTG-LDPC code $C_{ptg,qc}$.

The algebraic construction of a QC-PTG-LDPC code presented above is simply based on the expansion of the base matrix $\mathbf{B}_{ptg}$ of the protograph $\mathcal{G}_{ptg}$.

As we showed above, there are two forms of a QC-PTG-LDPC code, the block-cyclic form and the section-wise cyclic form. Each form consists of four construction steps. The construction of a QC-PTG-LDPC code $C_{ptg,cyc}$ in block-cyclic form consists of the following four steps: (1) Design an appropriate base matrix $\mathbf{B}_{ptg}$ for a desired code rate and choose a designed expansion factor k; (2) Decompose $\mathbf{B}_{ptg}$ into k constituent matrices $\mathbf{D}_0, \mathbf{D}_1, \ldots, \mathbf{D}_{k-1}$ which satisfy the RC- and the PW-RC-constraints; (3) Form a $k \times k$ array $\mathbf{H}_{ptg,cyc}(m,n)$ in the form of (4.1); and (4) Take the null space of $\mathbf{H}_{ptg,cyc}(m,n)$ to produce a QC-PTG-LDPC code $C_{ptg,cyc}$ in block-cyclic form.

To construct a QC-PTG-LDPC code in section-wise cyclic form, the first two and the fourth steps are the same as those for constructing a QC-PTG-LDPC code $C_{ptg,cyc}$ in block-cyclic form. In the third step, an $m \times n$ array $\mathbf{H}_{ptg,qc}(k,k) = [\mathbf{A}_{i,j}]_{0 \le i < m, 0 \le j < n}$ is constructed directly from the k constituent matrices $\mathbf{D}_0$, $\mathbf{D}_1$, ..., $\mathbf{D}_{k-1}$ formed in the second step (by skipping the row and column permutations of the array $\mathbf{H}_{ptg,cyc}(m,n)$). This is carried out by finding the set $\{\mathbf{g}_{i,j} : 0 \le i < m, 0 \le j < n\}$ of the generators of the set $\{\mathbf{A}_{i,j} : 0 \le i < m, 0 \le j < n\}$ of $k \times k$ circulants from $\mathbf{D}_0, \mathbf{D}_1, \ldots, \mathbf{D}_{k-1}$. Then, we replace the entries in the base matrix $\mathbf{B}_{ptg} = [b_{i,j}]_{0 \le i < m, 0 \le j < n}$ by their corresponding circulants in $\{\mathbf{A}_{i,j} : 0 \le i < m, 0 \le j < n\}$. So, the third step for constructing a QC-PTG-LDPC code $C_{ptg,qc}$ in section-wise cyclic form is simply a *replacement process* which is exactly the one used in the SP-construction method for constructing LDPC codes given in [76]. This correspondence will be discussed further in Chapter 5.

With the above algebraic construction of a QC-PTG-LDPC code, the design of a protograph is equivalent to the design of a base matrix, and the copy-and-permute process performed on a protograph is equivalent to the decomposition-and-replacement process performed on its associated base matrix.

4.2 Construction of RC-Constrained PTG Parity-Check Matrices

The girth and cycle distributions of the Tanner graph $\mathcal{G}_{ptg,cyc}(k,k)$ (or $\mathcal{G}_{ptg,qc}(k, k)$) and hence the performance of a QC-PTG-LDPC code $C_{ptg,cyc}$ (or $C_{ptg,qc}$) very much depend on the decomposition of the base matrix $\mathbf{B}_{ptg}$, i.e., the distribution of its nonzero entries and the *arrangement* (or *order*) of the nonzero constituent matrices in the generator row-block of the array $\mathbf{H}_{ptg,cyc}(m,n)$. The distribution of the nonzero entries in $\mathbf{B}_{ptg}$ and the order of the nonzero constituent matrices in the decomposition set $\Psi = \{\mathbf{D}_0, \mathbf{D}_1, \ldots, \mathbf{D}_{k-1}\}$ should be done in such a way that the array $\mathbf{H}_{ptg,cyc}(m,n)$ satisfies the RC-constraint (or the Tanner graph $\mathcal{G}_{ptg,cyc}(k,k)$ associated with the parity-check array $\mathbf{H}_{ptg,cyc}(m, n)$ has no cycle of length 4).

With iterative decoding of an LDPC code using a BP-algorithm, cycles of length 4 in the Tanner graph of the code cause serious decoding correlation after a few iterations and result in poor performance, in general. In the code design (or construction), effort must be made to remove these harmful cycles of length 4. The Tanner graph of an LDPC code whose parity-check matrix satisfies the RC-constraint has girth at least 6 and hence contains no cycle of length 4. In general, an LDPC code whose Tanner graph has a relatively large girth and a small number of short cycles performs well if it also has other good structural properties as described in Section 2.2. A girth of at least 6 is in general required for a code to achieve good error performance if it has no small trapping sets and has a large enough minimum distance, such as finite geometry LDPC codes and many other algebraic LDPC codes constructed based on finite fields and combinatorial designs.

Using the matrix decomposition method to construct a QC-PTG-LDPC code $C_{ptg,cyc}$ (or $C_{ptg,qc}$) whose parity-check array $\mathbf{H}_{ptg,cyc}(m,n)$ (or $\mathbf{H}_{ptg,qc}(k,k)$) satisfies the RC-constraint, the general guidelines to decompose the base matrix $\mathbf{B}_{ptg}$ of the protograph $\mathcal{G}_{ptg}$ are: (1) Each nonzero constituent matrix $\mathbf{D}_e$ in the decomposition set $\Psi = \{\mathbf{D}_0, \mathbf{D}_1, \ldots, \mathbf{D}_{k-1}\}$ satisfies the RC-constraint; and (2) Any two nonzero constituent matrices, $\mathbf{D}_e$ and $\mathbf{D}_j$, satisfy the PW-RC-constraint. These general guidelines are necessary but not sufficient to guarantee that the $k \times k$ array $\mathbf{H}_{ptg,cyc}(m,n)$ satisfies the RC-constraint. If the number of ZMs in the decomposition set Ψ is large enough and if we order the constituent matrices of Ψ in forming the generator row-block of the cyclic array $\mathbf{H}_{ptg,cyc}(m,n)$ in such a way that $\mathbf{H}_{ptg,cyc}(m,n)$ satisfies the 2×2 array RC-constraint, then $\mathbf{H}_{ptg,cyc}(m,n)$ satisfies the RC-constraint (as discussed in Section 2.1). A decomposition set Ψ of a base matrix whose nonzero constituent matrices satisfy the RC- and the PW-RC-constraints is called an *RC-constrained decomposition set*.

In forming an RC-constrained decomposition set Ψ of a base matrix $\mathbf{B}_{ptg}$, we first decompose $\mathbf{B}_{ptg}$ into a set of RC- and PW-RC-constrained nonzero constituent matrices. This can be achieved with computer aid. However, if each nonzero constituent matrix in the decomposition of $\mathbf{B}_{ptg}$ contains a single 1-entry, then all the nonzero constituent matrices in Ψ satisfy the RC- and the PW-RC-constraints. Another case when computer aid is not needed is when the base matrix $\mathbf{B}_{ptg}$ is binary and satisfies the RC-constraint. In this case, any decomposition of $\mathbf{B}_{ptg}$ will give a set of nonzero constituent matrices which satisfy both the RC- and the PW-RC-constraints. Construction of RC-constrained base matrices can be done using the PEG algorithm or using the algebraic methods presented in [58, 76, 110, 109, 35, 107, 3, 19, 74, 111, 102, 101, 112, 64, 65, 116, 97, 50, 99, 113, 114, 25, 46, 26, 70, 68, 77]. New construction methods of RC-constrained base matrices will be presented in later chapters. Once we obtain a set of RC- and PW-RC-constrained nonzero constituent matrices of the base matrix $\mathbf{B}_{ptg}$, we add ZMs to the set to form an RC-constrained decomposition set Ψ of the base matrix to meet the desired decomposition factor k.

Ordering the constituent matrices in an RC-constrained decomposition set Ψ of a base matrix $\mathbf{B}_{ptg}$ to form the generator row-block of a block-cyclic array $\mathbf{H}_{ptg,cyc}(m,n)$ such that $\mathbf{H}_{ptg,cyc}(m,n)$ satisfies the 2×2 array RC-constraint can also be achieved with computer aid. One approach to achieve this is to put ZMs between two nonzero constituent matrices in forming the generator row-block of $\mathbf{H}_{ptg,cyc}(m,n)$ such that any 2×2 subarray of $\mathbf{H}_{ptg,cyc}(m,n)$ contains *at least one ZM*. This ensures that $\mathbf{H}_{ptg,cyc}(m,n)$ satisfies the RC-constraint as described in Section 2.1.

Another approach to achieve the RC-constraint on $\mathbf{H}_{ptg,cyc}(m,n)$ is an algebraic approach. Let δ be the number of nonzero constituent matrices in Ψ. We first construct a $k \times k$ binary circulant $\mathbf{M}$ with weight δ which satisfies the RC-constraint. Note that any 2×2 submatrix of $\mathbf{M}$ contains at least one 0-entry. Next, we replace the δ 1-components of the generator (or top row) of the RC-constrained circulant $\mathbf{M}$ with the nonzero constituent matrices in Ψ and the $k-\delta$ 0-components with ZMs. (The replacement can be carried out arbitrarily.) This replacement results in a row-block that consists of all the δ constituent matrices in Ψ. Cyclically shifting this row-block to the right $k-1$ times, one constituent matrix per shift, we form a block-cyclic array $\mathbf{H}_{ptg,cyc}(m,n)$. Since any 2×2 submatrix of $\mathbf{M}$ contains at least one 0-entry, any 2×2 subarray of $\mathbf{H}_{ptg,cyc}(m,n)$ contains at least one ZM. Since the constituent matrices of $\mathbf{H}_{ptg,cyc}(m,n)$ are member matrices of the decomposition set Ψ which satisfy both the RC- and the PW-RC-constraints, every 2×2 subarray of $\mathbf{H}_{ptg,cyc}(m,n)$ satisfies the RC-constraint. Hence, the block-cyclic array $\mathbf{H}_{ptg,cyc}(m,n)$ satisfies the RC-constraint. Construction of RC-constrained circulants can be achieved by using finite geometries over finite fields [58, 74, 97]. This will be discussed in Chapter 6 and also in Appendix A.

It is clear that, if the block-cyclic array $\mathbf{H}_{ptg,cyc}(m,n)$ is RC-constrained, the array $\mathbf{H}_{ptg,qc}(k,k)$ is also RC-constrained.

4.3 Examples

In this section, we first use a simple example to demonstrate the decomposition process of a small base matrix and the construction of an array of circulants based on the decomposition of the base matrix. We then give three binary QC-PTG-LDPC codes constructed based on the proposed algebraic construction and present their error performances over the AWGNC using BPSK signaling and the BEC. We show that these codes perform very well over both the AWGNC and the BEC.

Example 4.1. Consider the protograph $\mathcal{G}_{ptg}$ shown in Fig. 4.1(a). The base matrix $\mathbf{B}_{ptg}$ of $\mathcal{G}_{ptg}$ is a 2×3 matrix given below:

$$\mathbf{B}_{ptg} = \begin{bmatrix} 2 & 0 & 1 \\ 1 & 1 & 1 \end{bmatrix}.$$

Note that the base matrix $\mathbf{B}_{ptg}$ does not satisfy the RC-constraint.

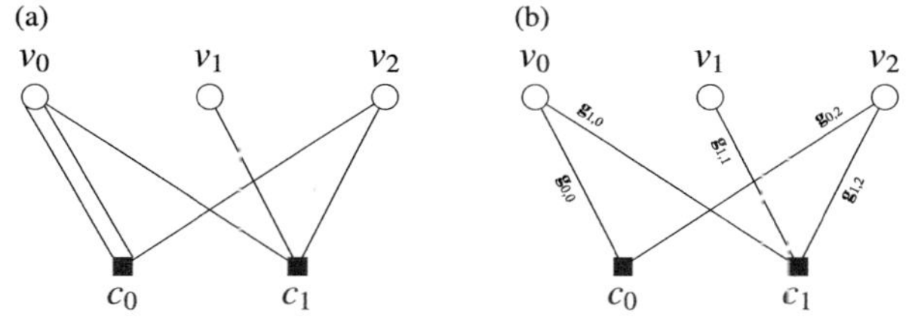

Figure 4.1. (a) The protograph $\mathcal{G}_{ptg}$ given in Example 4.1; and (b) the protograph $\mathcal{G}_{ptg}$ in Example 4.1 with vector labels, where $\mathbf{g}_{0,0} = (1,1,0)$, $\mathbf{g}_{0,2} = (0,0,1)$, $\mathbf{g}_{1,0} = (0,0,1)$, $\mathbf{g}_{1,1} = (0,1,0)$, and $\mathbf{g}_{1,2} = (0,0,1)$.

Choose the decomposition factor $k = 3$. Suppose we decompose $\mathbf{B}_{ptg}$ into the following three 2×3 matrices:

$$\mathbf{D}_0 = \begin{bmatrix} 1 & 0 & 0 \\ 0 & 0 & 0 \end{bmatrix}, \ \mathbf{D}_1 = \begin{bmatrix} 1 & 0 & 0 \\ 0 & 1 & 0 \end{bmatrix}, \ \mathbf{D}_2 = \begin{bmatrix} 0 & 0 & 1 \\ 1 & 0 & 1 \end{bmatrix}.$$

It is easy to check that the above constituent matrices of the decomposition of $\mathbf{B}_{ptg}$ satisfy the RC- and the PW-RC-constraints. Based on these three constituent matrices, we form the following 3×3 array $\mathbf{H}_{ptg,cyc}(2,3)$ of 2×3 matrices with a block-cyclic structure:

$$\mathbf{H}_{ptg,cyc}(2,3) = \left[\begin{array}{ccc|ccc|ccc} 1 & 0 & 0 & 1 & 0 & 0 & 0 & 0 & 1 \\ 0 & 0 & 0 & 0 & 1 & 0 & 1 & 0 & 1 \\ \hline 0 & 0 & 1 & 1 & 0 & 0 & 1 & 0 & 0 \\ 1 & 0 & 1 & 0 & 0 & 0 & 0 & 1 & 0 \\ \hline 1 & 0 & 0 & 0 & 0 & 1 & 1 & 0 & 0 \\ 0 & 1 & 0 & 1 & 0 & 1 & 0 & 0 & 0 \end{array}\right].$$

Note that $\mathbf{H}_{ptg,cyc}(2,3)$ satisfies the 2×2 array RC-constraint and hence the RC-constraint. The null space of $\mathbf{H}_{ptg,cyc}(2,3)$ gives a QC-PTG-LDPC code $C_{ptg,cyc}$ of length 9 and rate $1/3$, whose Tanner graph has girth 6. The code has a block-cyclic structure.

Permuting the columns and rows of $\mathbf{H}_{ptg,cyc}(2,3)$ based on the row and column permutations π_{row} and π_{col} defined by (4.3) and (4.5), respectively, we obtain the following 2×3 array $\mathbf{H}_{ptg,qc}(3,3)$ of circulants of size 3×3:

$$\mathbf{H}_{ptg,qc}(3,3) = \begin{bmatrix} \mathbf{A}_{0,0} & \mathbf{A}_{0,1} & \mathbf{A}_{0,2} \\ \mathbf{A}_{1,0} & \mathbf{A}_{1,1} & \mathbf{A}_{1,2} \end{bmatrix} = \left[\begin{array}{ccc|ccc|ccc} 1 & 1 & 0 & 0 & 0 & 0 & 0 & 0 & 1 \\ 0 & 1 & 1 & 0 & 0 & 0 & 1 & 0 & 0 \\ 1 & 0 & 1 & 0 & 0 & 0 & 0 & 1 & 0 \\ \hline 0 & 0 & 1 & 0 & 1 & 0 & 0 & 0 & 1 \\ 1 & 0 & 0 & 0 & 0 & 1 & 1 & 0 & 0 \\ 0 & 1 & 0 & 1 & 0 & 0 & 0 & 1 & 0 \end{array}\right].$$

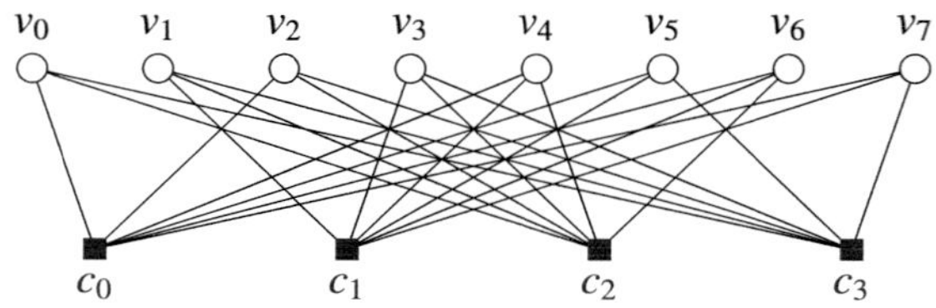

Figure 4.2. The Tanner graph $\mathcal{G}_{ptg,1}$ of the base matrix $\mathbf{B}_{ptg,1}$ given in Example 4.2.

From the structure of the array $\mathbf{H}_{ptg,qc}(3,3)$, we see that the array is obtained by replacing each entry in the base matrix $\mathbf{B}_{ptg}$ by a circulant of size 3×3 whose generator can be determined directly from the three constituent matrices $\mathbf{D}_0, \mathbf{D}_1$, and $\mathbf{D}_2$ of $\mathbf{B}_{ptg}$ (or the generator row-block of the block-cyclic array $\mathbf{H}_{ptg,cyc}(2,3)$) as explained above. Note that the entry 2 in $\mathbf{B}_{ptg}$ is replaced by a circulant of weight 2 and each 1-entry is replaced by a CPM. The protograph $\mathcal{G}_{ptg}$ with vector labels is shown in Fig. 4.1(b). The null space of $\mathbf{H}_{ptg,qc}(3,3)$ gives a $(9,3)$ QC-PTG-LDPC code $C_{ptg,qc}$ with a section-wise cyclic structure. $C_{ptg,cyc}$ and $C_{ptg,qc}$ are combinatorially equivalent.

In this example, we show that an RC-constrained array can be constructed by decomposing the base matrix into constituent matrices which satisfy the RC- and the PW-RC-constraints and ordering the constituent matrices properly.

Example 4.2. Suppose we choose the following 4×8 binary matrix as the base matrix for constructing a QC-PTG-LDPC code with rate 1/2:

$$\mathbf{B}_{ptg,1} = [b_{i,j}]_{0 \le i < 4, 0 \le j < 8} = \begin{bmatrix} 1 & 0 & 1 & 0 & 1 & 1 & 1 & 1 \\ 0 & 1 & 0 & 1 & 1 & 1 & 1 & 1 \\ 1 & 1 & 1 & 1 & 1 & 0 & 1 & 0 \\ 1 & 1 & 1 & 1 & 0 & 1 & 0 & 1 \end{bmatrix}. \tag{4.7}$$

This base matrix is a $(3,6)$-regular binary matrix with column and row weights 3 and 6, respectively. It has 24 nonzero entries which are all 1s. The protograph associated with this matrix is shown in Fig. 4.2. It has 8 VNs and 4 CNs but no parallel edges.

Suppose we choose the decomposition (or expansion) factor $k = 330$. The base matrix and the decomposition factor are designed to construct a QC-PTG-LDPC code of rate 1/2 whose Tanner graph has girth 8 and contains a small number of short cycles. The base matrix is not designed based on the decoding threshold criterion.

Decompose the base matrix $\mathbf{B}_{ptg,1}$ into 330 constituent matrices of size 4×8, $\mathbf{D}_0, \mathbf{D}_1, \ldots, \mathbf{D}_{329}$. Among these 330 constituent matrices, 23 of them are nonzero matrices and the other 307 matrices are ZMs. Among the 23 nonzero matrices, one

Table 4.1. *The nonzeros matric in the decomposition set* $\Psi = \{\mathbf{D}_0, \mathbf{D}_1, \ldots, \mathbf{D}_{329}\}$ *and the locations of their 1-entries constructed in Example 4.2*

The nonzero matrices	The locations of 1-entries	The nonzero matrices	The locations of 1-entries
$\mathbf{D}_{10}$	(3,5)	$\mathbf{D}_{209}$	(2,1)
$\mathbf{D}_{39}$	(0,0)	$\mathbf{D}_{271}$	(3,1)
$\mathbf{D}_{68}$	(0,7)	$\mathbf{D}_{274}$	(1,7)
$\mathbf{D}_{75}$	(1,3)	$\mathbf{D}_{275}$	(3,2)
$\mathbf{D}_{79}$	(1,1), (0,2)	$\mathbf{D}_{284}$	(3,7)
$\mathbf{D}_{143}$	(3,3)	$\mathbf{D}_{287}$	(2,2)
$\mathbf{D}_{148}$	(1,6)	$\mathbf{D}_{294}$	(0,6)
$\mathbf{D}_{158}$	(1,4)	$\mathbf{D}_{295}$	(1,5)
$\mathbf{D}_{163}$	(2,4)	$\mathbf{D}_{297}$	(0,6)
$\mathbf{D}_{181}$	(3,0)	$\mathbf{D}_{324}$	(2,3)
$\mathbf{D}_{202}$	(2,6)	$\mathbf{D}_{328}$	(0,5)
$\mathbf{D}_{206}$	(2,0)		

consists of two 1s and each of the other 22 consists of a single 1. These 23 nonzero constituent matrices of $\mathbf{B}_{ptg,1}$ satisfy the RC- and the PW-RC-constraints. The locations of the 1s in these 23 nonzero matrices are given in Table 4.1, where the pair (i, j) denotes the location intersecting at the i-th row and the j-th column of each nonzero constitute matrix. Based on the locations of 1s of these nonzero constituent matrices of $\mathbf{B}_{ptg,1}$, we construct a set of 24 CPMs of size 330×330 which correspond to the 24 1s in $\mathbf{B}_{ptg,1}$. Replacing each 1-entry in $\mathbf{B}_{ptg,1}$ by its corresponding CPM and each 0-entry by a ZM of size 330×330, we obtain the following 4×8 array $\mathbf{H}_{ptg,qc,1}(330, 330)$ of CPMs and ZMs of size 330×330 in which each CPM is represented by a positive integer less than 330:

$$\mathbf{H}_{ptg,qc,1}(330, 330) = \begin{bmatrix} 39 & z & 79 & z & 294 & 328 & 297 & 68 \\ z & 79 & z & 75 & 158 & 295 & 148 & 274 \\ 206 & 209 & 287 & 324 & 163 & z & 202 & z \\ 181 & 271 & 275 & 143 & z & 10 & z & 284 \end{bmatrix}. \quad (4.8)$$

A nonzero entry l in $\mathbf{H}_{ptg,qc,1}(330, 330)$ represents a CPM whose generator, a 330-tuple with components labeled from 0 to 329, has its single 1-entry at location l. The symbol z in $\mathbf{H}_{ptg,qc,1}(330, 330)$ represents a ZM of size 330×330. The array $\mathbf{H}_{ptg,qc,1}(330, 330)$ is a 1320×2640 matrix with column and row weights 3 and 6, respectively. It can be checked that $\mathbf{H}_{ptg,qc,1}(330, 330)$

satisfies the RC-constraint and hence its associated Tanner graph, denoted by $\mathcal{G}_{ptg,qc,1}(330,330)$, has girth at least 6. Using the cycle counting algorithm presented in [71], we find that the Tanner graph $\mathcal{G}_{ptg,qc,1}(330,330)$ of the code $C_{ptg,qc,1}$ has girth 8 and contains 990 cycles of length 8 and 8580 cycles of length 10. The connection number of each VN in $\mathcal{G}_{ptg,qc,1}(330,330)$ is 15.

The null space of $\mathbf{H}_{ptg,qc,1}(330,330)$ gives a $(3,6)$-regular $(2640,1320)$ QC-PTG-LDPC code, denoted by $C_{ptg,qc,1}$, with rate $1/2$. Suppose the code is used over the AWGNC with BPSK signaling. The bit error rate (BER) and block error rate (BLER) performances of the code $C_{ptg,qc,1}$ decoded using the MSA [97, 18] with a maximum of 50 iterations and the SPA [81, 74, 97] with a maximum of 100 iterations are shown in Fig. 4.3(a). We see that the code achieves a BER of 10^{-10} without a visible error-floor decoded using the MSA. The decoding threshold of this code is 1.1 dB. At BERs of 10^{-5} and 10^{-10}, the code performs 1 and 1.4 dBs away from the decoding threshold, respectively. At a BLER of 10^{-8}, the code performs within 1.25 dB of the SPB.

For comparison, in Fig. 4.3(a), we also include the BER and BLER performances of a $(2664,1334)$ QC-PTG-LDPC code $C^{\star}_{ptg,qc}$ with rate 0.5008 which was recently reported in [85]. This code was constructed based on the conventional PTG-based construction method using a pre-lifting technique. The protograph of this code is designed based on the decoding threshold criterion. The decoding threshold of this code is also 1.1 dB, the same as the decoding threshold of the $(2640,1320)$ QC-PTG-LDPC code $C_{ptg,qc,1}$. The QC-PTG-LDPC code $C^{\star}_{ptg,qc}$ is 24 bits longer than the code $C_{ptg,qc,1}$ constructed using the base matrix given by (4.7) and the algebraic decomposition method. In [85], the code $C^{\star}_{ptg,qc}$ was decoded using the SPA with 100 iterations. We see that the two codes perform basically the same above the BER of 10^{-7} decoded with 100 iteration of the SPA. Note that the protograph of the code $C^{\star}_{ptg,qc}$ was designed based on the decoding threshold criterion but the base matrix given by (4.7) is not. Surprisingly, it turns out that the two codes $C_{ptg,qc,1}$ and $C^{\star}_{ptg,qc}$ have the same decoding threshold and perform almost the same.

It has been shown in the literature (e.g., [101, 64, 65, 97]) that, if an LDPC code performs well over the AWGNC, it also performs well over the BEC. Fig. 4.3(b) shows that the $(2640,1320)$ code $C_{ptg,qc,1}$ indeed performs well over the BEC. At a unresolved erasure bit rate (UEBR) of 10^{-10} (an unresolved erasure block rate (UEBLR) of 10^{-9}), the code performs 0.14 from the Shannon limit (0.5) without a visible error-floor.

Example 4.3. In the previous example, a QC-PTG-LDPC code was constructed based on a binary base matrix, i.e., its associated Tanner graph contains no parallel edges. In this example, we construct a QC-PTG-LDPC code using a base matrix

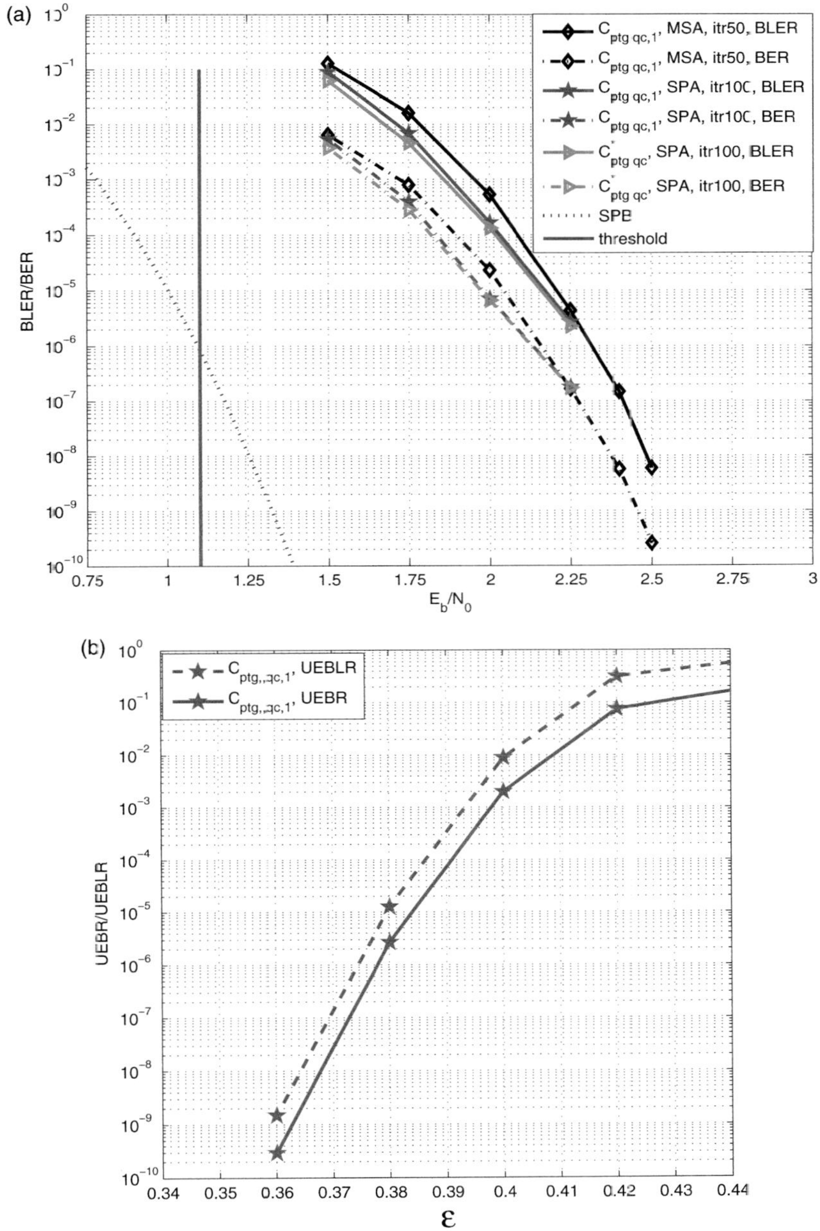

Figure 4.3. (a) The BER and BLER performances of the (2640, 1320) QC-PTG-LDPC code $C_{ptg,qc,1}$ given in Example 4.2 and the (2664, 1334) QC-PTG-LDPC code $C^{\star}_{ptg,qc}$ reported in [85] over the AWGNC; and (b) the performance of the (2640, 1320) QC-PTG-LDPC code $C_{ptg,qc,1}$ over the BEC.

whose Tanner graph contains parallel edges. The chosen base matrix is a 2×16 matrix whose entries are all equal to 2 as given below:

$$\mathbf{B}_{ptg,2} = \begin{bmatrix} 2 & 2 & 2 & 2 & 2 & 2 & 2 & 2 & 2 & 2 & 2 & 2 & 2 & 2 & 2 & 2 \\ 2 & 2 & 2 & 2 & 2 & 2 & 2 & 2 & 2 & 2 & 2 & 2 & 2 & 2 & 2 & 2 \end{bmatrix}. \quad (4.9)$$

Suppose we choose the decomposition factor $k = 511$ and decompose the base matrix $\mathbf{B}_{ptg,2}$ into 511 constituent matrices $\mathbf{D}_0, \mathbf{D}_1, \ldots, \mathbf{D}_{510}$ of size 2×16. In decomposing the base matrix $\mathbf{B}_{ptg,2}$, each 2-entry is split into two 1-entries which are put into two different constituent matrices. Furthermore, each nonzero constituent matrix contains only one 1-entry. Thus, among the 511 constituent matrices of $\mathbf{B}_{ptg,2}$, there are 64 nonzero matrices which satisfy both the RC- and the PW-RC-constraints. In forming the generator row-block of the block-cyclic array $\mathbf{H}_{ptg,cyc,2}(2,16)$ in the form of (4.1), we order the constituent matrices of $\mathbf{B}_{ptg,2}$ in such a way that $\mathbf{H}_{ptg,cyc,2}(2,16)$ satisfies the 2×2 array RC-constraint and hence $\mathbf{H}_{ptg,cyc,2}(2,16)$ satisfies the RC-constraint. Based on the decomposition of $\mathbf{B}_{ptg,2}$ and the order of the constituent matrices of the first row-block of $\mathbf{H}_{ptg,cyc,2}(2,16)$, we can construct a 2×16 array $\mathbf{H}_{ptg,qc,2}(511,511) = [\mathbf{A}_{i,j}]_{0 \le i < 2, 0 \le j < 16}$ of weight-2 circulants of size 511×511, which is a 1022×8176 matrix with constant column and row weights 4 and 32, respectively. The generators of the 32 circulants are given in Table 4.2, where the pair (a, b) denotes the locations of the two 1-components in the generator of each circulant in $\mathbf{H}_{ptg,qc,2}(511,511)$.

The Tanner graph $\mathcal{G}_{ptg,qc,2}(511,511)$ associated with $\mathbf{H}_{ptg,qc,2}(511,511)$ has girth 6 and contains 112,420 cycles of length 6. The connection number of each VN is 124. We see that $\mathcal{G}_{ptg,qc,2}(511,511)$ has a high degree of connectivity. The null space of $\mathbf{H}_{ptg,qc,2}(511,511)$ gives a (4, 32)-regular (8176, 7156) QC-PTG-LDPC code $C_{ptg,qc,2}$ with rate 0.875.

The BER and BLER performances of the code $C_{ptg,qc,2}$ decoded with 50, 10, and 5 iterations of the SPA over the AWGNC are shown in Fig. 4.4(a). We see that the decoding of the code converges very fast due to the high-degree connectivity of its Tanner graph. At a BER of 10^{-8}, the performance gap between 5 and 10 iterations is 0.4 dB and the gap between 10 and 50 iterations is about 0.2 dB. With 50 iterations, the code performs about 1.1 dB away from the Shannon limit at a BER of 10^{-8} and at a BLER of 10^{-6}, the code performs about 0.65 dB away from the SPB. The decoding threshold of $C_{ptg,qc,2}$ is 3.35 dB. At BERs of 10^{-5} and 10^{-8}, the code performs about 0.43 and 0.65 dBs from the decoding threshold, respectively, very close to the decoding threshold.

The QC-PTG-LDPC code $C_{ptg,qc,2}$ constructed above has the same length and the same rate as the NASA (CCSDS standard) (8176, 7156) QC-LDPC code C_{nasa}

[97, 87] for satellite and space communications. This code has been used in the Land Remote-Sensing Satellite (LANDSAT), the Interface Region Imaging Spectrograph (IRIS), and other NASA missions. The parity-check matrix of the NASA code C_{nasa} is also a 2×16 array of 511×511 circulants of weight 2. However, the compositions of the circulants of the parity-check arrays of the two codes are different as shown in Table 4.2. For comparison, the BER and BLER performances [97, Example 10.10, pp. 453–454] of the NASA code C_{nasa} decoded with 50 iterations of the SPA are also included in Fig. 4.4(a). We see that, in the simulation range, the two codes basically perform the same. Note that the NASA code C_{nasa} can reach down to a BER of 10^{-14} without a visible error-floor [97].

The error performance of the code $C_{ptg,qc,2}$ over the BEC is shown in Fig. 4.4(b). At a UEBR of 10^{-10}, it performs about 0.049 away from the Shannon limit (0.125) without a visible error-floor.

In the last two examples, we constructed two regular QC-PTG-LDPC codes using the proposed algebraic construction method. In the next example, we use the method to construct an *irregular* QC-PTG-LDPC code.

Example 4.4. Suppose it is desired to construct an irregular QC-PTG-LDPC code of length around 20 000 bits with rate close to 0.8. Applying density-evolution [96], we find that the following degree distributions [96, 97] of VNs and CNs of the Tanner graph give a good decoding threshold: $\gamma(X) = 0.1905X + 0.6074X^2 + 0.1466X^6 + 0.0555X^7$ and $\rho(X) = X^{18}$. Based on these two degree distributions, we construct a 12×63 base matrix $\mathbf{B}_{ptg,3}$ with only binary entries (i.e., the associated protograph has no parallel edges) whose row and column weight distributions are given in Table 4.3.

To achieve the desired code length of around 20 000 bits, we choose the decomposition factor $k = 330$. Decomposing the base matrix $\mathbf{B}_{ptg,3}$ into 330 constituent matrices, we construct a 12×63 array $\mathbf{H}_{ptg,qc,3}(330, 330)$ of CPMs and ZMs of size 330×330. The null space of $\mathbf{H}_{ptg,qc,3}(330, 330)$ gives an irregular $(20790, 16830)$ QC-PTG-LDPC code $C_{ptg,qc,3}$ with rate 0.8095. The Tanner graph of the code has girth 6 and contains 32 340 cycles of length 6. The BER and BLER performances of this code over the AWGNC decoded with 50 iterations of the MSA are shown in Fig. 4.5(a). The figure shows that the code $C_{ptg,qc,3}$ achieves a BER of 10^{-10} without a visible error-floor and performs about 1 dB away from the Shannon limit at a BER of 10^{-10}. At a BLER of 10^{-6}, it performs about 0.6 dB away from the SPB. The decoding threshold of this code is 2.66 dB. At BERs of 10^{-5} and 10^{-10}, the code performs 0.24 and 0.35 dBs away from its

Table 4.2. *The locations of 1-entries in the 32 circulants of parity-check matrices of the code $C_{ptg,qc,2}$ given in Example 4.3 and those for the NASA code C_{nasa} [97]*

Circulants	1's positions of generators	Circulants	1's positions of generators	Circulants (NASA)	1's positions of generators (NASA)	Circulants (NASA)	1's positions of generators (NASA)
$\mathbf{A}_{0,0}$	(209,284)	$\mathbf{A}_{1,0}$	(180,305)	$\mathbf{A}_{0,0}$	(0,176)	$\mathbf{A}_{1,0}$	(99,471)
$\mathbf{A}_{0,1}$	(45,469)	$\mathbf{A}_{1,1}$	(315,432)	$\mathbf{A}_{0,1}$	(12,239)	$\mathbf{A}_{1,1}$	(130,473)
$\mathbf{A}_{0,2}$	(111.482)	$\mathbf{A}_{1,2}$	(373,384)	$\mathbf{A}_{0,2}$	(0,352)	$\mathbf{A}_{1,2}$	(198,435)
$\mathbf{A}_{0,3}$	(55,317)	$\mathbf{A}_{1,3}$	(35,232)	$\mathbf{A}_{0,3}$	(24,431)	$\mathbf{A}_{1,3}$	(260,478)
$\mathbf{A}_{0,4}$	(466,467)	$\mathbf{A}_{1,4}$	(106,306)	$\mathbf{A}_{0,4}$	(0,392)	$\mathbf{A}_{1,4}$	(215,420)
$\mathbf{A}_{0,5}$	(149,229)	$\mathbf{A}_{1,5}$	(65,84)	$\mathbf{A}_{0,5}$	(151,409)	$\mathbf{A}_{1,5}$	(282,481)
$\mathbf{A}_{0,6}$	(93,364)	$\mathbf{A}_{1,6}$	(131,221)	$\mathbf{A}_{0,6}$	(0,351)	$\mathbf{A}_{1,6}$	(48,396)
$\mathbf{A}_{0,7}$	(330,337)	$\mathbf{A}_{1,7}$	(137,230)	$\mathbf{A}_{0,7}$	(9,359)	$\mathbf{A}_{1,7}$	(193,445)
$\mathbf{A}_{0,8}$	(162,219)	$\mathbf{A}_{1,8}$	(303,475)	$\mathbf{A}_{0,8}$	(0,307)	$\mathbf{A}_{1,8}$	(273,430)
$\mathbf{A}_{0,9}$	(244,314)	$\mathbf{A}_{1,9}$	(388,448)	$\mathbf{A}_{0,9}$	(53,329)	$\mathbf{A}_{1,9}$	(302,451)
$\mathbf{A}_{0,10}$	(406,441)	$\mathbf{A}_{1,10}$	(62,79)	$\mathbf{A}_{0,10}$	(0,207)	$\mathbf{A}_{1,10}$	(96,379)
$\mathbf{A}_{0,11}$	(273,419)	$\mathbf{A}_{1,11}$	(263,461)	$\mathbf{A}_{0,11}$	(18,281)	$\mathbf{A}_{1,11}$	(191,386)
$\mathbf{A}_{0,12}$	(86,133)	$\mathbf{A}_{1,12}$	(18,242)	$\mathbf{A}_{0,12}$	(0,399)	$\mathbf{A}_{1,12}$	(244,467)
$\mathbf{A}_{0,13}$	(16,359)	$\mathbf{A}_{1,13}$	(343,404)	$\mathbf{A}_{0,13}$	(202,457)	$\mathbf{A}_{1,13}$	(364,470)
$\mathbf{A}_{0,14}$	(190,268)	$\mathbf{A}_{1,14}$	(56,69)	$\mathbf{A}_{0,14}$	(0,247)	$\mathbf{A}_{1,14}$	(51,382)
$\mathbf{A}_{0,15}$	(324,338)	$\mathbf{A}_{1,15}$	(389,455)	$\mathbf{A}_{0,15}$	(36,261)	$\mathbf{A}_{1,15}$	(192,414)

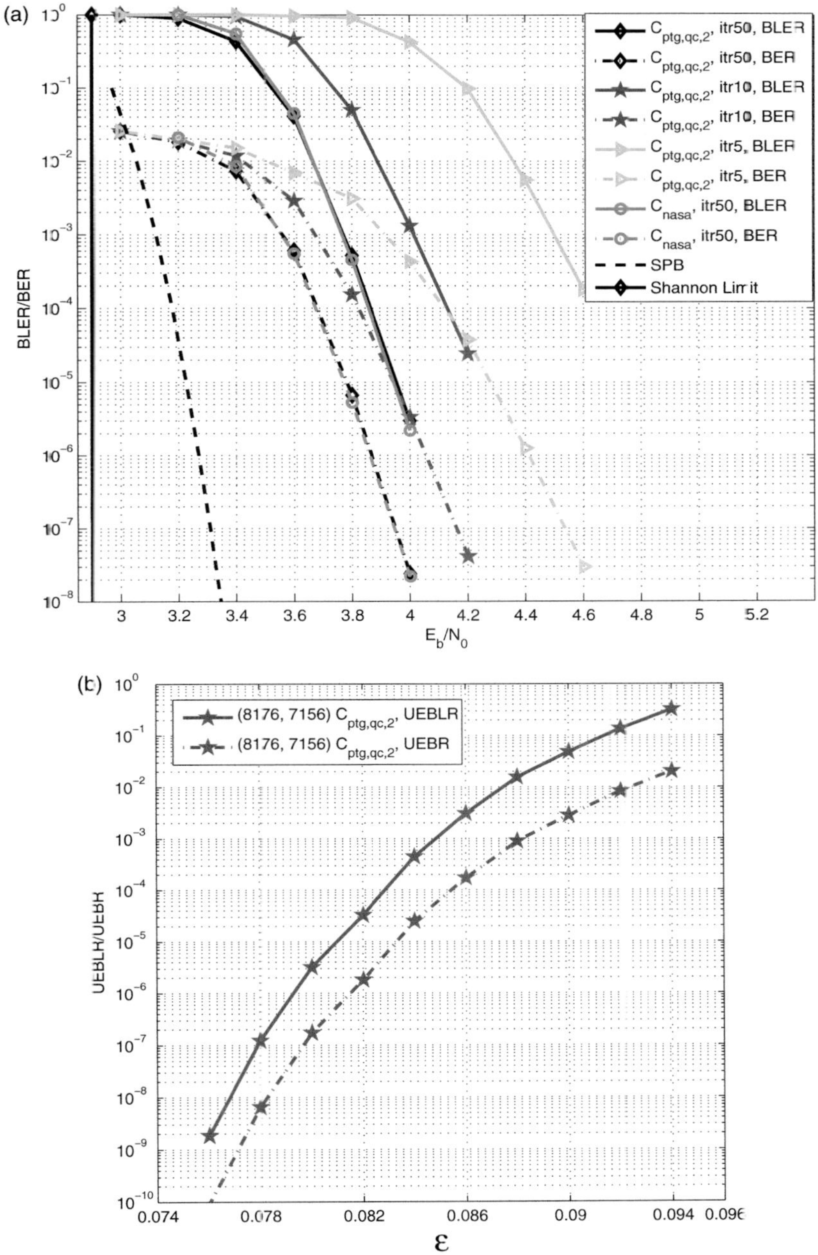

Figure 4.4. (a) The BER and BLER performances of the (8176, 7156) QC-PTG-LDPC code $C_{ptg,qc,2}$ decoded with 50, 10, and 5 iterations of the SPA given in Example 4.3 and the NASA code C_{nasa} [97, 87] over the AWGNC; and (b) the performance of the code $C_{ptg,qc,2}$ over the BEC.

Table 4.3. *Row and column weight distributions of the base matrix* $\mathbf{B}_{ptg,3}$ *given in Example 4.4*

Column weight distribution		Row weight distribution	
Column weight	No. of columns	Row weight	No. of rows
2	12	19	12
3	39		
7	9		
8	3		

decoding threshold, respectively. We see that the code performs very close to the decoding threshold.

For the BEC, the UEBR and UEBLR performances of the code $C_{ptg,qc,3}$ are shown in Fig. 4.5(b). We see that at a UEBR of 10^{-8}, the code performs only about 0.04 away from Shannon limit.

4.4 Construction of the Ensemble of PTG-LDPC Codes from an Algebraic Point of View

In this section, we create an ensemble of PTG-LDPC codes from an algebraic point of view.

For a given rate R_c, we choose the parameters m, n, and k such that $R_c = (n-m)/n$. Next, we choose, *at random*, an $m \times n$ matrix $\mathbf{B}_{ptg}$ over nonnegative integers. Then, we decompose this base matrix into an arbitrary number k of constituent matrices *randomly*, i.e., we split the integer entries in $\mathbf{B}_{ptg}$ that exceed one and distribute them among k constituent matrices randomly. This random decomposition of a base matrix results in an ensemble ξ_{cyc} of block-cyclic arrays in the form of (4.1), called a *block-cyclic ensemble*. Next, we permute the rows and columns of *each block-cyclic array* in the ensemble ξ_{cyc} based on the permutations π_{row} and π_{col} defined by (4.3) and (4.5) to obtain an ensemble ξ_{qc} of QC arrays in the form of (4.6), called a *QC ensemble*. Each constituent matrix in a QC array in the ensemble ξ_{qc} is either a $k \times k$ circulant or a $k \times k$ ZM. Note that the ensemble ξ_{cyc} is equivalent to the ensemble ξ_{qc}.

Now, we permute the rows and/or columns of *each* $k \times k$ nonzero constituent circulant of a QC array in the ensemble ξ_{qc} *randomly* to obtain a $k \times k$ regular matrix. These row and/or column permutations of *each* nonzero constituent circulant in *each* of the QC arrays in the ensemble ξ_{qc} result in an ensemble ξ_{ptg} of arrays of $k \times k$ regular matrices. The size of the ensemble ξ_{ptg} is much larger than that of the ensemble ξ_{qc} (or the ensemble ξ_{cyc}).

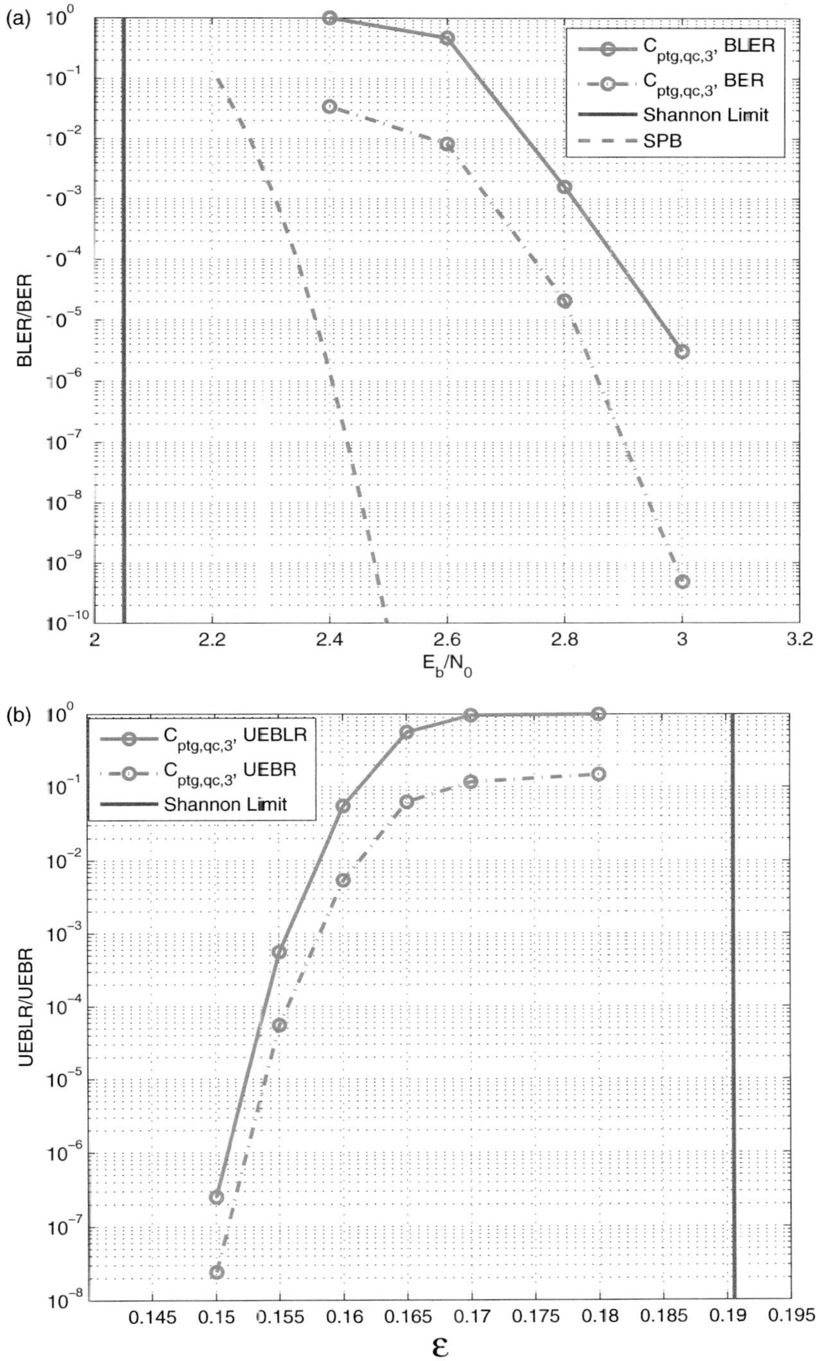

Figure 4.5. (a) The BER and BLER performances of the (20790, 16830) QC-PTG-LDPC code $C_{ptg,qc,3}$ given in Example 4.4 over the AWGNC; and (b) the performance of the code in (a) over the BEC.

The null spaces of the arrays in the ensemble ξ_{ptg} give an ensemble $\mathbf{E}_{ptg}$ of PTG-LDPC codes with rate R_c. This ensemble $\mathbf{E}_{ptg}$ of PTG-LDPC codes is equivalent to the ensemble of PTG-LDPC codes investigated in [34] from the graphical point of view, where its asymptotic properties were analyzed.

4.5 Discussion and Remarks

In this chapter, we presented an algebraic method for constructing binary QC-PTG-LDPC codes. The construction is based on the decomposition of a base matrix over nonnegative integers. The matrix decomposition process seems more convenient and flexible than the copy-and-permute process of a protograph. Examples showed that a well-designed base matrix gives a QC-PTG-LDPC code that performs very close to the decoding threshold. How to design good base matrices algebraically is an interesting research problem. This problem will be further discussed in Chapters 5, 7, and 11. Even though we have provided some general guidelines and constraints for decomposing a base matrix to avoid short cycles, a more specific method or algorithm is needed. Finding such a method may be a good research problem. How to order the constituent matrices in an RC-constrained decomposition set of a base matrix to form an RC-constrained parity-check array is also an interesting problem to investigate.

We showed that a QC-PTG-LDPC code constructed by decomposing a base matrix can be put either in a block-cyclic form or in a section-wise cyclic form. In a block-cyclic form, the code can be decoded with the iterative BP-decoding scheme proposed in [79] to reduce the decoder complexity significantly by using a small submatrix of the parity-check matrix of the code as the decoding matrix. This iterative BP-decoding scheme not only significantly reduces the number of CN-MPUs (and/or VN-MPUs) and the number of wires for message exchanges between the CN-MPUs and the VN-MPUs but also greatly reduces the size of memory required for storing messages for exchanges between the CN-MPUs and the VN-MPUs. In a section-wise cyclic form, we can use the reduced-complexity iterative BP-decoding scheme proposed in [69, 68, 78] to decode the code using a decoding matrix much smaller than the parity-check matrix of the code.

In two later chapters, we show that we can generalize base matrices over nonnegative integers for the algebraic construction of PTG-LDPC codes to base matrices over finite fields and matrices with set-entries over finite fields. This is used for constructing both binary and NB QC-PTG-LDPC codes with a matrix decomposition method similar to the method presented in this chapter.

5

Superposition Construction of LDPC Codes

The SP-construction, proposed in 2002 [76], is one of the earliest algebraic-based methods for constructing LDPC codes, especially QC-LDPC codes. In this chapter, we interpret this construction in a *broader way*, algebraically and graph-theoretically, and show that the PTG-based method of LDPC code construction is a *special case* of the SP-construction method.

Even though the PTG-construction of LDPC codes is a special case of the SP-construction of LDPC codes, it should be pointed out that design techniques for these two classes of codes have been different over the years since their discoveries in 2003 [105] and 2002 [76], respectively. Over the years, SP-LDPC codes and their related codes have employed algebraic approaches in their design with the goal of very low decoding error-floors and highly structured (lower-complexity) decoders. PTG-LDPC codes have instead been designed using ensemble approaches, with good ensembles found via density evolution or EXIT charts. The goal in this case is an optimum decoding threshold (waterfall region). Whether or not these ensemble approaches can be extended to general SP-LDPC codes will require further research.

5.1 SP-Construction of LDPC Codes and Its Graphical Interpretation

Let $\mathbf{B}_{sp} = [b_{i,j}]_{0 \leq i < m, 0 \leq j < n}$ be a small $m \times n$ binary matrix. Let $\boldsymbol{R} = \{\mathbf{A}_0, \mathbf{A}_1, \ldots, \mathbf{A}_{r-1}\}$ be a set of r $k \times t$ sparse matrices over GF(2). From $\mathbf{B}_{sp}$ and the member matrices of $\boldsymbol{R}$, we construct an $m \times n$ array $\mathbf{H}_{sp}(k,t)$ of $k \times t$ matrices by replacing each 1-entry in $\mathbf{B}_{sp}$ by a member matrix of $\boldsymbol{R}$ and each 0-entry by a $k \times t$ ZM. This replacement of entries in $\mathbf{B}_{sp}$ by the member matrices in $\boldsymbol{R}$ and/or ZMs expands the matrix $\mathbf{B}_{sp}$ into an $mk \times nt$ sparse matrix $\mathbf{H}_{sp}(k,t)$. As an $m \times n$ array, $\mathbf{H}_{sp}(k,t)$ consists of n column-blocks, denoted by $\mathbf{H}_0, \mathbf{H}_1, \ldots, \mathbf{H}_{n-1}$, and m row-blocks, denoted by $\mathbf{M}_0, \mathbf{M}_1, \ldots, \mathbf{M}_{m-1}$. Each column-block $\mathbf{H}_j$ consists of t consecutive columns of $\mathbf{H}_{sp}(k,t)$ and each row-block $\mathbf{M}_i$ consists of k consecutive rows of $\mathbf{H}_{sp}(k,t)$. The above operation of replacing the entries in $\mathbf{B}_{sp}$ by the member matrices in $\boldsymbol{R}$ and/or ZMs was called *SP-operation* in [76, 110, 109, 111, 97]. The SP-operation is simply an *expansion operation*. The null space of $\mathbf{H}_{sp}(k,t)$ gives an LDPC code C_{sp}, called a *superposition LDPC code* (or SP-LDPC code) of length nt. The rate R_c of the SP-LDPC code C_{sp} is at least $(nt - mk)/nt$ and depends on the parameters n, m, k, and t. For a given rate R_c, there are many possible choices of n, m, k, and t. In the SP-construction of an LDPC code, the matrix $\mathbf{B}_{sp}$ is called the *base matrix* (or *SP-base matrix*) and $\boldsymbol{R}$ is referred to as the *replacement set*.

Let $\mathcal{G}_{sp}$ be the Tanner graph associated with the SP-base matrix $\mathbf{B}_{sp}$, that consists of n VNs and m CNs, labeled by $v_0, v_1, \ldots, v_{n-1}$ and $c_0, c_1, \ldots, c_{m-1}$, respectively. The Tanner graph associated with the matrix $\mathbf{H}_{sp}(k,t)$, denoted by $\mathcal{G}_{sp}(k,t)$, consists of nt VNs and mk CNs. Now, we partition the nt VNs of $\mathcal{G}_{sp}(k,t)$ into n disjoint sets, denoted by $\Phi_0, \Phi_1, \ldots, \Phi_{n-1}$. For $0 \le j < n$, Φ_j consists of t VNs of $\mathcal{G}_{sp}(k,t)$, which correspond to the t columns of the j-th column-block $\mathbf{H}_j$ of $\mathbf{H}_{sp}(k,t)$. The t VNs in Φ_j are referred to as the type-j VNs of $\mathcal{G}_{sp}(k,t)$. The t type-j VNs in Φ_j are simply t copies of the VN v_j of the Tanner graph $\mathcal{G}_{sp}$ associated with the SP-base matrix $\mathbf{B}_{sp}$. Next, we partition the mk CNs of $\mathcal{G}_{sp}(k,t)$ into m disjoint sets, denoted by $\Omega_0, \Omega_1, \ldots, \Omega_{m-1}$. For $0 \le i < m$, Ω_i consists of k CNs of $\mathcal{G}_{sp}(k,t)$, which correspond to the k rows of the i-th row-block $\mathbf{M}_i$ of $\mathbf{H}_{sp}(k,t)$. The k CNs in Ω_i are referred to as the type-i CNs of $\mathcal{G}_{sp}(k,t)$. The k CNs in Ω_i are simply k copies of CN c_i of the Tanner graph $\mathcal{G}_{sp}$.

The SP-operation simply expands each VN in $\mathcal{G}_{sp}$ into t VNs in $\mathcal{G}_{sp}(k,t)$ and each CN in $\mathcal{G}_{sp}$ into k CNs in $\mathcal{G}_{sp}(k,t)$. In $\mathcal{G}_{sp}(k,t)$, the VNs in $\Phi_0, \Phi_1, \ldots, \Phi_{n-1}$ are connected to the CNs in $\Omega_0, \Omega_1, \ldots, \Omega_{m-1}$ based on the 1-entries in the member matrices of $\boldsymbol{R}$ that are used to replace the nonzero entries in $\mathbf{B}_{sp}$. The graph $\mathcal{G}_{sp}(k,t)$ may be viewed as a Tanner graph consisting of n super-VNs, $\Phi_0, \Phi_1, \ldots, \Phi_{n-1}$, and m super-CNs, $\Omega_0, \Omega_1, \ldots, \Omega_{m-1}$, as shown in Fig. 5.1. The t type-j VNs in Φ_j can only connect to the k type-i CNs in Ω_i if there is an edge (j,i) in $\mathcal{G}_{sp}$. The graph $\mathcal{G}_{sp}(k,t)$ is an *expansion* of the Tanner graph $\mathcal{G}_{sp}$ of the SP-base matrix $\mathbf{B}_{sp}$. In this expansion, if $t \neq k$, the VNs and CNs of $\mathcal{G}_{sp}$ are expanded by *two different factors* (or degrees), t and k, respectively, (*non-uniform expansion*). Therefore, from the graphical point of view, the SP-construction of the Tanner graph of an SP-LDPC code is carried out using the following three steps: (1) Design a small bipartite graph $\mathcal{G}_{sp}$ with n VNs and m CNs; (2) Duplicate each VN v_j of $\mathcal{G}_{sp}$ t times to form a set Φ_j of t type-j VNs and duplicate each CN c_i of $\mathcal{G}_{sp}$ k times to form a set Ω_i of k type-i CNs; and (3) Connect the type-j VNs and the type-i CNs by the type-(j,i) edges using edge permutations. Step 3 results

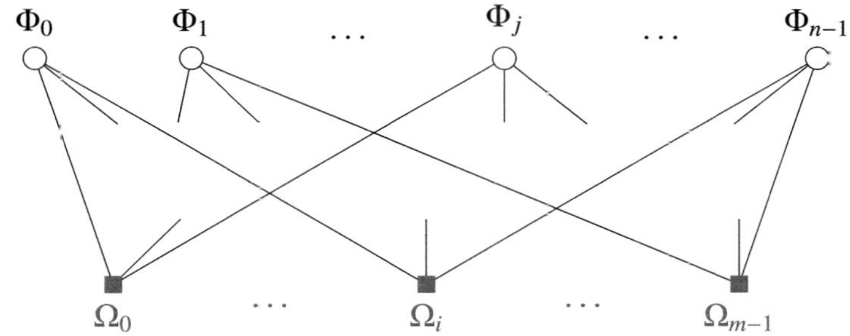

Figure 5.1. The Tanner graph $\mathcal{G}_{sp}(k,t)$ with super-VNs and super-CNs.

in an expanded and connected bipartite graph $\mathcal{G}_{sp}(k,t)$. Then, the null space of the adjacency matrix $\mathbf{H}_{sp}(k,t)$ of the bipartite graph $\mathcal{G}_{sp}(k,t)$ gives an SP-LDPC code. The connection of the t type-j VNs to the k type-i CNs can be specified by a $k \times t$ matrix, denoted by $\mathbf{A}_{i,j}$ and called a *connection matrix*. From the algebraic SP point of view, the connection matrix $\mathbf{A}_{i,j}$ is simply a member matrix of the replacement set $\boldsymbol{R}$.

5.2 Ensembles of SP-LDPC Codes

For a given rate R_c, we choose the parameters m, n, k, and t such that $R_c = (nt - mk)/nt$. Suppose we choose *at random* both an $m \times n$ SP-base matrix $\mathbf{B}_{sp}$ and a replacement set $\boldsymbol{R}$ of $k \times t$ sparse constituent matrices over GF(2) (i.e., the constituent matrices in $\boldsymbol{R}$ are chosen at random). Random replacements of the 1-entries of the SP-base matrix $\mathbf{B}_{sp}$ by the constituent matrices in $\boldsymbol{R}$ would result in an ensemble, denoted by $\xi_{sp}(k,t)$, of sparse arrays. This ensemble $\xi_{sp}(k,t)$ of sparse arrays gives an ensemble, denoted by $\mathbf{E}_{sp}(k,t)$, of SP-LDPC codes of rate R_c. The ensemble $\mathbf{E}_{sp}(k,t)$ of SP-LDPC codes of rate R_c defined above contains two major subensembles with different structures, one for the case $k = t$ and the other for the case $k \neq t$.

Consider the case for which $k = t$ and the member matrices in $\boldsymbol{R}$ are $k \times k$ regular matrices. For this case, from the graphical point of view presented above, the SP-operation duplicates each VN, each CN, and each edge of the associated Tanner graph $\mathcal{G}_{sp}$ of the SP-base matrix $\mathbf{B}_{sp}$ k times (*uniform expansion*). This amounts to duplicating the graph $\mathcal{G}_{sp}$ into k copies. These k copies of $\mathcal{G}_{sp}$ are then connected by permuting their edges based on the member matrices in $\boldsymbol{R}$ that are used to replace the 1-entries in $\mathbf{B}_{sp}$. The connection of the k copies of $\mathcal{G}_{sp}$ results in the Tanner graph $\mathcal{G}_{sp}(k,k)$ associated with the $m \times n$ array $\mathbf{H}_{sp}(k,k)$ which is obtained by replacing the 1-entries of the SP-base matrix $\mathbf{B}_{sp}$ by the $k \times k$ regular member matrices in the replacement set $\boldsymbol{R}$. This shows that, for the case with $k = t$, an SP-LDPC code is a PTG-LDPC code (or vice versa, a PTG-LDPC code is an SP-LDPC code). Hence, the ensemble $\mathbf{E}_{sp}(k,k)$ of SP-LDPC codes of rate R_c is an ensemble of PTG-LDPC codes of rate R_c.

From the algebraic construction of the ensemble $\mathbf{E}_{ptg}$ of PTG-LDPC codes given in Section 4.4 and the construction of the ensemble $\mathbf{E}_{sp}(k,t)$ of SP-LDPC codes given above, we readily see that the ensemble $\mathbf{E}_{ptg}$ of PTG-LDPC codes is equivalent to the *special ensemble* $\mathbf{E}_{sp}(k,k)$ of SP-LDPC codes. This, in turn, implies that the ensemble $\mathbf{E}_{sp}(k,k)$ of SP-LDPC codes is equivalent to the ensemble of PTG-LDPC codes investigated in [34] from the graphical point of view. Therefore, for a properly designed SP-base matrix, the ensemble $\mathbf{E}_{sp}(k,k)$ is asymptotically good.

Based on the above interpretations of PTG-LDPC and SP-LDPC codes from both the graph-theoretic and the algebraic (or matrix-theoretic) points of view and the constructions of their ensembles, we see that the PTG-LDPC code construction is a *special case* of the SP-LDPC code construction. The *unification* of the SP and the PTG-based constructions from both the algebraic and the graph-theoretic points of view may lead to better designs and constructions of LDPC codes with *good overall performance* in terms of the error performance in the waterfall region, the error-floor location, and the rate of decoding convergence.

For the case $k \neq t$, the member matrices in the replacement set $\boldsymbol{R}$ for the SP-construction of LDPC codes are not *regular*. In this case, the VNs and the CNs in the Tanner graph of an SP-LDPC code are expanded by two different factors, one by t and the other by k, i.e., the expansion of the Tanner graph is *not uniform*. This non-uniform expansion defines another ensemble, denoted by $\mathbf{E}'_{sp}(k,t)$, of the ensemble $\mathbf{E}_{sp}(k,t)$ of SP-LDPC codes. An analysis of the ensemble $\mathbf{E}'_{sp}(k,t)$ could be difficult, but it is a good research problem. In the following chapters, QC-SP-LDPC codes with $k \neq t$ are constructed and they perform well over both the AWGNC and the BEC. This may be an indication that the ensemble $\mathbf{E}'_{sp}(k,t)$ is also a good ensemble.

5.3 Constraints on the Construction of SP-LDPC Codes Free of Cycles of Length 4

In designing an SP-LDPC code to avoid short cycles in its Tanner graph, especially cycles of length 4, certain constraints must be imposed on the member matrices of the replacement set $\boldsymbol{R}$ and the method of replacing the nonzero entries in the SP-base matrix $\mathbf{B}_{sp}$ by the member matrices of $\boldsymbol{R}$. First, the member matrices in $\boldsymbol{R}$ must be designed to satisfy both the RC- and the PW-RC-constraints. Second, in the replacement process, all the 1-entries in a row and a column of $\mathbf{B}_{sp}$ must be replaced by *distinct* member matrices in $\boldsymbol{R}$ (unless they are CPMs).

This replacement rule is called the *replacement constraint* [76, 110, 109, 111, 97]. The RC- and the PW-RC-constraints on the member matrices of the replacement set $\boldsymbol{R}$ and the constraint on the replacement of the nonzero entries of the SP-base matrix $\mathbf{B}_{sp}$ by the member matrices of $\boldsymbol{R}$ are necessary but not sufficient to guarantee that the Tanner graph of the constructed SP-LDPC code is free of cycles of length 4, i.e., its girth is at least 6. However, if the base matrix $\mathbf{B}_{sp}$ also satisfies the RC-constraint, then the SP-construction under the replacement constraint will result in an SP-LDPC code whose the Tanner graph has girth at least 6 [76, 110, 109, 111, 97].

If the base matrix $\mathbf{B}_{sp}$ is a weight-w $n \times n$ circulant, the SP replacement constraint can be met easily. First, we replace the w 1-entries in the top row of $\mathbf{B}_{sp}$ by w distinct member matrices in $\boldsymbol{R}$. This replacement results in a row-block (generator row-block) that consists of n submatrices of size $k \times t$, including w nonzero matrices and $n - w$ ZMs. Then, we cyclically shift this generator row-block $n - 1$ times, each time one submatrix to the right. This results in an $n \times n$ array $\mathbf{H}_{sp}(k, t)$ with n row-blocks and n column-blocks. The w nonzero matrices (member matrices in $\boldsymbol{R}$) in each row-block (or column-block) are distinct. Therefore, the replacement constraint is satisfied. The above replacement is called *cyclic replacement*.

A replacement set $\boldsymbol{R}$ whose member matrices satisfy both the RC- and the PW-RC-constraints is called an *RC-constrained replacement set*. Constructions of RC-constrained SP-base matrices and RC-constrained replacement sets will be discussed further in Chapter 6.

From the above discussion, we see that the SP-construction is a very general algebraic construction of LDPC codes. It can be interpreted from either the matrix-theoretic point of view or the graph-theoretic point of view. From the perspective of simplicity, the matrix-theoretic approach to the SP-construction of LDPC codes is preferable in terms of designs of base matrices and replacement sets, as will be shown in later chapters.

In the following, we give a simple example to illustrate the SP-construction of an LDPC code and its graphical structure.

Example 5.1. Consider the following base matrix $\mathbf{B}_{sp}$ and the replacement set $\boldsymbol{R}$ for constructing an SP-LDPC code:

$$\mathbf{B}_{sp} = \begin{bmatrix} 1 & 0 & 1 \\ 1 & 1 & 0 \\ 0 & 1 & 1 \end{bmatrix},$$

$$\boldsymbol{R} = \left\{ \mathbf{A}_0 = \begin{bmatrix} 1 & 0 & 1 \\ 0 & 1 & 0 \end{bmatrix}, \mathbf{A}_1 = \begin{bmatrix} 1 & 0 & 0 \\ 0 & 0 & 1 \end{bmatrix} \right\}.$$

We see that the base matrix $\mathbf{B}_{sp}$ satisfies the RC-constraint and the two constituent matrices in $\boldsymbol{R}$ satisfy the RC- and the PW-RC-constraints. Note that $\mathbf{B}_{sp}$ is a circulant. Next, we replace the 1-entries and 0-entries of the base matrix $\mathbf{B}_{sp}$ by the 2×3 constituent matrices in $\boldsymbol{R}$ and 2×3 ZMs, respectively. The replacement results in a 3×3 array $\mathbf{H}_{sp}(2,3)$ of 2×3 matrices. Under the replacement constraint, there are several possible replacements. One such a replacement (a cyclic replacement as described above) gives the following array:

$$\mathbf{H}_{sp}(2,3) = \begin{bmatrix} \mathbf{A}_0 & \mathbf{0} & \mathbf{A}_1 \\ \mathbf{A}_1 & \mathbf{A}_0 & \mathbf{0} \\ \mathbf{0} & \mathbf{A}_1 & \mathbf{A}_0 \end{bmatrix} = \left[\begin{array}{ccc|ccc|ccc} 1 & 0 & 1 & 0 & 0 & 0 & 1 & 0 & 0 \\ 0 & 1 & 0 & 0 & 0 & 0 & 0 & 0 & 1 \\ \hline 1 & 0 & 0 & 1 & 0 & 1 & 0 & 0 & 0 \\ 0 & 0 & 1 & 0 & 1 & 0 & 0 & 0 & 0 \\ \hline 0 & 0 & 0 & 1 & 0 & 0 & 1 & 0 & 1 \\ 0 & 0 & 0 & 0 & 0 & 1 & 0 & 1 & 0 \end{array}\right].$$

It is easy to check that the array $\mathbf{H}_{sp}(2,3)$, a 6×9 matrix over GF(2), satisfies the RC-constraint. The null space of $\mathbf{H}_{sp}(2,3)$ gives a $(9,3)$ QC-SP-LDPC code C_{sp} in block-cyclic form. The Tanner graph $\mathcal{G}_{sp}$ associated with the base matrix $\mathbf{B}_{sp}$ and the Tanner graph $\mathcal{G}_{sp}(2,3)$ associated with the array $\mathbf{H}_{sp}(2,3)$ are shown in Fig. 5.2(a) and Fig. 5.2(b), respectively. We see that $\mathcal{G}_{sp}(2,3)$ is an expansion of $\mathcal{G}_{sp}$. In $\mathcal{G}_{sp}$, each VN is expanded by a factor of 3 and each CN is expanded by a factor of 2. The Tanner graph $\mathcal{G}_{sp}$ is non-uniformly expanded and their VNs and CNs are connected based on the two constituent matrices $\mathbf{A}_0$ and $\mathbf{A}_1$ of $\boldsymbol{R}$ serving as the connection matrices. Both $\mathcal{G}_{sp}$ and $\mathcal{G}_{sp}(2,3)$ have girth 6.

Now suppose we add (integer sum) the three constituent matrices of the generator row-block of the 3×3 array $\mathbf{H}_{sp}(2,3)$. We obtain the following 2×3 matrix over nonnegative integers:

$$\mathbf{B}_{ptg} = \begin{bmatrix} b_{0,0} & b_{0,1} & b_{0,2} \\ b_{1,0} & b_{1,1} & b_{1,2} \end{bmatrix} = \begin{bmatrix} 2 & 0 & 1 \\ 0 & 1 & 1 \end{bmatrix}.$$

The matrix $\mathbf{B}_{ptg}$ is simply the base matrix of the protograph $\mathcal{G}_{ptg}$ shown in Fig. 5.2(c). From the generator row-block of $\mathbf{H}_{sp}(2,3)$, we find the generators of the six circulants of size 3×3 as described in Section 4.1, which are:

$$\mathbf{g}_{0,0} = (1,0,1), \qquad \mathbf{g}_{0,1} = (0,0,0), \qquad \mathbf{g}_{0,2} = (1,0,0),$$
$$\mathbf{g}_{1,0} = (0,0,0), \qquad \mathbf{g}_{1,1} = (1,0,0), \qquad \mathbf{g}_{1,2} = (0,0,1).$$

For $0 \le i < 2$ and $0 \le j < 3$, we replace the entry $b_{i,j}$ by the circulant $\mathbf{A}_{i,j}$ generated by $\mathbf{g}_{i,j}$. This replacement results in the following 2×3 array of circulants

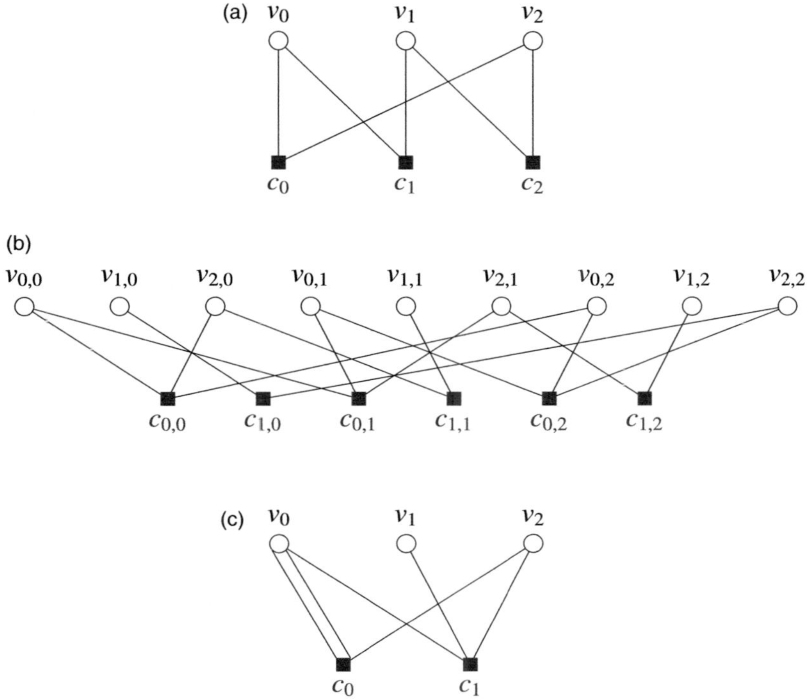

Figure 5.2. (a) The Tanner graph $\mathcal{G}_{sp}$ associated with the base matrix $\mathbf{B}_{sp}$ given in Example 5.1; (b) the Tanner graph $\mathcal{G}_{sp}(2,3)$ associated with the array $\mathbf{H}_{sp}(2,3)$ in Example 5.1; and (c) the protograph $\mathcal{G}_{ptg}$ associated with the base matrix $\mathbf{B}_{ptg}$ given in Example 5.1.

and ZMs of size 3×3:

$$\mathbf{H}_{ptg,qc}(3,3) = \left[\begin{array}{ccc|ccc|ccc} 1 & 0 & 1 & 0 & 0 & 0 & 1 & 0 & 0 \\ 1 & 1 & 0 & 0 & 0 & 0 & 0 & 1 & 0 \\ 0 & 1 & 1 & 0 & 0 & 0 & 0 & 0 & 1 \\ \hline 0 & 0 & 0 & 1 & 0 & 0 & 0 & 0 & 1 \\ 0 & 0 & 0 & 0 & 1 & 0 & 1 & 0 & 0 \\ 0 & 0 & 0 & 0 & 0 & 1 & 0 & 1 & 0 \end{array}\right].$$

The array $\mathbf{H}_{ptg,qc}(3,3)$ is the adjacency matrix of the Tanner graph $\mathcal{G}_{ptg}(3,3)$ obtained by expanding the protograph $\mathcal{G}_{ptg}$ by a factor of 3. Note that $b_{0,0} = 2$, and it is replaced by a weight-2 circulant. The null space of $\mathbf{H}_{ptg,qc}(3,3)$ gives a $(9,6)$ QC-PTG-LDPC code $C_{ptg,qc}$ with a section-wise cyclic structure. Here, we see that the QC-PTG-LDPC code $C_{ptg,qc}$ is constructed algebraically using the SP-construction and is an SP-LDPC code.

Note that the 2×3 array $\mathbf{H}_{ptg,qc}(3,3)$ can also be obtained from the 3×3 array $\mathbf{H}_{sp}(2,3)$ by permuting the rows and columns using permutations similar to π_{row} and π_{col} defined by (4.3) and (4.5). The Tanner graphs $\mathcal{G}_{sp}(2,3)$ and $\mathcal{G}_{ptg}(3,3)$ are structurally identical, i.e., one can be obtained from the other by permuting the VNs and the CNs. The above shows again the connection between a PTG-LDPC code and an SP-LDPC code.

5.4 SP-Construction of QC-LDPC Codes

Suppose the member matrices of the replacement set $\boldsymbol{R}$ are sparse binary circulants of the same size $k \times k$, not necessarily of the same weight. Replacing the 1-entries in the binary $m \times n$ SP-base matrix $\mathbf{B}_{sp}$ by the member circulants in $\boldsymbol{R}$ and the 0-entries in $\mathbf{B}_{sp}$ by ZMs of size $k \times k$, we obtain an $m \times n$ array $\mathbf{H}_{sp,qc}(k,k) = [\mathbf{A}_{i,j}]_{0 \leq i < m, 0 \leq j < n}$ of circulants and/or ZMs of size $k \times k$ in the form given by (4.6). The null space of $\mathbf{H}_{sp,qc}(k,k)$ gives a binary QC-SP-LDPC code $C_{sp,qc}$ of length nk. Typically, in binary QC-SP-LDPC code construction, the member matrices in $\boldsymbol{R}$ are binary weight-1 circulants (i.e., CPMs) of size $k \times k$.

Using the inverses of the row and column permutations defined by (4.3) and (4.5), we can put the array $\mathbf{H}_{sp,qc}(k,k)$ into a $k \times k$ array $\mathbf{H}_{sp,cyc}(m,n) = [\mathbf{D}_{i,j}]_{0 \leq i,j, < k}$ of $m \times n$ matrices in the form given in (4.1) with a block-cyclic structure. The integer sum of the constituent matrices of the generator row-block (or any row- or column-block) of $\mathbf{H}_{sp,cyc}(m,n)$ gives the base matrix $\mathbf{B}_{ptg}$ of the protograph $\mathcal{G}_{ptg}$ for the PTG-construction of the QC-PTG-LDPC code $C_{ptg,qc}$ (or $C_{ptg,cyc}$). This shows that the special case of the SP-construction with $k = t$ and the replacement set $\boldsymbol{R}$ being a set of circulants of size $k \times k$ is exactly the algebraic construction of a QC-PTG-LDPC code presented in Chapter 4.

Alternatively, if we combine the k constituent matrices of the top row-block of the array $\mathbf{H}_{sp,cyc}(m,n)$ with corresponding entries added using the logic-OR operation, we obtain an $m \times n$ binary matrix which is simply the SP-base matrix $\mathbf{B}_{sp}$ for the SP-construction of the QC-SP-LDPC code $C_{sp,qc}$.

Example 5.2. In this example, we show that, from the the SP-construction point of view, the $(2640, 1320)$ QC-PTG-LDPC code given in Example 4.2 is a QC-SP-LDPC code. The matrix used for constructing the QC-PTG-LDPC code given by (4.7) is now taken as the SP-base matrix $\mathbf{B}_{sp}$. The replacement set $\boldsymbol{R}$ consists of a set of binary CPMs of size 330×330. In this set, there are 23 CPMs whose generators are specified by the locations of their single 1-entries which are 10, 39, 68, 75, 79, 143, 148, 158, 163, 181, 202, 206, 209, 271, 274, 275, 284, 287, 294, 295, 297, 324, and 328. Replacing the 24 1-entries in the base matrix $\mathbf{B}_{sp}$ by these CPMs in the pattern shown in (4.8), we obtain a

4×8 array $\mathbf{H}_{sp,qc}(330, 330)$ of CPMs and ZMs of size 330×330. The null space of $\mathbf{H}_{sp,qc}(330, 330)$ gives a $(2640, 1320)$ QC-SP-LDPC code $C_{sp,qc}$ which is exactly the $(2640, 1320)$ QC-PTG-LDPC code $C_{ptg,qc,1}$ given in Example 4.2.

Example 5.3. From the SP-construction point of view, the $(8176, 7156)$ QC-PTG-LDPC code given in Example 4.3 is a QC-SP-LDPC code with the following base matrix:

$$\mathbf{B}_{sp} = \begin{bmatrix} 1 & 1 & 1 & 1 & 1 & 1 & 1 & 1 & 1 & 1 & 1 & 1 & 1 & 1 & 1 & 1 \\ 1 & 1 & 1 & 1 & 1 & 1 & 1 & 1 & 1 & 1 & 1 & 1 & 1 & 1 & 1 & 1 \end{bmatrix}. \tag{5.1}$$

The 32 weight-2 circulants of size 511×511 whose generators are given in Table 4.2 form the replacement set $\boldsymbol{R}$. The $(8176, 7156)$ NASA-CCSDS standard QC-LDPC code C_{nasa} cited in Example 4.3 is also a QC-SP-LDPC code, with the matrix given by (5.1) as its SP-base matrix $\mathbf{B}_{sp}$. The replacement set $\boldsymbol{R}$ for constructing this code consists of 32 weight-2 circulants given in Table 4.2 and labeled with NASA, which are constructed based on the 3-dimensional Euclidean geometry EG$(3, 2^3)$ over GF(2^3) [97].

Note that, in constructing a QC-SP-LDPC code, each member matrix in the replacement set $\boldsymbol{R}$ can be a $b \times c$ array of CPMs of the same size. In this case, the replacement of the 1-entries in the SP-base matrix by member arrays in $\boldsymbol{R}$ results in a parity-check array of CPMs and ZMs of the same size. Constructions of replacement sets whose member matrices are arrays of CPMs and/or ZMs will be discussed in Chapters 6–11.

5.5 SP-Base Matrices over Nonnegative Integers

The base matrix $\mathbf{B}_{sp}$ for the SP-construction of LDPC codes can be a matrix over nonnegative integers. Similar to the PTG-based construction, a nonzero entry represents the number of parallel edges connecting a VN and a CN in its associated Tanner graph. In this case, the constituent matrices in the replacement set $\boldsymbol{R}$ are designed (or constructed) as regular matrices with weights in agreement with integer entries in $\mathbf{B}_{sp}$. In the construction of a parity-check array, an integer entry w in the SP-base matrix $\mathbf{B}_{sp}$ is replaced by a member matrix in $\boldsymbol{R}$ with weight w. As an example, suppose we take the base matrix given by (4.9) as the base matrix for the SP-construction of a QC-SP-LDPC code. Since all of its entries are 2s, then the member matrices of the replacement set $\boldsymbol{R}$ are circulants with weight-2. Therefore, the $(4, 32)$-regular $(8176, 7156)$ QC-PTG-LDPC code given in Example 4.3 (also the NASA CCSDS standard $(8176, 7156)$ code C_{nasa}) may be regarded as a QC-SP-LDPC code whose SP-base matrix contains only 2-entries.

In Chapter 7, we will present methods for constructing SP-LDPC codes using base matrices over NB fields, which are not considered in the conventional PTG-based construction. So the base matrix for constructing an SP-LDPC code can be a binary matrix, a matrix over nonnegative integers, or a matrix over a finite field.

5.6 Discussion and Remarks

As presented in Section 5.2, two major ensembles of SP-LDPC codes for a given rate can be formed. One ensemble is equivalent to the ensemble of PTG-LDPC codes. In forming this ensemble, the member matrices in the replacement set $\boldsymbol{R}$ for the SP-construction of LDPC codes are regular. Since this ensemble is equivalent to the ensemble of PTG-LDPC codes, SP-LDPC codes in this ensemble have asymptotically good performance and structural properties. For the other ensemble, the member matrices in the replacement set $\boldsymbol{R}$ for the SP-construction of LDPC codes are not regular. Whether this ensemble of SP-LDPC codes has good asymptotic properties and performance is unknown at this point. However, some examples given in later chapters show that SP-LDPC codes constructed using non-regular matrices for replacement do perform very well. This may be an indication that this ensemble is a good ensemble. An analysis of this ensemble is a good research problem.

It is known that the EXIT chart approach can be applied to PTG-LDPC codes [80, 2]. Whether the EXIT chart approach can also be applied to SP-LDPC codes is another interesting problem to look at.

In the SP-construction of an LDPC code with an $m \times n$ base matrix $\mathbf{B}_{sp}$ and a replacement set $\boldsymbol{R}$, suppose the member matrices of $\boldsymbol{R}$ are $k \times t$ arrays of $l \times l$ regular matrices (e.g., PMs or CPMs). Then, the replacement of 1-entries of the base matrix $\mathbf{B}_{sp}$ by member arrays in $\boldsymbol{R}$ and 0-entries by $kl \times tl$ ZMs results in a parity-check matrix $\mathbf{H}_{sp}(kl, tl)$ which is an $mk \times nt$ array of regular matrices and/or ZMs of size $l \times l$. In this case, the null space of $\mathbf{H}_{sp}(kl, tl)$ may be regarded as an SP-LDPC code constructed from an $mk \times nt$ base matrix and a replacement set whose constituent matrices are regular matrices of size $l \times l$. From the PTG point of view, the code is also a PTG-LDPC code whose protograph has nt VNs and mk CNs. Various constructions of RC-constrained replacement sets whose member matrices are arrays of regular matrices, especially PMs or CPMs, will be discussed in later chapters and in Appendix A.

6

Construction of Base Matrices and RC-Constrained Replacement Sets for SP-Construction

The two key components in the SP-construction of an LDPC code are an SP-base matrix $\mathbf{B}_{sp}$ and a replacement set $\boldsymbol{R}$ of sparse member matrices. To ensure that the Tanner graph of an SP-LDPC code has girth at least 6, it is, in general, required that the SP-base matrix $\mathbf{B}_{sp}$ satisfies the RC-constraint and the member matrices in the replacement set $\boldsymbol{R}$ satisfy both the RC- and the PW-RC-constraints. In this chapter, we present several algebraic constructions of RC-constrained SP-base matrices and replacement sets whose member matrices satisfy both the RC- and the PW-RC-constraints. More constructions of RC-constrained SP-base matrices and replacement sets will be presented in Chapters 7 to 11 and Appendix A.

6.1 RC-Constrained Base Matrices

The algebraic methods presented in [58, 35, 3, 101, 112, 64, 65, 50, 113, 114, 46, 70, 68] can be used to construct RC-constrained SP-base matrices. All these methods are based on finite geometries, finite fields, and combinatorial designs (such as balanced incomplete block designs (BIBDs) or Latin squares). In this section, we use the construction based on finite Euclidean geometries. In Chapter 7, we will present a very flexible and powerful construction based on finite fields.

Consider the two-dimensional Euclidean geometry (EG) over the field GF(q) [84, 74, 97] (see Appendix A), denoted by EG$(2, q)$. This geometry consists of $q^2 + q$ lines, each consisting of q points. Among these lines, $q + 1$ lines pass through the origin of the geometry. Based on the $q^2 - 1$ lines of EG$(2, q)$ not passing through the origin, it is possible to construct a $(q + 1) \times (q + 1)$ array $\mathbf{H}_{EG}$ of CPMs and ZMs of size $(q - 1) \times (q - 1)$ [46, 26] which satisfies the RC-constraint [58, 97, 46] (see Appendix A). This array $\mathbf{H}_{EG}$ contains $q + 1$ ZMs which can be put on the main diagonal of the array. Since $\mathbf{H}_{EG}$ satisfies the RC-constraint, any subarray of $\mathbf{H}_{EG}$ also satisfies the RC-constraint and hence can be used as an SP-base matrix for constructing an SP-LDPC code. The Tanner graph of the SP-LDPC code constructed based on this SP-base matrix has girth at least 6.

If we replace some CPMs by ZMs in a subarray of $\mathbf{H}_{EG}$, it is possible to obtain an SP-LDPC code whose Tanner graph has girth at least 8, or a smaller number of short cycles of lengths 6, 8, and 10. The replacement of a CPM by a ZM is referred to as *masking* [112, 64, 97, 68]. Masking a subarray of $\mathbf{H}_{EG}$ may result in an SP-LDPC code with good error performance. Masking will be discussed further in Chapter 7.

There is another method for constructing SP-base matrices based on the two-dimensional geometry EG$(2, q)$ over GF(q). Using the $q^2 - 1$ lines of EG$(2, q)$ not passing through the origin, we can construct a weight-q binary circulant, denoted by $\mathbf{G}_{EG}$, of size $(q^2 - 1) \times (q^2 - 1)$, which satisfies the RC-constraint [58, 74, 97, 46, 26]. This circulant can be used as the base matrix for constructing SP-LDPC codes. If the weight q is too large, we can replace l 1-entries in its generator by 0-entries with $l < q$. This gives a new generator with $q - l$ nonzero components. The circulant generated by this new generator has weight $(q - l)$. The new circulant is called a *descendant* of $\mathbf{G}_{EG}$. Clearly, it also satisfies the RC-constraint and can be used as an SP-base matrix for constructing SP-LDPC codes. This construction gives SP-base matrices with a cyclic structure.

The circulant $\mathbf{G}_{EG}$ can be decomposed into a group of descendants of the same size to form SP-base matrices for the SP-construction of LDPC codes. Let e be a positive integer less than q. We decompose the generator $\mathbf{g}$ of the circulant $\mathbf{G}_{EG}$ into e new generators $\mathbf{g}_0, \mathbf{g}_1, \ldots, \mathbf{g}_{e-1}$ of the same length with the q nonzero components in $\mathbf{g}$ distributed among the e new generators. Each nonzero component in $\mathbf{g}$ is put in one and only one new generator $\mathbf{g}_i$ at the same position as its original position in $\mathbf{g}$. Let $w_0, w_1, \ldots, w_{e-1}$ be e positive integers such that $w_0 + w_1 + \ldots + w_{e-1} = q$. For $0 \leq i < e$, let w_i be the number of nonzero components in $\mathbf{g}$ that are put in $\mathbf{g}_i$. Let $\mathbf{G}_0, \mathbf{G}_1, \ldots, \mathbf{G}_{e-1}$ be the circulants generated by $\mathbf{g}_0, \mathbf{g}_2, \ldots, \mathbf{g}_{e-1}$, respectively. These e circulants $\mathbf{G}_0, \mathbf{G}_1, \ldots, \mathbf{G}_{e-1}$ are descendants of $\mathbf{G}_{EG}$ with weights $w_0, w_1, \ldots, w_{e-1}$, respectively. It follows from the construction of their generators that these e descendants of $\mathbf{G}_{EG}$ are mutually disjoint and their logic-OR sum $\mathbf{G}_0 + \mathbf{G}_1 + \ldots + \mathbf{G}_{e-1}$ gives $\mathbf{G}_{EG}$. Since $\mathbf{G}_{EG}$ satisfies the RC-constraint, its descendants must satisfy both the RC- and the PW-RC-constraints. Each descendant of $\mathbf{G}_{EG}$ can be used as an SP-base matrix. For $1 \leq r < e$, if we take r of these descendants of $\mathbf{G}_{EG}$ and arrange them as a row (or a column) to form a $(q^2 - 1) \times r(q^2 - 1)$ (or an $r(q^2 - 1) \times (q^2 - 1)$) matrix, then this matrix also satisfies the RC-constraint and can be used as an SP-base matrix for the SP-construction of LDPC codes. With properly chosen e and weights $w_0, w_1, \ldots, w_{e-1}$, we can form regular or irregular base matrices for the

SP-construction of LDPC codes. An EG circulant $\mathbf{G}_{EG}$ can also be decomposed as an array of circulants of the same size and this array (or its subarray) can be used as an SP-base matrix [19, 74, 97].

As an example, suppose we use the two-dimensional Euclidean geometry EG(2, 7) over GF(7) for constructing SP-base matrices. Based on the 48 lines not passing through the origin of EG(2, 7), we can construct a 48×48 circulant $\mathbf{G}_{EG}$ with weight 7. This 48×48 circulant can be used as an SP-base matrix. Suppose we decompose this circulant into three descendant circulants $\mathbf{G}_0$, $\mathbf{G}_1$, and $\mathbf{G}_2$, the first two with weight 3 and the third one with weight 1. Arranging the first two descendant circulants, $\mathbf{G}_0$ and $\mathbf{G}_1$, in a row, we obtain a 48×96 SP-base matrix $\mathbf{B}_{sp}$ with column and row weights 3 and 6, respectively. The Tanner graph of the base matrix $\mathbf{B}_{sp}$ has girth 6.

Besides finite geometries, large classes of RC-constrained arrays of CPMs and ZMs can be constructed from finite fields. One very flexible construction is based on two *arbitrary subsets* of a finite field reported recently in [70, 68]. This general construction includes several other known constructions of LDPC codes based on finite fields as special cases. This construction method will be presented in Chapter 7 to construct QC-SP-LDPC codes. The PEG algorithm [43, 44] can also be used to design base matrices for the SP-construction of LDPC codes, especially irregular LDPC codes.

Note that the RC-constrained circulant (or its descendant) constructed based on a two-dimensional Euclidean geometry EG$(2, q)$ can be used as a base matrix for constructing an RC-constrained parity-check array of a QC-PTG-LDPC code using the matrix decomposition method presented in Section 4.1.

6.2 Construction of RC-Constrained Replacement Sets Based on Hamming Codes

Methods for constructing RC-constrained replacement sets for constructing SP-LDPC codes were first discussed in [76, 111]. In this and the next two sections, we present three new methods for constructing RC-constrained replacement sets for the SP-construction of LDPC codes, especially QC-SP-LDPC codes. More discussion on the construction of RC-constrained replacement sets will be given in Chapter 7 and Appendix A.

Let m be a positive integer with $m \geq 3$. Consider the $(2^m - 1, 2^m - m - 1)$ cyclic Hamming code C generated by a primitive polynomial $p(X)$ of degree m over GF(2) [74, 89]. The minimum weight, denoted by m-w, of this code is 3. Two different m-w codewords cannot have more than one 1-component in common. Otherwise, the sum of two m-w codewords would give a nonzero codeword with weight less than 3, which contradicts the fact that the m-w of the code C is 3. Let $n = 2^m - 1$. Let $\mathbf{v} = (v_0, v_1, \ldots, v_{n-1})$ be an m-w codeword of C and $\mathbf{v}^{(i)}$ be its i-th cyclic shift (cyclically shifting $\mathbf{v}$ to the right i positions). It is clear that $\mathbf{v}^{(0)} = \mathbf{v}^{(n)} = \mathbf{v}$. Let ℓ be the smallest positive integer less than or equal to n such that $\mathbf{v}^{(\ell)} = \mathbf{v}$. We call ℓ the *cycle span* of $\mathbf{v}$. It is clear that ℓ divides n. If $\ell = n$, then $\mathbf{v}$ is said to be *primitive*. In this case, $\mathbf{v}$ and all its $n-1$ cyclic shifts are different. If $\ell < n$, then $\mathbf{v}$ is said to be *nonprimitive*. If $\mathbf{v}$ is nonprimitive, then its cycle span ℓ divides n and is equal to 3ℓ since the m-w of the code is 3. If 3 is not a factor of n or n is a prime, then every m-w codeword of C is primitive.

Since the weight distribution of a Hamming code is completely known [74, 89, 8], the number of primitive m-w codewords can be determined and these m-w codewords can be constructed [74]. The primitive m-w codewords in C can be partitioned into *cyclic classes*, say k of them, denoted by $\mathbf{Q}_0, \mathbf{Q}_1, \ldots, \mathbf{Q}_{k-1}$. Each cyclic class contains n m-w codewords of length n which are cyclic shifts of each other. For each cyclic class $\mathbf{Q}_i$, we can construct an $n \times n$ circulant $\mathbf{G}_i$ with its n m-w codewords as rows. This is achieved by using any m-w codeword in $\mathbf{Q}_i$ as the first row and cyclically shifting it $n-1$ times to form the other $n-1$ rows. The weight of the circulant $\mathbf{G}_i$ is 3. Note that any column of $\mathbf{G}_i$ is also an m-w codeword in the Hamming code C. Based on the k cyclic classes $\mathbf{Q}_0, \mathbf{Q}_1, \ldots, \mathbf{Q}_{k-1}$, we can construct k circulants, $\mathbf{G}_0, \mathbf{G}_1, \ldots, \mathbf{G}_{k-1}$, each of size $n \times n$. Since no two m-w codewords have more than one 1-component in common, these circulants have the following properties: no two rows (or two columns) either from the same circulant or from two different circulants, $\mathbf{G}_i$ and $\mathbf{G}_j$, have more than one 1-component in common. Hence, the circulants $\mathbf{G}_0, \mathbf{G}_1, \ldots, \mathbf{G}_{k-1}$ satisfy both the RC- and the PW-RC-constraints and they can be used as the member matrices of a replacement set $\boldsymbol{R}_{Ham}$ for the SP-construction of LDPC codes, where the subscript "*Ham*" in $\boldsymbol{R}_{Ham}$ stands for "*Hamming*."

For $0 \leq i < k$, we decompose the circulant $\mathbf{G}_i$ into 3 descendant CPMs, $\mathbf{G}_{i,0}$, $\mathbf{G}_{i,1}$, and $\mathbf{G}_{i,2}$, each of size $n \times n$. The decomposition is done in exactly the same manner as the decomposition of an EG circulant described in Section 6.1, i.e., we split the generator $\mathbf{g}_i$ of $\mathbf{G}_i$ into 3 generators, $\mathbf{g}_{i,0}$, $\mathbf{g}_{i,1}$, and $\mathbf{g}_{i,2}$, each with a single 1-component, and then use each of these 3 generators to construct a CPM. Clearly, these 3 CPMs satisfy the PW-RC-constraint. For $0 \leq i < k$, we form the following two matrices from $\mathbf{G}_{i,0}$, $\mathbf{G}_{i,1}$, and $\mathbf{G}_{i,2}$: $\mathbf{A}_i^0 = [\mathbf{G}_{i,0}\mathbf{G}_{i,1}]$ and $\mathbf{A}_i^1 = [\mathbf{G}_{i,0}\mathbf{G}_{i,1}\mathbf{G}_{i,2}]$, where $\mathbf{A}_i^0$ is an $n \times 2n$ matrix with constant column and row weights 1 and 2, respectively, and $\mathbf{A}_i^1$ is an $n \times 3n$ matrix with constant column and row weights 1 and 3, respectively. Next, we form the following two replacement sets: $\boldsymbol{R}_{Ham,0} = \{\mathbf{A}_i^0 : 0 \leq i < k\}$ and $\boldsymbol{R}_{Ham,1} = \{\mathbf{A}_i^1 : 0 \leq i < k\}$. Both of these two replacement sets satisfy the RC- and the PW-RC-constraints and can be used for the SP-construction of QC-SP-LDPC codes.

Let r be an integer with $1 \leq r \leq k$. Suppose we take r circulants from $\mathbf{G}_0, \mathbf{G}_1, \ldots, \mathbf{G}_{k-1}$, say the first r of them, and form an $n \times nr$ matrix $\mathbf{B}_{sp} = [\mathbf{G}_0\ \mathbf{G}_1\ \ldots\ \mathbf{G}_{r-1}]$ which satisfies the RC-constraint. The matrix $\mathbf{B}_{sp}$ has column and row weights, 3 and $3r$, respectively, and can be used as an SP-base matrix for the SP-construction of LDPC codes. Using this SP-base matrix and the replacement set, either $\boldsymbol{R}_{Ham,0}$ or $\boldsymbol{R}_{Ham,1}$ given above, we can construct SP-QC-LDPC codes whose Tanner graphs have girth at least 6.

Suppose we use $\mathbf{B}_{sp} = [\mathbf{G}_0\ \mathbf{G}_1\ \ldots\ \mathbf{G}_{r-1}]$ as the base matrix for the SP-construction of QC-SP-LDPC codes. Let all the binary CPMs of size $l \times l$ be the constituent matrices of a replacement set $\boldsymbol{R}_{cpm}$. Since these CPMs satisfy both the RC- and the PW-RC-constraints, $\boldsymbol{R}_{cpm}$ is an RC-constrained replacement set. Based on $\mathbf{B}_{sp} = [\mathbf{G}_0\ \mathbf{G}_1\ \ldots\ \mathbf{G}_{r-1}]$ and $\boldsymbol{R}_{cpm}$, we can construct a QC-LDPC code of length nrl and rate at least $(r-1)/r$ whose Tanner graph has girth at least 6. For $r = 2, 3, 4, \ldots, k$, we can construct a sequence of QC-SP-LDPC codes with rates at least $1/2, 2/3, 3/4, \ldots, (k-1)/k$.

What we have shown above is that, using the m-w codewords of a Hamming code, we can form both the SP-base matrices as well as replacement sets for the SP-construction of QC-SP-LDPC codes.

Example 6.1. Let $m = 5$. Consider the $(31, 26)$ cyclic Hamming code C generated by the primitive polynomial $p(X) = 1 + X^2 + X^5$ over GF(2). Since 31 is a prime, then every m-w codeword of C is primitive. The code C has 155 m-w codewords which can be used to form 5 circulants $\mathbf{G}_0$, $\mathbf{G}_1$, $\mathbf{G}_2$, $\mathbf{G}_3$, and $\mathbf{G}_4$, of size 31×31, each with weight 3. Decompose each of these 5 circulants into 3 descendant CPMs of size 31×31. Based on these 5 groups of descendant CPMs, we can form an RC-constrained replacement set $\boldsymbol{R}_{Ham,1}$ using the method described above. Each member matrix in $\boldsymbol{R}_{Ham,1}$ is a 31×93 matrix with three CPMs in a row. Based on this replacement set $\boldsymbol{R}_{Ham,1}$ and a chosen base matrix, we can construct a family of QC-SP-LDPC codes.

Suppose we construct a 15×15 circulant $\mathbf{B}_{sp}$ using the 15 lines not passing through the origin of the two-dimensional Euclidean geometry EG$(2, 2^2)$ over GF(2^2). The generator of the circulant $\mathbf{B}_{sp}$ has weight 4, i.e., $\mathbf{B}_{sp}$ is a weight-4 circulant. In the SP-construction, we use $\mathbf{B}_{sp}$ as the SP-base matrix and the set $\boldsymbol{R}_{Ham,1}$ as the replacement set. Replacing the 60 1-entries in $\mathbf{B}_{sp}$ with member matrices in $\boldsymbol{R}_{Ham,1}$ following the replacement constraint (see Section 5.3), we obtain a 15×45 array $\mathbf{H}_{sp,qc}(31, 31)$ of CPMs and ZMs of size 31×31. The array $\mathbf{H}_{sp,qc}(31, 31)$ is a 465×1359 matrix with constant column and row weights 4 and 12, respectively. The null space of this matrix gives a (4, 12)-regular $(1395, 937)$ QC-SP-LDPC code $C_{sp,qc,Ham}$ with rate 0.6717. The Tanner graph of this code has girth 6 and contains 6200 cycles of length 6 and 164 951 cycles of length 8.

The BLER and BER performances of this code over the AWGNC decoded with 50 iterations of the MSA are shown in Fig. 6.1(a). We see that this code performs well, at about 1.25 dB away from the SPB at a BLER of 10^{-8}. The decoding threshold of this code is 1.95 dB. At BERs 10^{-5} and 10^{-9}, the code performs 1.0 and 1.55 dBs from the decoding threshold, respectively.

Fig. 6.1(b) shows the performance of the code $C_{sp,qc,Ham}$ over the BEC. For the BEC, it performs about 0.14 from the Shannon limit at a UEBR of 10^{-8}.

If we put the array $\mathbf{H}_{sp,qc}(31,31)$ into the form of (4.1), we obtain a 31×31 array $\mathbf{H}_{ptg,cyc}(15,45)$ of submatrices of size 15×45 with a block-cyclic structure. The integer sum of the constituent matrices in the top row-block of $\mathbf{H}_{ptg,cyc}(15,45)$ gives the 15×45 base matrix $\mathbf{B}_{ptg}$ for constructing the above code using the algebraic construction presented in Chapter 4. In the code construction by using this base matrix, we can choose any decomposition factor. This would result in a family of QC-PTG-LDPC codes.

Example 6.2. In Example 6.1, we used the $(31,26)$ cyclic Hamming code C to construct a replacement set for the SP-construction of a QC-SP-LDPC code, but used a base matrix constructed based on a Euclidean geometry. Suppose we use one of the five 31×31 circulants $\mathbf{G}_0$, $\mathbf{G}_1$, $\mathbf{G}_2$, $\mathbf{G}_3$, and $\mathbf{G}_4$ constructed from the m-w codewords in C as the base matrix, denoted by $\mathbf{B}_{sp,Ham}$ and the same replacement set $\boldsymbol{R}_{Ham,1}$ constructed in Example 6.1 for the SP-construction. The replacement of the 1-entries of the base matrix $\mathbf{B}_{sp,Ham}$ by the member matrices in $\boldsymbol{R}_{Ham,1}$ results in the 31×93 array $\mathbf{H}_{sp,qc}(31,31)$ of CPMs and ZMs of size 31×31 which is a 961×2883 matrix with column and row weights 3 and 9, respectively.

The null space of $\mathbf{H}_{sp,qc}(31,31)$ gives a $(3,9)$-regular $(2883,1927)$ QC-SP-LDPC code $C_{sp,qc,Ham}$ with rate 0.6684. The Tanner graph of the code $C_{sp,qc,Ham}$ has girth 6 and contains 2387 cycles of length 6 and 20 088 cycles of length 8. The BLER and BER performances of this code decoded with 50 iterations of the MSA over the AWGNC are shown in Fig. 6.2(a). We see that the code performs about 1.0 dB away from the SPB at a BLER of 10^{-4} and 1.4 dB away from the Shannon limit at a BER of 10^{-5}. Its performance over the BEC is shown in Fig. 6.2(b). At a UEBR of 10^{-7}, it performs within 0.1 of the Shannon limit.

There is another way to construct SP-base matrices from the circulants $\mathbf{G}_0, \mathbf{G}_1, \ldots, \mathbf{G}_{k-1}$ constructed from a Hamming code. First, we use the k circulants

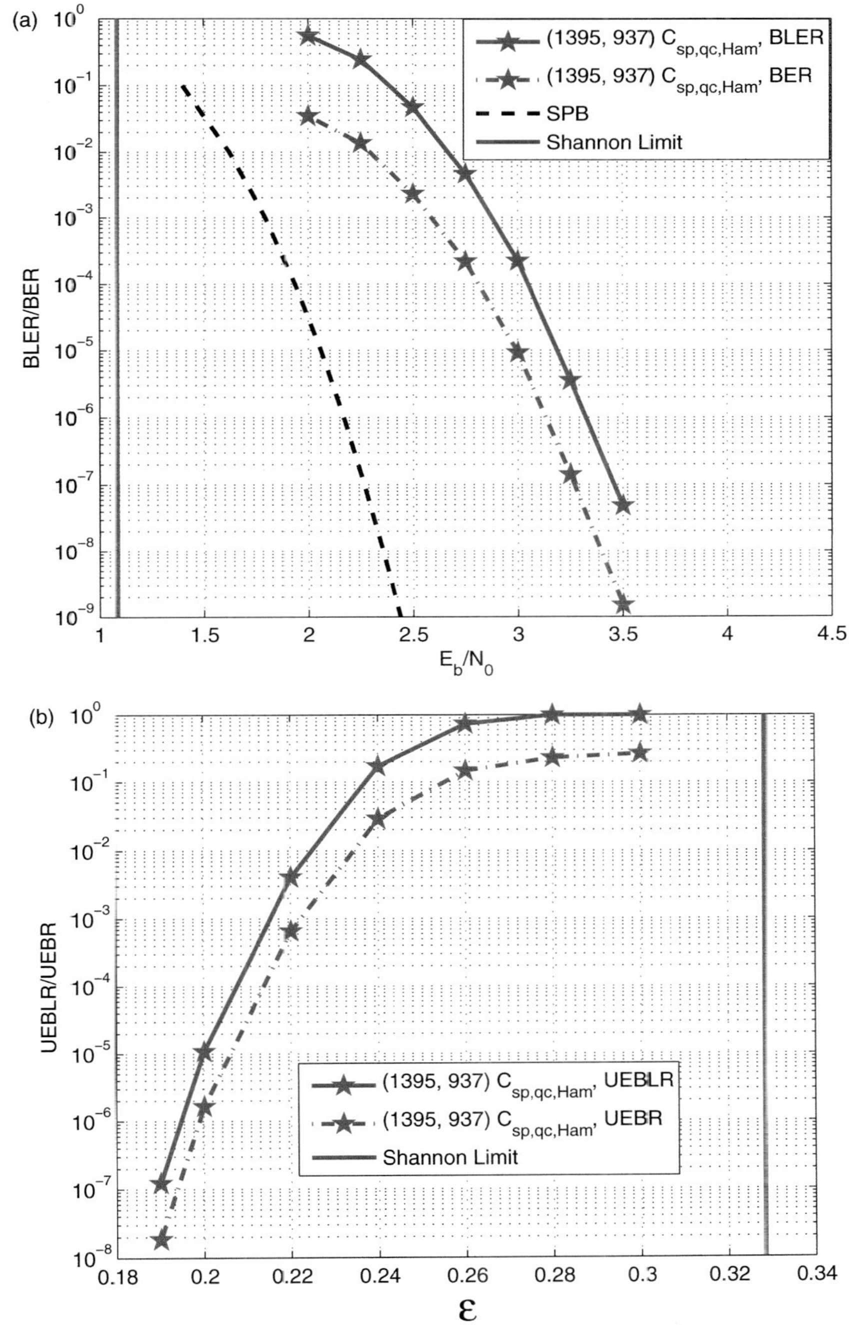

Figure 6.1. (a) The BER and BLER performances of the (1395, 937) QC-SP-LDPC code $C_{sp,qc,Ham}$ given in Example 6.1 over the AWGNC; and (b) performance of the code in (a) over the BEC.

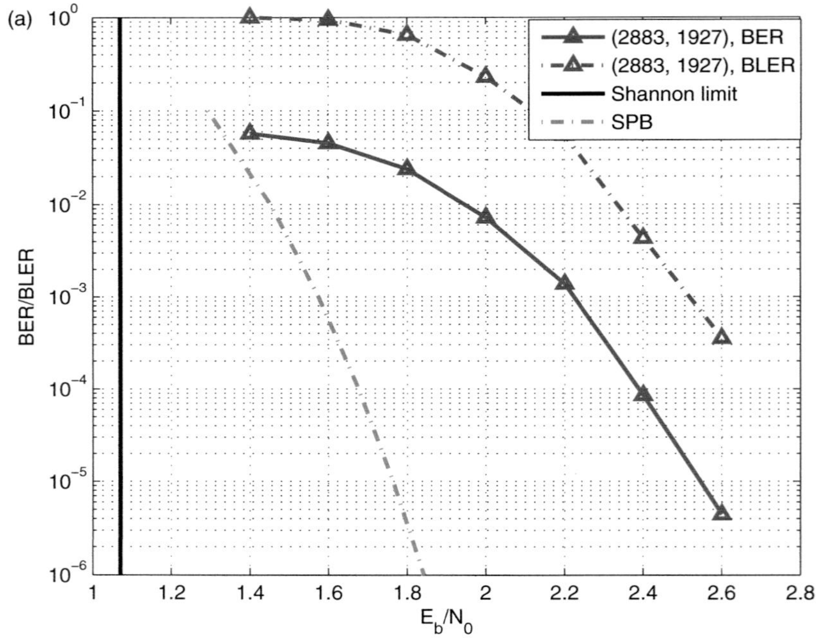

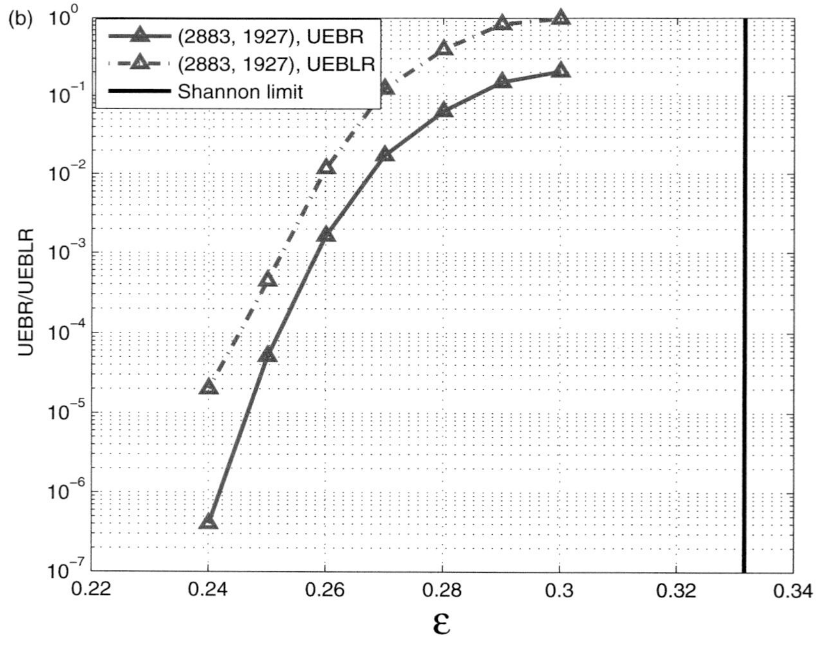

Figure 6.2. (a) The BER and BLER performances of the (2883, 1927) QC-SP-LDPC codes $C_{sp,qc,Ham}$ given in Example 6.2 over the AWGNC; and (b) performance of the code in (a) over the BEC.

$\mathbf{G}_0, \mathbf{G}_1, \ldots, \mathbf{G}_{k-1}$ to form a $k \times 1$ column array as follows:

$$\mathbf{G}_{Ham} = \begin{bmatrix} \mathbf{G}_0 \\ \mathbf{G}_1 \\ \cdots \\ \mathbf{G}_{k-1} \end{bmatrix}. \tag{6.1}$$

Since $\mathbf{G}_0, \mathbf{G}_1, \ldots, \mathbf{G}_{k-1}$ satisfy both the RC- and PW-RC-constraints, the array $\mathbf{G}_{Ham}$ satisfies the RC-constraint. For $0 \le i < k$, we decompose $\mathbf{G}_i$ into a row of three CPMs of size $n \times n$, denoted by $\mathbf{G}_{i,0}$, $\mathbf{G}_{i,1}$, and $\mathbf{G}_{i,2}$. Next, we form the following $k \times 3$ array of CPMs of size $n \times n$:

$$\mathbf{H}_{Ham}(n,n) = \begin{bmatrix} \mathbf{G}_{0,0} & \mathbf{G}_{0,1} & \mathbf{G}_{0,2} \\ \mathbf{G}_{1,0} & \mathbf{G}_{1,1} & \mathbf{G}_{1,2} \\ \vdots & \vdots & \vdots \\ \mathbf{G}_{k-1,0} & \mathbf{G}_{k-1,1} & \mathbf{G}_{k-1,2} \end{bmatrix}. \tag{6.2}$$

It is clear that the array $\mathbf{H}_{Ham}(n,n)$ satisfies the RC-constraint. Any subarray of $\mathbf{H}_{Ham}(n,n)$ can be used as an SP-base matrix for the SP-construction of LDPC codes. Suppose we take the transpose of the array $\mathbf{H}_{Ham}(n,n)$. We obtain the following $3 \times k$ RC-constrained array:

$$\mathbf{H}^{\star}_{Ham}(n,n) = \begin{bmatrix} \mathbf{G}_{0,0} & \mathbf{G}_{1,0} & \cdots & \mathbf{G}_{k-1,0} \\ \mathbf{G}_{0,1} & \mathbf{G}_{1,1} & \cdots & \mathbf{G}_{k-1,1} \\ \mathbf{G}_{0,2} & \mathbf{G}_{1,2} & \cdots & \mathbf{G}_{k-1,2} \end{bmatrix}. \tag{6.3}$$

Any subarray of $\mathbf{H}^{\star}_{Ham}(n,n)$ can be used as an SP-base matrix for the SP-construction of LDPC codes. In fact, for $4 \le \ell < k$, if we take a $3 \times \ell$ subarray from $\mathbf{H}^{\star}_{Ham}(n,n)$, the null space of $\mathbf{H}^{\star}_{Ham}(n,n)$ gives a $(3,\ell)$-regular QC-SP-LDPC code of length ℓn with rate at least $(\ell-3)/\ell$, whose Tanner graph has girth at least 6. For $\ell = 4, 5, 6, \ldots, k$, we can construct a sequence of regular QC-SP-LDPC codes with rates $1/4$, $2/5$, $1/2, \ldots, (k-3)/k$.

6.3 Construction of RC-Constrained Replacement Sets Based on m-dimensional Euclidean Geometry EG(m, 2) over GF(2)

For any positive integer $m \ge 2$, the m-dimensional Euclidean (or affine) geometry EG(m, 2) over the binary field GF(2) can be used to construct an RC-constrained replacement set. The geometry EG(m, 2) consists of $(2^{m-1}-1) \times (2^m-1)$ lines not passing through the origin of the geometry [74, 97]. Each of these lines consists of 2 points. Based on the incidence vectors of these lines, we can form

$2^{m-1} - 1$ circulants of size $(2^m - 1) \times (2^m - 1)$ with weight 2 [74, 97], denoted by $\mathbf{G}_0, \mathbf{G}_1, \ldots, \mathbf{G}_{s-1}$, where $s = 2^{m-1} - 1$. Since two lines in EG$(m, 2)$ are either parallel or intersect at one and only one point, all these circulants satisfy the RC- and the PW-RC-constraints. Therefore, these circulants form an RC-constrained replacement set for the SP-construction of QC-SP-LDPC codes. Furthermore, any subset of the circulants $\mathbf{G}_0, \mathbf{G}_1, \ldots, \mathbf{G}_{s-1}$ can be arranged in a row (or a column) to form an SP-base matrix $\mathbf{B}_{sp}$.

If we decompose each of the circulants $\mathbf{G}_0, \mathbf{G}_1, \ldots, \mathbf{G}_{s-1}$ into two CPMs of size $(2^m - 1) \times (2^m - 1)$ and arrange them in a row to form a $(2^m - 1) \times 2(2^m - 1)$ matrix, we obtain $2^{m-1} - 1$ matrices of size $(2^m - 1) \times 2(2^m - 1)$. These matrices also satisfy the RC- and the PW-RC-constraints and they form an RC-constrained replacement set for the SP-construction of QC-SP-LDPC codes as well as base matrices.

Example 6.3. Consider the geometry EG$(5, 2)$. Based on the 465 lines not passing through the origin of the geometry, we can form 15 circulants $\mathbf{G}_0, \mathbf{G}_1, \ldots, \mathbf{G}_{14}$ of size 31×31 which satisfy the RC- and the PW-RC-constraints. Using these 15 circulants of weight 2, we can form both SP-base matrices and RC-constrained replacement sets for the SP-construction of QC-SP-LDPC codes.

Suppose we take two circulants $\mathbf{G}_0$ and $\mathbf{G}_1$ to form a 31×62 SP-base matrix $\mathbf{B}_{sp}$ with column and row weights 2 and 4, respectively, and use $\mathbf{G}_0, \mathbf{G}_1, \ldots, \mathbf{G}_{14}$ as the replacement matrices to replace the 1-entries in $\mathbf{B}_{sp}$. Then, we can form an RC-constrained 31×62 array $\mathbf{H}_{sp,qc}(31, 31)$ of circulants of size 31×31. This array is a 961×1922 matrix with column and row weights 4 and 8, respectively. The null space of this array gives a (4, 8)-regular (1922, 997) QC-SP-LDPC code whose Tanner graph has girth at least 6. By using the cycle counting method in [71], we find that its Tanner graph has girth 6 and contains 7595 cycles of length 6 and 107 260 cycles of length 8. The BER and BLER performances of this code over the AWGNC decoded with 50 iterations of the MSA and the UEBR and UEBLR performances over the BEC are shown in Fig. 6.3(a) and Fig. 6.3(b), respectively. At a BLER of 10^{-5}, the code performs about 2.4 dB away from the SPB. At a UEBR of 10^{-7}, the code performs within 0.17 of the Shannon limit (0.52).

For a large m, a large number of circulants of weight 2 that satisfy the RC- and the PW-RC-constraints can be constructed to form replacement sets for constructing QC-SP-LDPC codes. Each of these weight-2 circulants can be decomposed into two circulants of weight 1 (CPMs). By properly grouping circulants of weights 1 and 2, we can form both RC-constrained replacement sets with member matrices of various weights, and SP-base matrices.

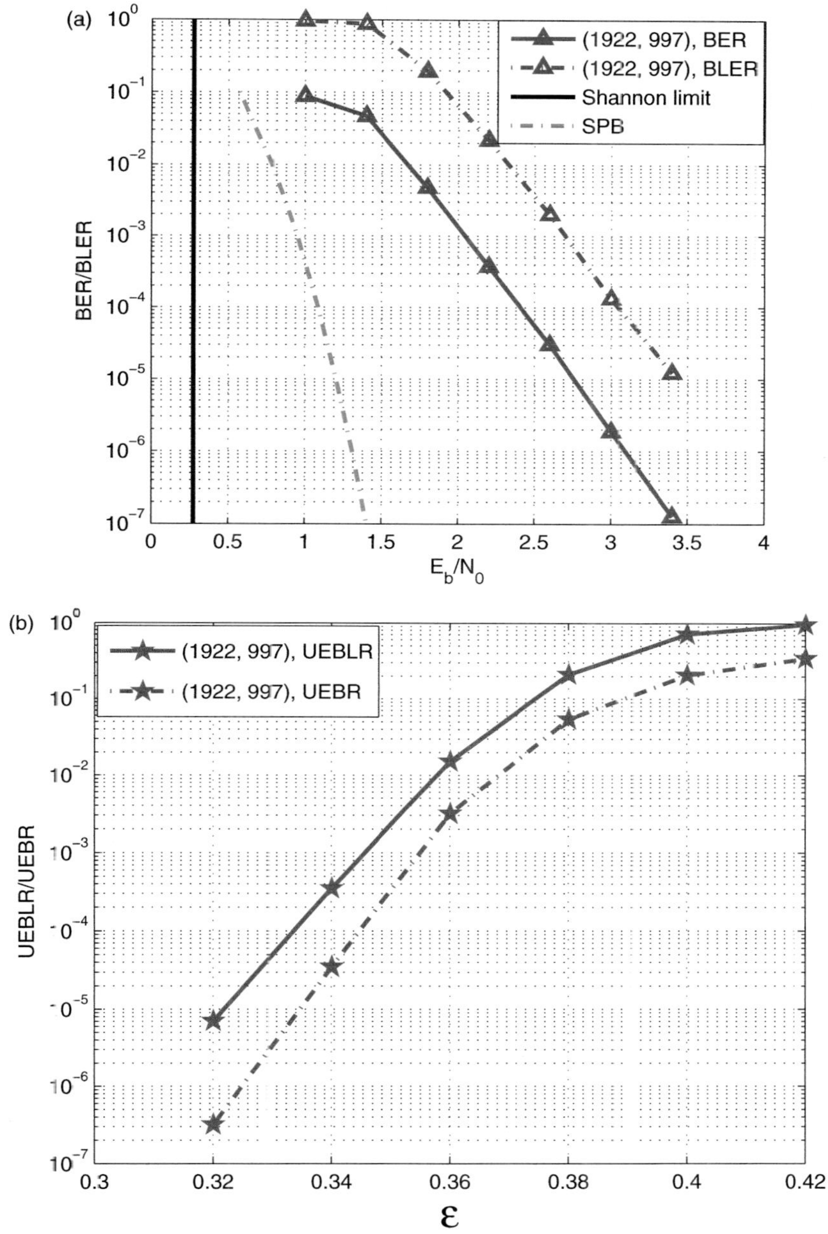

Figure 6.3. (a) The BER and BLER performances of the (1922, 997) QC-SP-LDPC code given in Example 6.3 over the AWGNC; and (b) performance of the code in (a) over the BEC.

The class of m-dimensional Euclidean geometries EG(m, 2) over GF(2) is well known for its application in constructing *Reed–Muller* (RM) codes which have good trellis structure [74, 89, 75] and can be decoded with the BCJR (or MAP) decoding algorithm to achieve very good error performance [75, 89, 4]. It was recently shown that the ensemble of RM codes achieves the capacity of the BEC [61].

6.4 Construction of RC-Constrained Replacement Sets Based on RC-Constrained Arrays of CPMs

RC-constrained replacement sets for the SP-construction of LDPC codes can be constructed by decomposing an RC-constrained array of CPMs.

Let $\mathbf{H}_{qc}(l,l)$ be an $m \times n$ RC-constrained array of CPMs of size $l \times l$. This array consists of m row-blocks, denoted by $\mathbf{A}_0, \mathbf{A}_1, \ldots, \mathbf{A}_{m-1}$, each consisting of n CPMs of size $l \times l$. Since each of these row-blocks of $\mathbf{H}_{qc}(l,l)$ is an $l \times ln$ matrix with column weight 1, it certainly satisfies the RC-constraint. If we arrange any two of these row-blocks in a row, we obtain an $l \times 2ln$ matrix which also has column weight 1 and hence satisfies the RC-constraint. Note that this is true even if the two row-blocks are identical. If we arrange any two of the m row-blocks of $\mathbf{H}_{qc}(l,l)$ one on the top of the other, we obtain a $2 \times n$ subarray of $\mathbf{H}_{qc}(l,l)$. Since $\mathbf{H}_{qc}(l,l)$ satisfies the RC-constraint, this subarray of $\mathbf{H}_{qc}(l,l)$ must also satisfy the RC-constraint. In fact, for $1 \leq k < m$, any $k \times n$ subarray that consists of k row-blocks of $\mathbf{H}_{qc}(l,l)$ satisfies the RC-constraint. Based on these facts, the set $\boldsymbol{R} = \{\mathbf{A}_0, \mathbf{A}_1, \ldots, \mathbf{A}_{m-1}\}$ of $l \times ln$ matrices forms an RC-constrained replacement set for the SP-construction of QC-SP-LDPC codes. In the SP-construction, if the SP-base matrix $\mathbf{B}_{sp}$ satisfies the RC-constraint, replacing the 1-entries in each column of $\mathbf{B}_{sp}$ with distinct member matrices in $\boldsymbol{R}$ will result in a QC-SP-LDPC code whose Tanner graph has girth at least 6.

Major methods for constructing RC-constrained arrays of CPMs and/or ZMs can be found in [35, 3, 101, 112, 64, 65, 50, 113, 114, 46, 70, 68]. In Section 6.1, it was pointed out that an RC-constrained $(q+1) \times (q+1)$ array $\mathbf{H}_{EG}$ of CPMs and ZMs of size $(q-1) \times (q-1)$ can be constructed based on the lines not passing through the origin of the two-dimensional Euclidean geometry EG(2, q) over GF(q) [46, 26] (also see Appendix A). This array contains exactly $q+1$ ZMs which can be put on the main diagonal of the array. Hence, for $1 \leq m, n \leq q+1$, any $m \times n$ subarray $\mathbf{H}_{EG}(q-1, q-1)$ of $\mathbf{H}_{EG}$, avoiding the ZMs, can be used to construct an RC-constrained replacement set of m member matrices of size $(q-1) \times n(q-1)$ for the SP-construction of QC-SP-LDPC codes. Constructions of arrays of CPMs based on two types of finite geometries are presented in Appendix A.

The array $\mathbf{H}_{Ham}(n,n)$ given by (6.2) constructed based on the m-w codewords of a Hamming code can be used to construct RC-constrained replacement sets as described above.

Another method for constructing RC-constrained array of CPMs and/or ZMs is based on a *Latin square* [73, 113]. A Latin square of order q is a $q \times q$ matrix for which each row and each column contains every element of a set of q elements exactly once [73]. For every finite field GF(q), a Latin square of order q can be constructed [113]. Let α be a primitive element of GF(q). Then, $\alpha^{-\infty} = 0, \alpha^0 = 1, \alpha, \alpha^2, \ldots, \alpha^{q-2}$ give all the q elements of GF(q). Let η be a nonzero element of GF(q). Then, the $q \times q$ matrix $\mathbf{L} = [\eta\alpha^i - \alpha^j]$ over GF(q), with the row index i and column index j running through $-\infty, 0, 1, \ldots, q-2$, is a Latin square of order q. If we disperse each nonzero entry in $\mathbf{L}$ into a CPM of size $(q-1) \times (q-1)$ (see Chapter 2, Section 2.1) and each zero entry into a ZM of size $(q-1) \times (q-1)$, we obtain an RC-constrained $q \times q$ array $\mathbf{H}_{Lat}$ of CPMs and ZMs of size $(q-1) \times (q-1)$ [113], where the subscript "*Lat*" in $\mathbf{H}_{Lat}$ stands for "*Latin square*." There are exactly q ZMs in $\mathbf{H}_{Lat}$. If we choose $\eta = 1$, then the q ZMs lie on the main diagonal of $\mathbf{H}_{Lat}$. For $1 \leq m, n \leq q$, any $m \times n$ subarray of $\mathbf{H}_{Lat}$, avoiding the ZMs, can be used to construct an RC-constrained replacement set of m constituent matrices of size $(q-1) \times n(q-1)$ for the SP-construction of QC-LDPC codes.

Example 6.4. Using the Latin square over GF(2^5), we can construct a 32×32 RC-constrained array $\mathbf{H}_{Lat}$ of CPMs and ZMs of size 31×31 with the 32 ZMs lying on the main diagonal of $\mathbf{H}_{Lat}$. Choose a 29×3 subarray from $\mathbf{H}_{Lat}$, avoiding the ZMs. Based on this subarray, we can construct an RC-constrained replacement set $\boldsymbol{R}_{Lat}$ with 29 matrices of size 31×93, each with column and row weights 1 and 3, respectively.

Suppose we use the base matrix $\mathbf{B}_{sp}$ given in Example 6.1 as the base matrix for the SP-construction. This matrix is a 15×15 circulant with weight 4 constructed based on the Euclidean geometry EG($2, 2^2$) over GF(2^2). Replacing the four 1-entries in each column of $\mathbf{B}_{sp}$ by 4 distinct arbitrary member matrices of $\boldsymbol{R}_{Lat}$, we obtain a 15×45 RC-constrained array $\mathbf{H}_{sp,qc,Lat}(31,31)$ of CPMs and ZMs of size 31×31. It is a 465×1395 matrix with column and row weights 4 and 12, respectively. The null space of $\mathbf{H}_{sp,qc,Lat}(31,31)$ gives a (4, 12)-regular $(1395, 937)$ QC-SP-LDPC code $C_{sp,qc,Lat}$ with rate 0.6717. The Tanner graph $\mathcal{G}_{sp,qc,Lat}(31,31)$ of this code has girth 6 and contains 6 200 cycles of length 6 and 164 951 cycles of length 8.

The BLER and BER performances of this code over the AWGNC decoded with 50 iterations of the MSA are shown in Fig. 6.4(a). Its performance over the BEC is shown in Fig. 6.4(b). The code $C_{sp,qc,Lat}$ given above has the same length, dimension, and rate as the code $C_{sp,qc,Ham}$ constructed in Example 6.1 but with

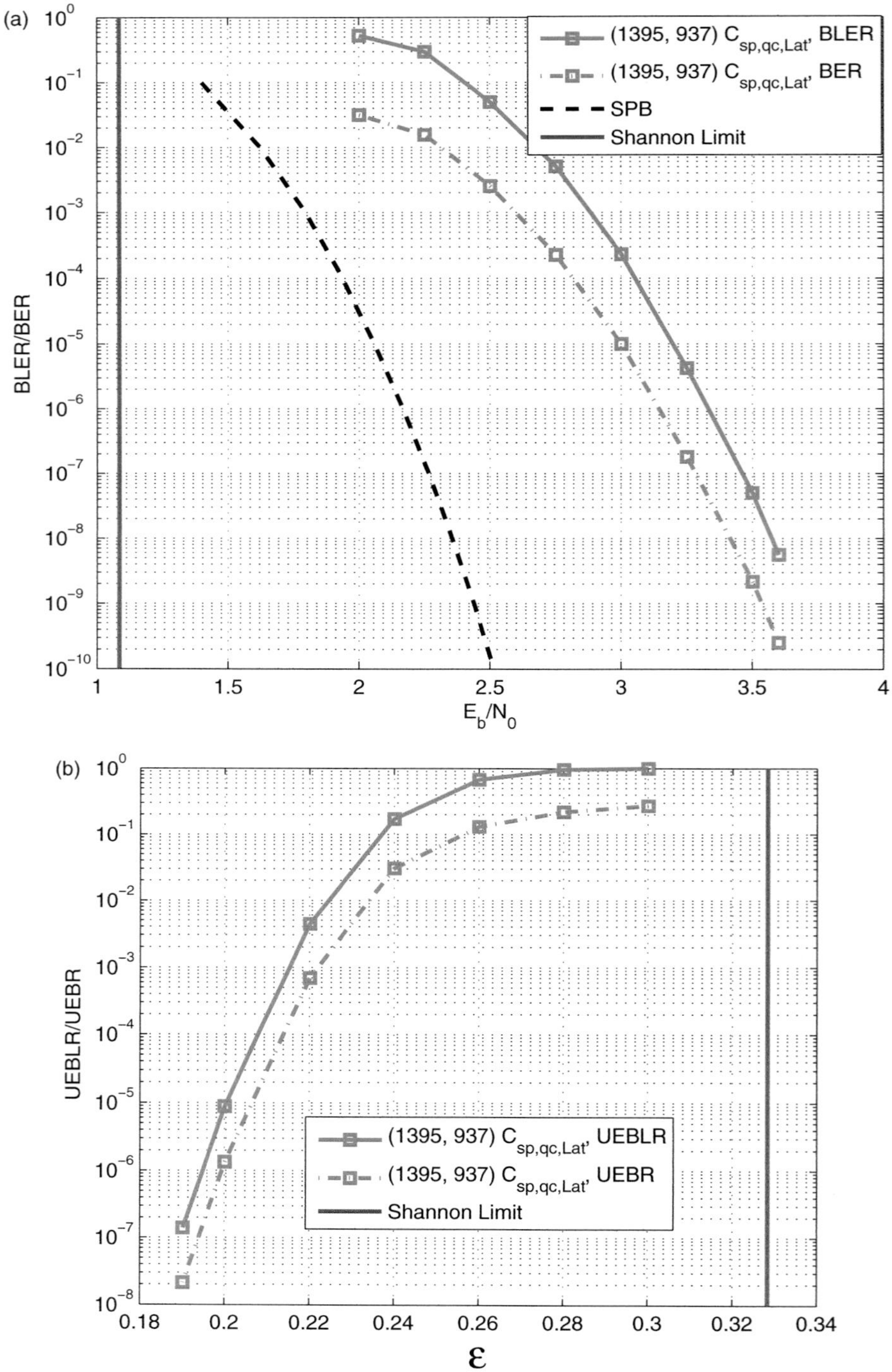

Figure 6.4. (a) The BER and BLER performances of the (1395, 937) QC-SP-LDPC code $C_{sp,qc,Lat}$ given in Example 6.4 over the AWGNC; and (b) performance of the code in (a) over the BEC.

Table 6.1. *The code parameters of the QC-SP-LDPC codes constructed in Example 6.5 with different lengths and rates*

Code	$m \times n$	(Length, dimension)	Rate	(w_c, w_r)	Girth	No. of cycle-6	No. of cycle-8
C_0	30×2	(930, 472)	0.5075	(4, 8)	6	1736	25668
C_1	29×3	(1395, 976)	0.6715	(4, 12)	6	5888	151552
C_2	28×4	(1860, 1402)	0.7539	(4, 16)	6	15252	526318
C_3	27×5	(2325, 1867)	0.8030	(4, 20)	6	30721	1345927
C_4	26×6	(2790, 2332)	0.8358	(4, 24)	6	53320	2903553
C_5	25×7	(3255, 2797)	0.8569	(4, 28)	6	86304	5469175
C_6	24×8	(3720, 3262)	0.8769	(4, 32)	6	128774	9535817
C_7	23×9	(4185, 3727)	0.8906	(4, 36)	6	185535	15491010
C_8	22×10	(4650, 4192)	0.9015	(4, 40)	6	256886	23901310
C_9	17×15	(6975, 6517)	0.9343	(4, 60)	6	881144	125243782
C_{10}	12×20	(9300, 8842)	0.9508	(4, 80)	6	2113673	402691271

different structure. From Fig. 6.1 and Fig. 6.4, we see that the two codes perform almost the same.

Example 6.5. Based on the above example, we can construct a family of QC-SP-LDPC codes with different lengths and rates by choosing different m and n. (Here, we only consider the case where the $m \times n$ subarray does not contain any ZMs.) Based on the base matrix $\mathbf{B}_{sp}$ used in the above example and the 32×32 RC-constrained array $\mathbf{H}_{Lat}$, we construct a family of QC-SP-LDPC codes by choosing different m and n. The code parameters including length, dimension, rate, column and row weights, and cycle distributions are given in Table 6.1. The BER performances of these codes over the AWGNC decoded with 50 iterations of the MSA are shown in Fig. 6.5. This figure shows that all the codes have no visible error-floor down to a BER of 10^{-9} or 10^{-10}.

In the above construction of an RC-constrained replacement set $\boldsymbol{R}$ for the SP-construction, we put a restriction on the $m \times n$ array $\mathbf{H}_{qc}(l, l)$ that it does not contain ZMs. With this restriction, every member matrix in $\boldsymbol{R}$ has constant column weight 1 and constant row weight n. This restriction can be removed. In this case, some of the member matrices in $\boldsymbol{R}$ may contain ZMs. With this replacement set $\boldsymbol{R}$, the replacement operation may result in an irregular array, i.e., a matrix with multiple column and/or row weights. An irregular array with good column and row weight distributions may result in an LDPC code with good waterfall error performance.

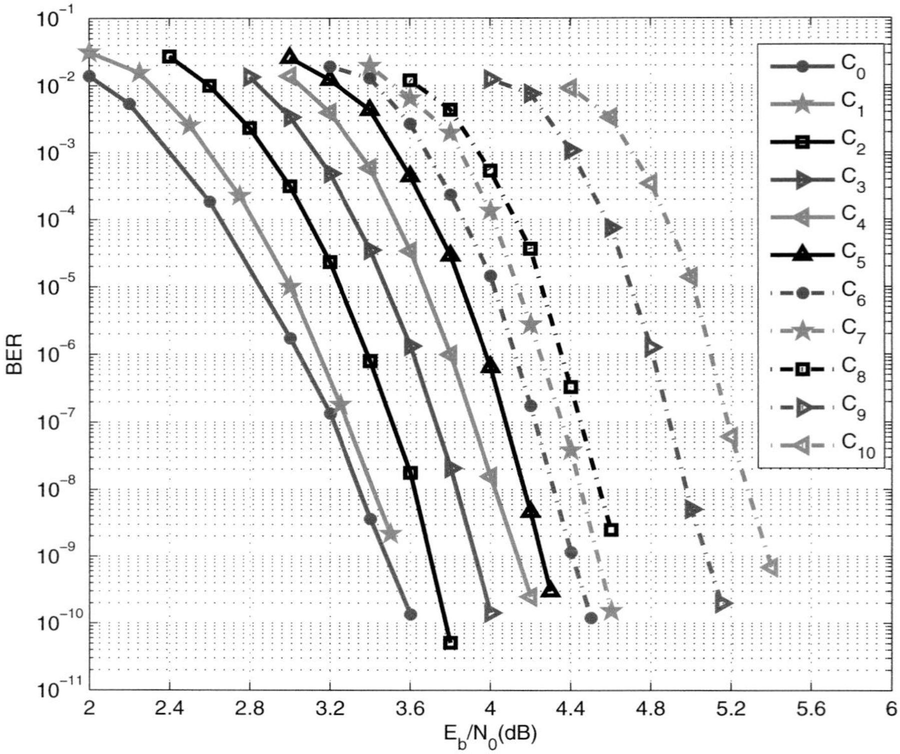

Figure 6.5. The BER performances of a family of QC-SP-LDPC codes given in Example 6.5 with different code lengths and rates over the AWGNC.

Example 6.6. In Example 6.4, instead of choosing a 29×3 subarray, we can choose a 32×3 subarray and use it to construct an RC-constrained replacement set $\boldsymbol{R}_{Lat}$ with 32 member matrices of size 31×93, in which there are 3 matrices with two column weights (0 and 1) and constant row weight 2, and 29 matrices with constant column weight 1 and constant row weight 3. Based on such a replacement set and the same base matrix $\mathbf{B}_{sp}$, we can construct an irregular 15×45 RC-constrained array $\mathbf{H}^*_{sp,qc,Lat}(31,31)$ of CPMs and ZMs of size 31×31. This array $\mathbf{H}^*_{sp,qc,Lat}(31,31)$ has 155 weight-3 columns, 1240 weight-4 columns, and 31 weight-10 rows, 93 weight-11 rows, and 341 weight-12 rows. The null space of $\mathbf{H}^*_{sp,qc,Lat}(31,31)$ gives an irregular $(1395,932)$ QC-SP-LDPC code $C^*_{sp,qc,Lat}$ with rate 0.6681. The rate of the code $C^*_{sp,qc,Lat}$ is slightly lower than that of the $(1395,937)$ code $C_{sp,qc,Lat}$ constructed in Example 6.4. The BLER and BER performances of this code over the AWGNC decoded with 50 iterations of the MSA are included in Fig. 6.6(a). Its performance over the BEC is shown in

Fig. 6.6(b). From Fig. 6.6, we see that the code $C^*_{sp,qc,Lat}$ also performs well over both the AWGNC and the BEC.

In constructing irregular QC-SP-LDPC codes with the SP-construction, it is simpler to design a base matrix with the desired column and row weight distributions (which are also the degree distributions of the VNs and the CNs of the Tanner graph of the base matrix) and then replace its 1-entries with matrices in an RC-constrained set $\boldsymbol{R}$ whose member matrices all have the same column and row weights. Then, the replacement operation results in an array which has the same column and row weight distributions as the base matrix. In the following example, we use this method to construct an irregular QC-SP-LDPC code.

Example 6.7. Consider the following VN and CN degree distributions (calculated using density evolution): $\gamma(X) = 0.25X + 0.625X^2 + 0.125X^8$ and $\rho(X) = 0.0451X^3 + 0.1076X^4 + 0.2257X^5 + 0.1597X^6 + 0.3611X^7 + 0.0868X^8 + 0.0139X^9$, which are designed for constructing a rate-1/2 irregular QC-SP-LDPC code.

Using the lines not passing through the origin of the two-dimensional Euclidean geometry EG(2, 17), we can construct a 288×288 RC-constrained circulant $\mathbf{G}_{EG}$ of weight 17. Based on the degree distributions given above and the circulant $\mathbf{G}_{EG}$, we can then construct a 288×576 SP-base matrix $\mathbf{B}_{sp}$ with column weights and row weights in agreement with $\gamma(X)$ and $\rho(X)$. The construction of $\mathbf{B}_{sp}$ is done as follows. First, we decompose the circulant $\mathbf{G}_{EG}$ into 4 descendant circulants, $\mathbf{G}_0$, $\mathbf{G}_1$, $\mathbf{G}_2$, and $\mathbf{G}_3$, with weight 2, 3, 3, and 9, respectively. The decomposition is done as described in Section 6.1. The 4 descendant circulants $\mathbf{G}_0$, $\mathbf{G}_1$, $\mathbf{G}_2$, and $\mathbf{G}_3$ of $\mathbf{G}_{EG}$ satisfy both the RC- and the PW-RC-constraints. The SP-base matrix $\mathbf{B}_{sp}$ consists of a 288×144 submatrix of $\mathbf{G}_0$ (the first 144 columns of $\mathbf{G}_0$), the entire circulant $\mathbf{G}_1$, a 288×72 submatrix of $\mathbf{G}_2$ (the first 72 columns of $\mathbf{G}_2$), and a 288×72 submatrix of $\mathbf{G}_3$ (the first 72 columns of $\mathbf{G}_3$). Then, the base matrix $\mathbf{B}_{sp}$ is a 288×576 binary matrix whose associated Tanner graph has the degree distributions $\gamma(X)$ and $\rho(X)$. The row and column weight distributions (in numbers of rows and columns) of this base matrix $\mathbf{B}_{sp}$ are given in Table 6.2. Since the circulants $\mathbf{G}_0$, $\mathbf{G}_1$, $\mathbf{G}_2$, and $\mathbf{G}_3$ satisfy both the RC- and the PW-RC-constraints, $\mathbf{B}_{sp}$ also satisfies the RC-constraint and its associated Tanner graph has girth at least 6. Using the cycle counting algorithm proposed in [71], we find that the Tanner graph associated with $\mathbf{B}_{sp}$ has girth 6 and contains 2937 cycles of length 6.

Using the Latin square over GF(2^4), we can construct a 16×16 array $\mathbf{H}_{Lat}$ of CPMs and ZMs of size 15×15 which satisfies the RC-constraint. Choose a 15×1 subarray (a single column-block of $\mathbf{H}_{Lat}$) from $\mathbf{H}_{Lat}$ avoiding the ZMs. Based on this subarray, we can construct an RC-constrained replacement set $\boldsymbol{R}_{Lat}$ with 15 CPMs of size 15×15 as member matrices. Based on the irregular base matrix $\mathbf{B}_{sp}$

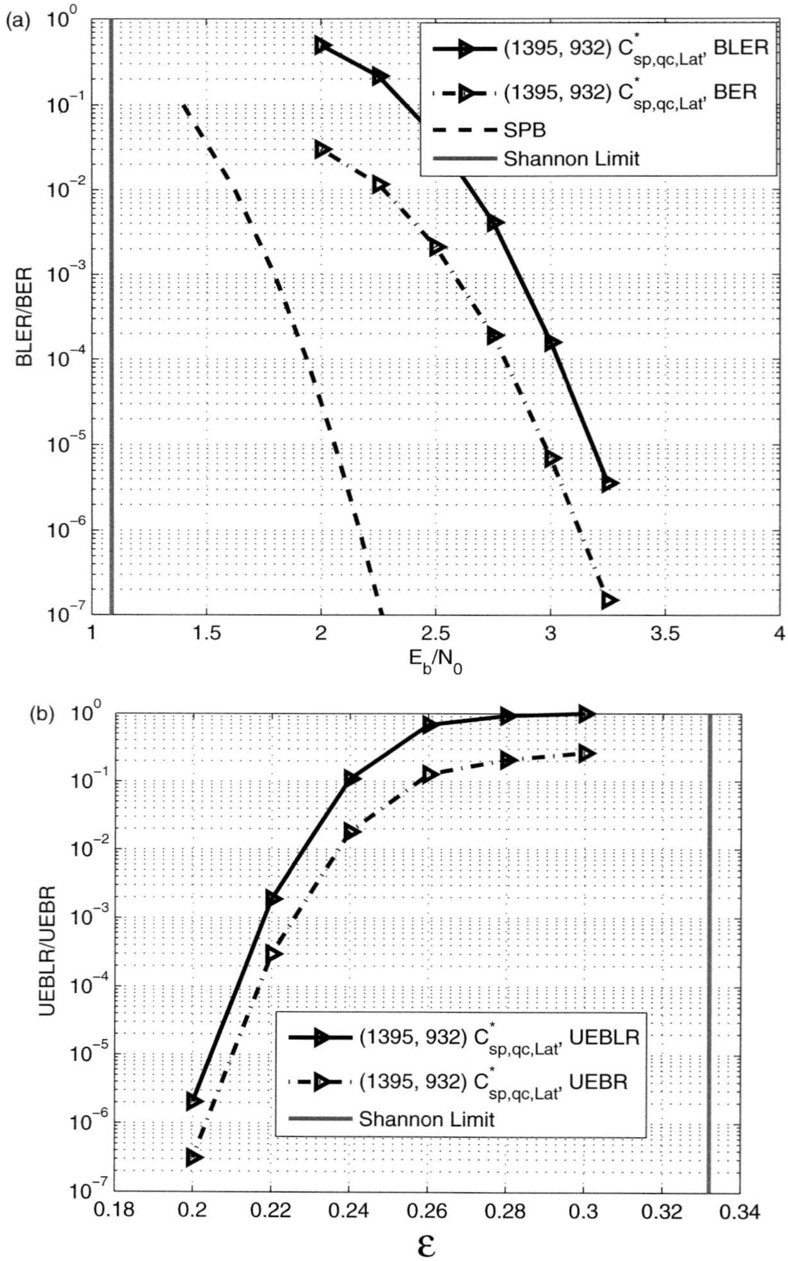

Figure 6.6. (a) The BER and BLER performances of the irregular (1395, 932) QC-SP-LDPC code $C^*_{sp,qc,Lat}$ given in Example 6.6 over the AWGNC; and (b) performance of the code in (a) over the BEC.

Table 6.2. *Row and column weight distributions of the SP-base matrix* $\mathbf{B}_{sp}$ *given in Example 6.7*

Column weight distribution		Row weight distribution	
Column weight	No. of columns	Row weight	No. of rows
2	144	4	13
3	360	5	31
9	72	6	65
		7	46
		8	104
		9	25
		10	4

and the replacement set $\boldsymbol{R}_{Lat}$ constructed above and using the SP-construction, we can construct an irregular 288×576 array $\mathbf{H}_{sp,qc,Lat}(15, 15)$ of CPMs and ZMs of size 15×15. The array $\mathbf{H}_{sp,qc,Lat}(15, 15)$ is a 4320×8640 matrix over GF(2) whose associated Tanner graph $\mathcal{G}_{sp,qc,Lat}(15, 15)$ has the same node degree distributions as the associated Tanner graph $\mathcal{G}_{sp}$ of the base matrix $\mathbf{B}_{sp}$, i.e., $\gamma(X)$ and $\rho(X)$.

The null space of $\mathbf{H}_{sp,qc,Lat}(15, 15)$ gives an irregular (8640, 4320) QC-SP-LDPC code $C_{sp,qc,Lat}$ with rate 0.5. The decoding threshold is 0.65 dB. The BER and BLER performances of this code over the AWGNC decoded with 50 iterations of the MSA are shown in Fig. 6.7(a). At BERs of 10^{-5} and 10^{-8}, the code performs 0.81 and 1.11 dBs away from the decoding threshold, respectively. At a BLER of 10^{-6}, the code performs 1.1 dB away from the SPB and it performs 1.6 dB away from the Shannon limit at a BER of 10^{-8}.

Its UEBR and UEBLR performances over the BEC are shown in Fig. 6.7(b). At a UEBR of 10^{-8}, the code performs about 0.1 away from the Shannon limit (0.5).

6.5 Discussion and Remarks

The methods presented or mentioned for constructing RC-constrained base matrices and replacement sets in this chapter are mostly based on finite fields or finite geometries. However, combinatorial designs, such as Latin squares and certain BIBDs, can also be used for constructing RC-constrained base matrices and replacement sets. Construction of a Latin square over a finite field GF(q) will be further discussed in the next chapter. The CPM-dispersion of such a Latin square gives an RC-constrained array $\mathbf{H}_{Lat}$ of CPMs and ZMs of size

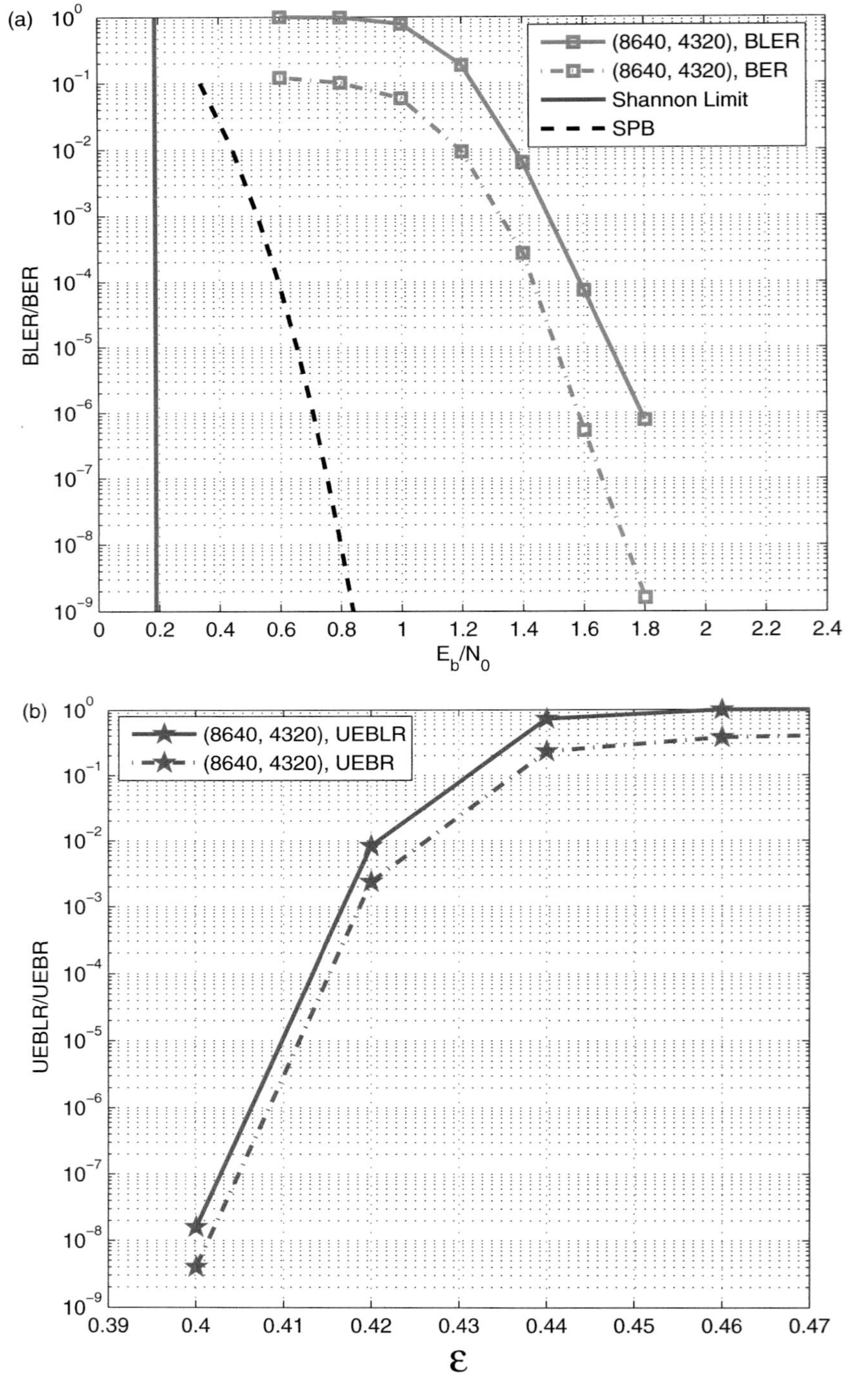

Figure 6.7. (a) The BER and BLER performances of the (8640, 4320) QC-SP-LDPC code $C_{sp,qc,Lat}$ given in Example 6.7 over the AWGNC; and (b) performance of the code in (a) over the BEC.

$(q-1) \times (q-1)$. Using this array, we can construct both RC-constrained base matrices and replacement sets for the SP-construction of QC-SP-LDPC codes. Furthermore, the null space of any subarray of $\mathbf{H}_{Lat}$ gives a QC-LDPC code whose Tanner graph has girth at least 6. Construction of QC-LDPC codes based on Latin squares can be found in [113]. These QC-LDPC codes have good trapping set structure [46].

Let $\mathbf{X} = \{x_0, x_1, \ldots, x_{\vartheta-1}\}$ be a set of ϑ objects and let $\mathbf{X}_0, \mathbf{X}_1, \ldots, \mathbf{X}_{b-1}$ be b distinct subsets of $\mathbf{X}$. These subsets are said to form a BIBD, denoted by $\boldsymbol{B}$, provided that they satisfy the following requirements [21, 98]: (1) For $0 \leq j < b$, each $\mathbf{X}_j$ is a k-subset of $\mathbf{X}$ (i.e., consisting of k objects in $\mathbf{X}$); (2) Each object appears in exactly r of the b subsets $\mathbf{X}_0, \mathbf{X}_1, \ldots, \mathbf{X}_{b-1}$; (3) Every two objects appear together in exactly λ of the b subsets $\mathbf{X}_0, \mathbf{X}_1, \ldots, \mathbf{X}_{b-1}$; and (4) The parameters ϑ, k and λ satisfy $0 < \lambda$ and $k < \vartheta - 1$. The subsets $\mathbf{X}_0, \mathbf{X}_1, \ldots, \mathbf{X}_{b-1}$ are called *blocks* of $\boldsymbol{B}$. A BIBD with parameters ϑ, b, k, r and λ is commonly referred to as a $(\vartheta, b, r, k, \lambda)$-BIBD (or a configuration). Form a $\vartheta \times b$ matrix $\mathbf{H}_{BIBD} = [h_{i,j}]_{0 \leq i < \vartheta, 0 \leq j < b}$ with rows corresponding to the ϑ objects and columns corresponding to the b blocks of $\boldsymbol{B}$, where $h_{i,j} = 1$ if the i-th object of $\mathbf{X}$ is in the j-th block of $\boldsymbol{B}$, otherwise $h_{i,j} = 0$. This matrix is called the *incidence matrix* of $\boldsymbol{B}$. It has constant column weight k and constant row weight r. For the special case with $\lambda = 1$, any two rows (or two columns) of the matrix $\mathbf{H}_{BIBD}$ have exactly one location where they both have 1-components. Therefore, $\mathbf{H}_{BIBD}$ satisfies the RC-constraint. Any submatrix of this RC-constrained matrix $\mathbf{H}_{BIBD}$ or $\mathbf{H}_{BIBD}$ itself can be used as a base matrix for the SP-construction of LDPC codes. For several classes of BIBDs with $\lambda = 1$, the incidence matrix $\mathbf{H}_{BIBD}$ of such a BIBD consists of a row of circulants of weight k [21, 98, 97]. These circulants satisfy both the RC- and the PW-RC-constraints and hence they can be used to form an RC-constrained replacement set. Of course, a subset of these circulants arranged in a row or a column can be used to form a base matrix for the SP-construction of QC-SP-LDPC codes. Constructions of BIBDs with $\lambda = 1$ can be found in [9, 21]. A good coverage of constructions of LDPC codes using BIBDs with $\lambda = 1$ can be found in [97, 3, 65].

Experimental results show that LDPC codes constructed based on BIBDs have low error-floors. This is an indication that these codes may have good trapping set structure, i.e., no small trapping sets. Investigation of the trapping set structure of LDPC codes constructed based on BIBDs may be an interesting research problem.

A final remark is that any RC-constrained base matrix constructed in this chapter can be used as the base matrix for constructing PTG-LDPC codes using the matrix decomposition method presented in Chapter 4.

7

SP-Construction of QC-LDPC Codes Using Matrix Dispersion and Masking

Since the discovery of SP-construction in 2002, many algebraic methods have been developed for constructing QC-LDPC codes [58, 76, 110, 109, 35, 107, 3, 19, 39, 74, 111, 102, 101, 112, 64, 65, 116, 97, 100, 50, 99, 113, 114, 25, 46, 26, 70, 68, 77, 78]. These methods basically evolved from SP-construction and they all have the same two ingredients as SP-construction, *a base matrix* and *matrix expansion*. The base matrix is constructed over a *NB field* satisfying certain constraints, in contrast to the base matrix over nonnegative integers for the PTG-construction of LDPC codes. These constraints ensure that the Tanner graph of the constructed QC-LDPC code has girth at least 6 or 8. All these methods are *special cases* of the SP-method for constructing QC-LDPC codes. Many codes with excellent performance, especially very low error-floors, have been constructed. For example, QC-LDPC codes with a BER as low as 10^{-15} without visible error-floors have recently been reported in [70, 68, 77, 78].

In this chapter, we unify all these algebraic constructions under the framework of the SP-construction. Also presented in this chapter is a technique, known as *masking* [112, 64, 97], to shape a chosen base matrix such that the produced QC-LDPC code has a better error performance. The masking technique is also used to construct QC-LDPC codes for correcting bursts of erasures over the BEC. Again, we only focus on binary LDPC codes.

7.1 A Deterministic SP-Construction of QC-LDPC Codes

Let α be a primitive element of GF(q). Then, $\alpha^0 = 1, \alpha^1, \alpha^2, \ldots, \alpha^{q-2}$ give all the nonzero elements of GF(q). Let $\mathbf{B}_{q,sp} = [b_{i,j}]_{0\le i<m, 0\le j<n}$ be an $m \times n$ matrix over GF(q), where the subscript "q" stands for "*q-ary*." Disperse each nonzero entry of $\mathbf{B}_{q,sp}$ into a binary CPM of size $(q-1) \times (q-1)$ as described in Section 2.1 and each 0-entry of $\mathbf{B}_{q,sp}$ into a ZM of size $(q-1) \times (q-1)$. This

results in an $m \times n$ array $\mathbf{H}_{b,sp,qc}(q-1,q-1)$ of binary CPMs and/or ZMs of size $(q-1) \times (q-1)$, where the subscript "b" stands for "*binary*." The array $\mathbf{H}_{b,sp,qc}(q-1,q-1)$ is called the *binary CPM-dispersion* of the q-ary matrix $\mathbf{B}_{q,sp}$. The array $\mathbf{H}_{b,sp,qc}(q-1,q-1)$ is an $m(q-1) \times n(q-1)$ matrix over GF(2). Note that the array is a *binary expansion* of the NB matrix $\mathbf{B}_{q,sp}$ (*a nonbinary-to-binary (NB-to-B) mapping*). The null space of $\mathbf{H}_{b,sp,qc}(q-1,q-1)$ gives a binary QC-LDPC code $C_{b,sp,qc}$ of length $n(q-1)$.

The above method for constructing binary QC-LDPC codes can be viewed as an SP-construction with a *deterministic* CPM replacement for each nonzero entry in the base matrix $\mathbf{B}_{q,sp}$, i.e., each nonzero entry of $\mathbf{B}_{q,sp}$ is replaced by a *specific* binary CPM of size $(q-1) \times (q-1)$. We call this SP-construction the *CPM-D-SP-construction*, where the symbol "*D*" stands for "*dispersion*," and the matrix $\mathbf{B}_{q,sp}$ is the base matrix. The code $C_{b,sp,qc}$ given by the null space of $\mathbf{H}_{b,sp,qc}(q-1,q-1)$ is called a *CPM-QC-SP-LDPC code*. Most of the algebraic methods for constructing QC-LDPC codes presented in [19, 39, 74, 111, 102, 101, 112, 64, 65, 116, 97, 100, 50, 99, 113, 114, 25, 46, 26, 70, 68, 77, 78] are CPM-D-SP-construction and hence special cases of the SP-construction, where the SP-base matrix in this case is constructed over an NB finite field.

From the graph-theoretic point of view, the CPM-D-SP-construction is a *specific* PTG-based construction. The Tanner graph $\mathcal{G}_{q,sp}$ associated with the NB base matrix $\mathbf{B}_{q,sp}$ is the protograph for the construction of the binary code $C_{b,sp,qc}$. Since $\mathbf{B}_{q,sp}$ is a matrix over GF(q), the edges of $\mathcal{G}_{q,sp}$ are labeled with nonzero elements in GF(q). In the PTG-based construction, the protograph $\mathcal{G}_{q,sp}$ is copied $q-1$ times and the copies are connected by permuting their edges based on the binary CPM-dispersions of the edge labels in $\mathcal{G}_{q,sp}$. This copy-and-permute operation results in a connected bipartite graph whose adjacency matrix is the binary parity-check array $\mathbf{H}_{b,sp,qc}(q-1,q-1)$ of a binary QC-PTG-LDPC code $C_{b,ptg,qc}$ which is exactly the binary CPM-QC-SP-LDPC code $C_{b,sp,qc}$ described above.

The nonzero entries in the base matrix $\mathbf{B}_{q,sp}$ can also be dispersed into binary CPMs of size equal to, smaller than, or larger than $q-1$ as described in Section 2.1.

7.2 Conditions on Girth of CPM-QC-SP-LDPC Codes

The above CPM-D-SP-construction of a binary CPM-QC-SP-LDPC code reduces to the construction of the base matrix $\mathbf{B}_{q,sp}$ over an NB field GF(q). For such a CPM-QC-SP-LDPC code, the girth of the Tanner graph of the code is related to the choice of the base matrix $\mathbf{B}_{q,sp}$. The following two theorems give the necessary and sufficient conditions on the base matrix $\mathbf{B}_{q,sp}$ for the Tanner graph of a CPM-QC-SP-LDPC code to have girth at least 6 and 8, respectively. These two theorems were proved in [25, 26]. Here, we rephrase them in terms of the CPM-D-SP-construction.

Theorem 7.1. *Let $\mathbf{B}_{q,sp}$ be the base matrix of a binary CPM-QC-SP-LDPC code $C_{b,sp,qc}$ whose parity-check matrix $\mathbf{H}_{b,sp,qc}(q-1,q-1)$ is the binary CPM-dispersion of $\mathbf{B}_{q,sp}$. A necessary and sufficient condition for the Tanner graph of $C_{b,sp,qc}$ to have girth at least 6 is that every 2×2 submatrix in the base matrix $\mathbf{B}_{q,sp}$ contains at least one zero entry or is non-singular.*

Theorem 7.2. *A necessary and sufficient condition for the Tanner graph of a binary CPM-QC-SP-LDPC code $C_{b,sp,qc}$ with base matrix $\mathbf{B}_{q,sp}$ to have girth at least 8 is that no 2×2 or 3×3 submatrix in the base matrix $\mathbf{B}_{q,sp}$ has two identical nonzero terms in its determinant expansion.*

For convenience, the conditions given in Theorems 7.1 and 7.2 are called the 2×2 and the 3×3 *submatrix (SM) constraints*, respectively. In the CPM-D-SP-construction of a binary CPM-QC-SP-LDPC code, the 2×2 and the 3×3 SM-constraints are used to construct a base matrix. Note that the conditions given in Theorems 7.1 and 7.2 imply that the parity-check matrix $\mathbf{H}_{b,sp,qc}(q-1,q-1)$ obtained by dispersing the base matrix $\mathbf{B}_{q,sp}$ satisfies the RC-constraint [25, 26]. Hence, the Tanner graph of the CPM-QC-SP-LDPC code given by the null space of $\mathbf{H}_{b,sp,qc}(q-1,q-1)$ has girth at least 6.

7.3 A Finite Field Construction of 2 x 2 SM-Constrained SP-Base Matrices and Their Associated CPM-QC-SP-LDPC Codes

There are many algebraic methods for constructing NB base matrices that satisfy the 2×2 SM-constraint given in Theorem 7.1 [19, 39, 74, 111, 102, 101, 64, 65, 116, 100, 50, 114, 70, 68, 78]. Among these methods, the most flexible one is the method proposed in [70, 68]. The construction is based on *two arbitrary sets* of a chosen NB field. In the following, we describe this method and its special cases for later use of constructing doubly QC-, QC spatially coupled, and QC globally coupled LDPC codes in Chapters 8 to 10, and NB LDPC codes in Chapter 11.

Let α be a primitive element of the field GF(q). For $1 \leq m,n \leq q$, let $\boldsymbol{S}_0 = \{\alpha^{i_0}, \alpha^{i_1}, \ldots, \alpha^{i_{m-1}}\}$ and $\boldsymbol{S}_1 = \{\alpha^{j_0}, \alpha^{j_1}, \ldots, \alpha^{j_{n-1}}\}$ be two arbitrary subsets of elements in GF(q) with i_k and j_l in $L \triangleq \{-\infty, 0, 1, 2, \ldots, q-2\}$, $i_0 < i_1 < \ldots < i_{m-1}$ and $j_0 < j_1 < \ldots < j_{n-1}$. Let η be any nonzero element in GF(q). We then form the following $m \times n$ matrix over GF(q):

$$\mathbf{B}_{q,sp,s} = \begin{bmatrix} \eta\alpha^{i_0} + \alpha^{j_0} & \eta\alpha^{i_0} + \alpha^{j_1} & \cdots & \eta\alpha^{i_0} + \alpha^{j_{n-1}} \\ \eta\alpha^{i_1} + \alpha^{j_0} & \eta\alpha^{i_1} + \alpha^{j_1} & \cdots & \eta\alpha^{i_1} + \alpha^{j_{n-1}} \\ \vdots & \vdots & \ddots & \vdots \\ \eta\alpha^{i_{m-1}} + \alpha^{j_0} & \eta\alpha^{i_{m-1}} + \alpha^{j_1} & \cdots & \eta\alpha^{i_{m-1}} + \alpha^{j_{n-1}} \end{bmatrix}. \quad (7.1)$$

Note that the entries of the above matrix $\mathbf{B}_{q,sp,s}$ are the *linear sums* of elements from the two subsets $\boldsymbol{S}_0$ and $\boldsymbol{S}_1$. This matrix $\mathbf{B}_{q,sp,s}$ has the following structural properties [68]: (1) All the entries in a row (or a column) of $\mathbf{B}_{q,sp,s}$ are distinct elements of GF(q); (2) Each row (or each column) of $\mathbf{B}_{q,sp,s}$ contains at most one zero element; and (3) No two rows (or two columns) in $\mathbf{B}_{q,sp,s}$ have identical entries at any position. It was proved in [68] that the matrix $\mathbf{B}_{q,sp,s}$ satisfies the 2×2 SM-constraint.

If we disperse each entry in $\mathbf{B}_{q,sp,s}$ into a $(q-1) \times (q-1)$ binary CPM or ZM, then the CPM-dispersion of $\mathbf{B}_{q,sp,s}$ results in an RC-constrained $m \times n$ array $\mathbf{H}_{b,sp,qc,s}(q-1, q-1)$ of binary CPMs and/or ZMs of size $(q-1) \times (q-1)$. The null space of $\mathbf{H}_{b,sp,qc,s}(q-1, q-1)$ gives a binary CPM-QC-SP-LDPC code $C_{b,sp,qc}$ whose Tanner graph has girth of at least 6. The subscript "s" in $\mathbf{B}_{q,sp,s}$ and $\mathbf{H}_{b,sp,qc,s}(q-1, q-1)$ represents "*sum-form*" of $\mathbf{B}_{q,sp,s}$ given by (7.1).

There are two special cases of the base matrix given in (7.1). The first case is that $\boldsymbol{S}_0 = \boldsymbol{S}_1 = \mathrm{GF}(q) = \{0, 1, \alpha, \alpha^2, \ldots, \alpha^{q-2}\}$. In this case, $\mathbf{B}_{q,sp,s}$ is a $q \times q$ matrix over GF(q) in which each row and each column contains every element of GF(q) exactly once. Thus, $\mathbf{B}_{q,sp,s}$ is simply a Latin square over GF(q) defined in Section 6.4, where we use Latin squares over GF(q) to construct RC-constrained replacement sets for the SP-construction of QC-SP-LDPC codes.

The second case is that $\boldsymbol{S}_0$ and $\boldsymbol{S}_1$ are two additive subgroups of the additive group of GF(q) with orders m and n, respectively, such that $m + n \leq q$ and $\boldsymbol{S}_0 \cap \boldsymbol{S}_1 = \{0\}$. Let $\alpha^{i_0} = \alpha^{j_0} = 0$. In this case, the first row of $\mathbf{B}_{q,sp,s}$ contains the elements of the subgroup $\boldsymbol{S}_1$ and all the rows of $\mathbf{B}_{q,sp,s}$ are *cosets* of $\boldsymbol{S}_1$, with elements in the set $\eta\boldsymbol{S}_0 = \{\eta\alpha^{i_0}, \eta\alpha^{i_1}, \ldots, \eta\alpha^{i_{m-1}}\}$ as *coset leaders*. The construction of QC-LDPC codes based on two additive subgroups of the additive group of GF(q) with orders m and $n = q - m$, respectively, and $\eta = 1$, was first proposed in [100] and later generalized in [50].

The construction of a 2×2 SM-constrained base matrix based on the two subsets $\boldsymbol{S}_0$ and $\boldsymbol{S}_1$ of GF(q) can also be put in *product-form* as follows:

$$\mathbf{B}_{q,sp,p} = \begin{bmatrix} \alpha^{i_0}\alpha^{j_0} - \eta & \alpha^{i_0}\alpha^{j_1} - \eta & \cdots & \alpha^{i_0}\alpha^{j_{n-1}} - \eta \\ \alpha^{i_1}\alpha^{j_0} - \eta & \alpha^{i_1}\alpha^{j_1} - \eta & \cdots & \alpha^{i_1}\alpha^{j_{n-1}} - \eta \\ \vdots & \vdots & \ddots & \vdots \\ \alpha^{i_{m-1}}\alpha^{j_0} - \eta & \alpha^{i_{m-1}}\alpha^{j_1} - \eta & \cdots & \alpha^{i_{m-1}}\alpha^{j_{n-1}} - \eta \end{bmatrix}. \quad (7.2)$$

The null space of the binary CPM-dispersion of the base matrix $\mathbf{B}_{q,sp,p}$, denoted by $\mathbf{H}_{b,sp,qc,p}(q-1,q-1)$, gives a CPM-QC-SP-LDPC code. The subscript "p" in $\mathbf{B}_{q,sp,p}$ and $\mathbf{H}_{b,sp,qc,p}(q-1,q-1)$ stands for the "*product-form*" of the base matrix $\mathbf{B}_{q,sp,p}$ given by (7.2).

There are four special cases of the base matrix $\mathbf{B}_{q,sp,p}$ in the product-form given in (7.2), which are summarized as follows. The first case is that $S_0 = S_1 = \text{GF}(q)\backslash\{0\} = \left\{1, \alpha, \alpha^2, \ldots, \alpha^{q-3}, \alpha^{q-2}\right\}$ and $\eta = 1$. In this case, the base matrix $\mathbf{B}_{q,sp,p}$ defined by (7.2) becomes the base matrix given by (4) in [64, p. 2432] shown as follows:

$$\mathbf{B}_{q,sp,p} = \begin{bmatrix} \alpha^0 - 1 & \alpha - 1 & \alpha^2 - 1 & \cdots & \alpha^{q-3} - 1 & \alpha^{q-2} - 1 \\ \alpha - 1 & \alpha^2 - 1 & \alpha^3 - 1 & \cdots & \alpha^{q-2} - 1 & \alpha^0 - 1 \\ \vdots & \vdots & \vdots & \ddots & \vdots & \vdots \\ \alpha^{q-2} - 1 & \alpha^0 - 1 & \alpha^1 - 1 & \cdots & \alpha^{q-4} - 1 & \alpha^{q-3} - 1 \end{bmatrix}. \tag{7.3}$$

It is a $(q-1) \times (q-1)$ matrix over GF(q). Note that each row of the base matrix $\mathbf{B}_{q,sp,p}$ given by (7.3) is the cyclic-shift of the row above it one position to the left and the top row is the cyclic-shift of the last row one position to the left. Each column of $\mathbf{B}_{q,sp,p}$ is the cyclic-shift of the column on its left one place upward and the first column is the cyclic-shift of the last column upward one place. Hence, it has a cyclic structure. We also note that for $0 \le i < q-1$, the i-th column is the transpose of the i-th row and vice versa.

The second case is that S_0 and S_1 are two cyclic subgroups of the multiplicative group of GF(q) with orders m and n, respectively, such that $S_0 \cap S_1 = \{1\}$, and m and n are relatively prime factors of $q-1$. Let $\alpha^{i_0} = \alpha^{j_0} = 1$ and $\eta = 1$. Then, all the entries in the matrix $\mathbf{B}_{q,sp,p}$ defined by (7.2) are distinct elements in GF(q). The construction of QC-LDPC codes based on two cyclic subgroups of the multiplicative group of GF(q) with orders m and n, respectively, such that $mn = q-1$ was first proposed in [100] and later extended in [114].

The third case is that $S_0 = \{\alpha^{-j_1}, \alpha^{-j_2}, \ldots, \alpha^{-j_k}\}$, $S_1 = \{\alpha^{j_1}, \alpha^{j_2}, \ldots, \alpha^{j_k}\}$, and $\eta = 1$, where α^{j_1}, α^{j_2}, ..., α^{j_k} are all the primitive elements of GF(q) and $k = (q-1)\prod_{i=1}^{t}(1 - \frac{1}{p_i})$ in which $p_1, p_2, \ldots, p_t$ are all the prime factors of $q-1$. Using these two sets of elements of GF(q), the base matrix $\mathbf{B}_{q,sp,p}$ constructed from (7.2) is exactly the base matrix given by equation (4) of [100, p. 86] or equation (11.17) of [97, p. 511].

For the fourth case, let GF(q^2) be an extension field of GF(q). Then, GF(q) is a subfield of GF(q^2). Let α and β be two primitive elements of GF(q) and GF(q^2), respectively. We then have $\text{GF}(q) = \{0, 1, \alpha, \alpha^2, \ldots, \alpha^{q-2}\}$ and $\text{GF}(q^2) = \{0, 1, \beta, \beta^2, \ldots, \beta^{q^2-2}\}$. There exists q distinct elements, β^{j_0}, β^{j_1}, ..., $\beta^{j_{q-1}}$ in GF(q^2) with $j_0, j_1, \ldots, j_{q-1} \in \{0, 1, 2, \ldots, q^2-2\}$, in which $\beta^{j_k} \neq \alpha^l \beta^0$ and for $i \neq k$, $\beta^{j_k} \neq 1 + \alpha^l \beta^{j_i}$ with $0 \le l < q-1$. Let $S_0 = \{\beta^{j_0}, \beta^{j_1}, \dots, \beta^{j_{q-1}}\}$, $S_1 = \text{GF}(q) = \{0, 1, \alpha, \alpha^2, \dots, \alpha^{q-2}\}$, and $\eta = -1$. In this case, the matrix $\mathbf{B}_{q,sp,p}$ constructed based on these two subsets S_0 and S_1 of GF(q^2) becomes the base matrix given by equation (11.20) of [97, Chapter 11, p. 514], constructed based on the two-dimensional Euclidean geometry EG($2, q$) [97].

The above six special cases of the base matrices defined by (7.1) and (7.2) in the sum- and product-forms constructed based on two subsets of an NB field GF(q) are the major constructions of base matrices given in [35, 101, 64, 65, 100, 50, 113, 114]. In Chapters 8, 9, and 10, we will use the special case of the product-from given in the form of (7.3) for constructing doubly, spatially coupled, and globally coupled QC-LDPC codes, respectively.

7.4 Masking

Given an NB base matrix $\mathbf{B}_{q,sp}$ that satisfies the 2×2 SM-constraint, a shaping operation can be performed on $\mathbf{B}_{q,sp}$ to obtain a new base matrix which can also satisfy the 3×3 SM-constraint. The matrix shaping operation is known as *masking* [112, 64, 97, 68].

Let $\mathbf{B}_{q,sp} = [b_{i,j}]_{0 \le i < m, 0 \le j < n}$ be an $m \times n$ 2×2 SM-constrained base matrix over GF(q). Suppose we replace a nonzero entry in $\mathbf{B}_{q,sp}$ by the zero-element of GF(q). In the CPM-dispersion $\mathbf{H}_{b,sp,qc}(q-1, q-1)$ of $\mathbf{B}_{q,sp}$, this replacement results in replacing a $(q-1) \times (q-1)$ CPM in $\mathbf{H}_{b,sp,qc}(q-1, q-1)$ by a $(q-1) \times (q-1)$ ZM which is referred to as *masking*. Let μ be a nonnegative integer less than the number of nonzero entries in $\mathbf{B}_{q,sp}$. The replacement of μ nonzero entries in $\mathbf{B}_{q,sp}$ by μ zeros amounts to replacing μ CPMs by μ ZMs at the locations in $\mathbf{H}_{b,sp,qc}(q-1, q-1)$ corresponding to the locations of the μ nonzero entries in $\mathbf{B}_{q,sp}$ which are replaced by zeros. Masking μ CPMs in $\mathbf{H}_{b,sp,qc}(q-1, q-1)$ amounts to removing $\mu(q-1)$ edges from the Tanner graph $\mathcal{G}_{b,sp,qc}(q-1, q-1)$ associated with the array $\mathbf{H}_{b,sp,qc}(q-1, q-1)$. Removing these edges from $\mathcal{G}_{b,sp,qc}(q-1, q-1)$ may break many short cycles. As a result, the resultant Tanner graph, denoted by $\mathcal{G}_{b,sp,qc,mask}(q-1, q-1)$, may have a smaller number of short cycles, or a larger girth (larger than 6), or both. The subscript "*mask*" in $\mathcal{G}_{b,sp,qc,mask}(q-1, q-1)$ stands for "*masking*." In choosing the nonzero entries in $\mathbf{B}_{q,sp}$ to be masked, we have to avoid disconnecting the Tanner graph of $\mathbf{H}_{b,sp,qc}(q-1, q-1)$ into disjoint graphs.

Masking the base matrix $\mathbf{B}_{q,sp} = [b_{i,j}]_{0 \le i < m, 0 \le j < n}$ can be modeled mathematically as a *Hadamard matrix product* [42]. Let $\mathbf{Z}$ be an $m \times n$ matrix with the zero and unit elements of GF(q) as entries in the form of $\mathbf{Z} = [z_{i,j}]_{0 \le i < m, 0 \le j < n}$. Define the following product of $\mathbf{Z}$ and $\mathbf{B}_{q,sp}$: $\mathbf{B}_{q,sp,mask} =$

$\mathbf{Z} \odot \mathbf{B}_{q,sp} = [z_{i,j}b_{i,j}]_{0 \le i < m, 0 \le j < n}$ where $z_{i,j}b_{i,j} = b_{i,j}$ if $z_{i,j} = 1$ and $z_{i,j}b_{i,j} = 0$ if $z_{i,j} = 0$. In this matrix product, the nonzero entries in $\mathbf{B}_{q,sp}$ at the locations corresponding to the zero entries in $\mathbf{Z}$ are replaced (or masked) by zeros. The binary CPM-dispersion of $\mathbf{B}_{q,sp,mask}$ gives an $m \times n$ masked array, denoted by $\mathbf{H}_{b,sp,qc,mask}(q-1, q-1)$, of CPMs and ZMs of size $(q-1) \times (q-1)$. We call $\mathbf{Z}$ and $\mathbf{B}_{q,sp,mask}$ the *masking matrix* and the *masked base matrix*, respectively. The null space of $\mathbf{H}_{b,sp,qc,mask}(q-1, q-1)$ also gives a binary CPM-QC-SP-LDPC code (called masked LDPC code), denoted by $C_{b,sp,qc,mask}$. Since $\mathbf{B}_{q,sp}$ already satisfies the 2×2 SM-constraint, the masking matrix $\mathbf{Z}$ should be designed to obtain a masked base matrix $\mathbf{B}_{q,sp,mask}$ which satisfies the 3×3 SM-constraint, called a 3×3 *SM-constrained base matrix*. In this case, the Tanner graph $\mathcal{G}_{b,sp,qc,mask}(q-1, q-1)$ associated with the masked array $\mathbf{H}_{b,sp,qc,mask}(q-1, q-1)$ has girth at least 8, as well as a small number of short cycles. Binary CPM-dispersions of properly chosen base matrices in conjunction with masking can result in CPM-QC-SP-LDPC codes with *good error probability as well as low error-floors*.

Given an $m \times n$ base matrix $\mathbf{B}_{q,sp}$, the problem of how to design a masking matrix $\mathbf{Z}$ to produce a masked base matrix $\mathbf{B}_{q,sp,mask}$ which results in a CPM-QC-SP-LDPC code whose Tanner graph not only has a large girth but also contains a small number of short cycles, is challenging. If the base matrix $\mathbf{B}_{q,sp}$ satisfies the 2×2 SM-constraint, a general guideline is to construct a masking matrix which yields a masked based matrix $\mathbf{B}_{q,sp,mask}$ in which each 3×3 submatrix contains at least one zero entry. This may produce a masked base matrix that satisfies the 3×3 SM-constraint given in Theorem 7.2. For constructing an irregular CPM-QC-SP-LDPC code based on designed degree distributions of VNs and CNs of its Tanner graph, the masking matrix must be constructed with column and row weight distributions matching the VN and CN degree distributions of the optimized Tanner graph. This can be achieved by using the PEG algorithm with computer search. In the PTG-based construction of LDPC codes, the PEG algorithm is often used to design a protograph.

In the following, we illustrate the CPM-D-SP-construction of LDPC codes using several examples, and demonstrate the effectiveness of the masking technique.

Example 7.1. Let α be a primitive element of the prime field GF(331). Set $m = 4$, $n = 8$, and $\eta = 1$. We choose the following two subsets of GF(331) for constructing a base matrix using the sum-form given by (7.1): $S_0 = \{\alpha^{112}, \alpha^{115}, \alpha^{148}, \alpha^{317}\}$ and $S_1 = \{\alpha^{24}, \alpha^{112}, \alpha^{115}, \alpha^{234}, \alpha^{236}, \alpha^{274}, \alpha^{316}, \alpha^{320}\}$. The elements of these two sets are chosen *at random* from the field GF(331). Using the construction given by

(7.1), we obtain the following 4×8 base matrix $\mathbf{B}_{q,sp,s}$ over GF(331):

$$\mathbf{B}_{q,sp,s} = \begin{bmatrix} \alpha^{39} & \alpha^{123} & \alpha^{79} & \alpha^{190} & \alpha^{294} & \alpha^{328} & \alpha^{297} & \alpha^{68} \\ \alpha^{139} & \alpha^{79} & \alpha^{126} & \alpha^{75} & \alpha^{158} & \alpha^{295} & \alpha^{148} & \alpha^{274} \\ \alpha^{206} & \alpha^{209} & \alpha^{287} & \alpha^{324} & \alpha^{163} & \alpha^{129} & \alpha^{202} & \alpha^{325} \\ \alpha^{181} & \alpha^{271} & \alpha^{275} & \alpha^{143} & \alpha^{208} & \alpha^{10} & \alpha^{173} & \alpha^{284} \end{bmatrix}. \quad (7.4)$$

This base matrix satisfies the 2×2 SM-constraint but not the 3×3 SM-constraint given in Theorem 7.2. All the entries in this matrix are nonzero elements in GF(331). The binary CPM-dispersion of this base matrix gives a 4×8 array $\mathbf{H}_{b,sp,qc,s}(330, 330)$ of binary CPMs of size 330×330. The Tanner graph $\mathcal{G}_{b,sp,qc,s}(330, 330)$ associated with the array $\mathbf{H}_{b,sp,qc,s}(330, 330)$ has girth 6 and contains 990 cycles of length six, 24 750 cycles of length 8, and 389 400 cycles of length 10.

Suppose we mask the base matrix $\mathbf{B}_{q,sp,s}$ with the following masking matrix:

$$\mathbf{Z} = \begin{bmatrix} 1 & 0 & 1 & 0 & 1 & 1 & 1 & 1 \\ 0 & 1 & 0 & 1 & 1 & 1 & 1 & 1 \\ 1 & 1 & 1 & 1 & 1 & 0 & 1 & 0 \\ 1 & 1 & 1 & 1 & 0 & 1 & 0 & 1 \end{bmatrix}. \quad (7.5)$$

Note that every 3×3 submatrix of $\mathbf{Z}$ contains at least one 0-entry. Masking the base matrix $\mathbf{B}_{q,sp,s}$ given by (7.4) with the masking matrix $\mathbf{Z}$ given by (7.5), we obtain the following masked base matrix:

$$\mathbf{B}_{q,sp,s,mask} = \begin{bmatrix} \alpha^{39} & 0 & \alpha^{79} & 0 & \alpha^{294} & \alpha^{328} & \alpha^{297} & \alpha^{68} \\ 0 & \alpha^{79} & 0 & \alpha^{75} & \alpha^{158} & \alpha^{295} & \alpha^{148} & \alpha^{274} \\ \alpha^{206} & \alpha^{209} & \alpha^{287} & \alpha^{324} & \alpha^{163} & 0 & \alpha^{202} & 0 \\ \alpha^{181} & \alpha^{271} & \alpha^{275} & \alpha^{143} & 0 & \alpha^{10} & 0 & \alpha^{284} \end{bmatrix}. \quad (7.6)$$

By checking, we find that the masked base matrix $\mathbf{B}_{q,sp,s,mask}$ satisfies the 3×3 SM-constraint. The binary CPM-dispersion of $\mathbf{B}_{q,sp,s,mask}$ gives a 4×8 array $\mathbf{H}_{b,sp,qc,s,mask}(330, 330)$ of binary CPMs and ZMs of size 330×330. It is a 1320×2640 matrix with column and row weights 3 and 6, respectively. Since $\mathbf{B}_{q,sp,s,mask}$ satisfies both the 2×2 and the 3×3 SM-constraints, it follows from Theorem 7.2 that the Tanner graph $\mathcal{G}_{b,sp,qc,s,mask}(330, 330)$ associated with the array $\mathbf{H}_{b,sp,qc,s,mask}(330, 330)$ has girth at least 8. Using the cycle counting algorithm given in [71], we find that $\mathcal{G}_{b,sp,qc,s,mask}(330, 330)$ has girth 8 and contains 990 cycles of length 8 and 8 580 cycles of length 10. We see that masking the base matrix $\mathbf{B}_{q,sp,s}$ with $\mathbf{Z}$ not only increases the girth of the Tanner graph associated with its CPM-dispersion $\mathbf{H}_{b,sp,qc,s}(330, 330)$ (from 6 to 8) but also greatly reduces the number of short cycles (lengths 6, 8, and 10).

The null space of $\mathbf{H}_{b,sp,qc,s,mask}(330, 330)$ gives a (3, 6)-regular (2640, 1320) CPM-QC-SP-LDPC code $C_{b,sp,qc,mask}$ with rate 0.5. The BER and BLER performances of this code decoded with 5, 10, and 50 iterations of the MSA are shown in Fig. 7.1(a). The decoding of the code converges quickly. At a BER of 10^{-10}, the gap between 5 and 10 iterations is about 1.8 dB and the gap between 10 and 50 iterations is within 0.8 dB. The code $C_{b,sp,qc,mask}$ is the same code given in Example 4.2 (the (2640, 1320) QC-PTG-LDPC code $C_{ptg,qc,1}$), where we constructed the code using the algebraic method with the masking matrix given by (7.5) as the base matrix for decomposition. Here, we construct the code using the CPM-D-SP-construction in conjunction with the matrix given by (7.5) as the masking matrix. This again shows the strong connection of the PTG-based construction to the algebraic SP-construction.

The UEBR and UEBLR performances of the code $C_{b,sp,qc,mask}$ over the BEC are shown in Fig. 7.1(b). At a UEBR of 10^{-10}, the code performs about 0.14 away from the Shannon limit.

In Examples 4.2 and 7.1, we use the same 4×8 matrix given by (4.7) and (7.5) in two different ways, one as a decomposition base matrix for the algebraic construction of a PTG-LDPC code and the other as a masking matrix for the SP-construction of an SP-LDPC code. Both ways result in a code with very good performance, especially in the error-floor region. Thus, it would be interesting to know whether a good binary masking matrix is also a good decomposition base matrix for algebraic construction of PTG-LDPC codes. Another interesting question is whether a large girth and a small number of short cycles result in a good decoding threshold. These two problems deserve further investigation.

Example 7.2. In this example, we use the same field GF(331) given in Example 7.1 to construct another CPM-QC-SP-LDPC code of length 2640 and rate 0.5 as the (2640, 1320) code given in Example 7.1. In Example 7.1, the elements of the two subsets of GF(331) for constructing the code were chosen at random. However, in this example, the elements of two subsets of GF(331) are chosen specifically to construct a CPM-QC-SP-LDPC code whose Tanner graph has girth 10.

The two sets of GF(331) chosen for constructing the base matrix are: $S_0^* = \{\alpha^0, \alpha^1, \alpha^2, \alpha^3\}$ and $S_1^* = \{\alpha^{130}, \alpha^{131}, \alpha^{132}, \alpha^{133}, \alpha^{134}, \alpha^{135}, \alpha^{136}, \alpha^{137}\}$. The first subset consists of 4 consecutive powers of α, a primitive element of GF(331), and the second subset consists of 8 consecutive powers of α. Based on these two subsets with $m = 4$, $n = 8$, and $\eta = -1$, we construct a 4×8 base matrix $\mathbf{B}_{q,sp,s}^*$ over GF(331). The CPM-dispersion of the base matrix $\mathbf{B}_{q,sp,s}^*$ gives a 4×8 array $\mathbf{H}_{b,sp,qc,s}^*(330, 330)$ of CPMs of size 330×330 (no ZM), a 1320×2640 matrix

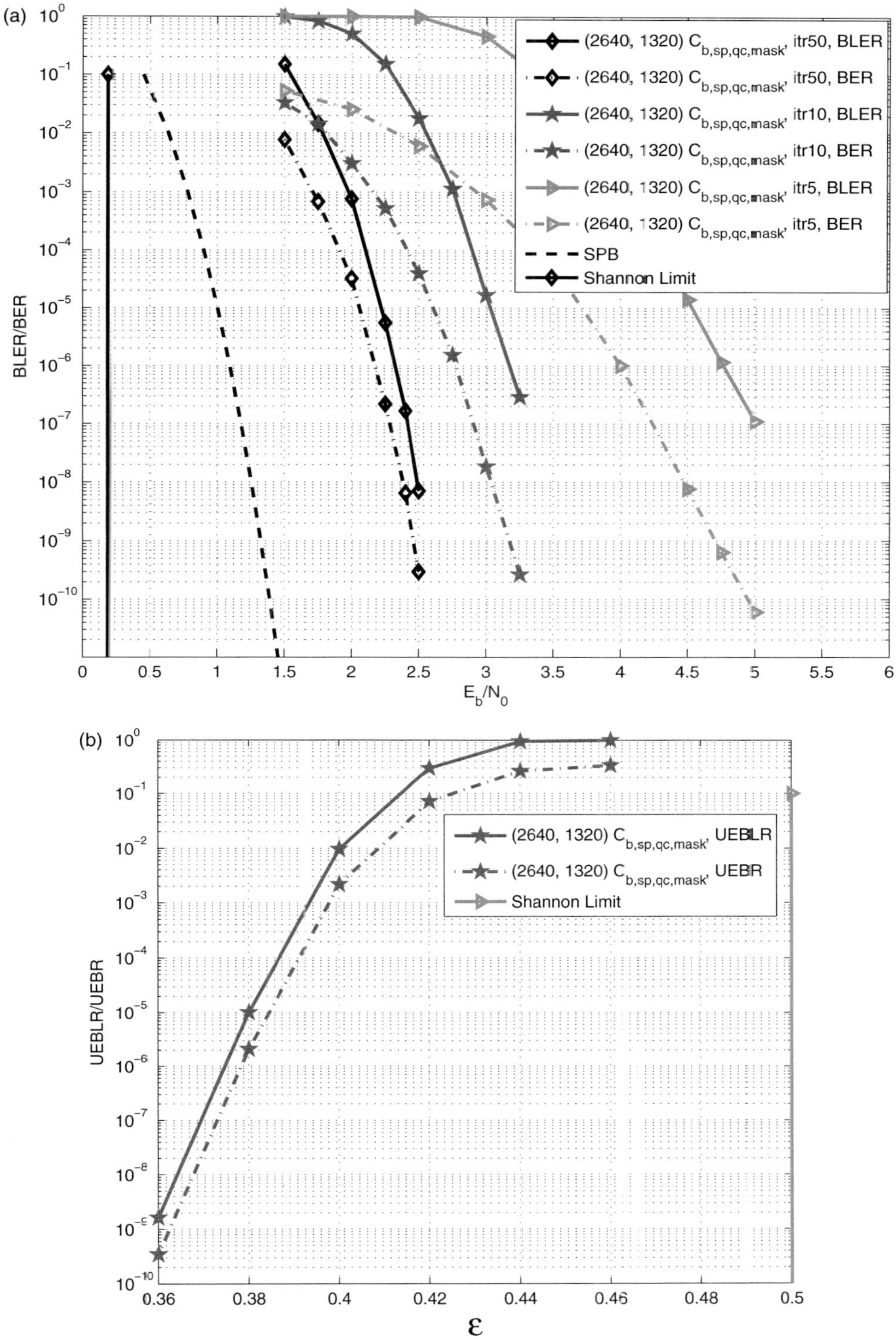

Figure 7.1. (a) The BER and BLER performances of the (2640, 1320) code $C_{b,sp,qc,mask}$ given in Example 7.1 decoded with 5, 10, and 50 iterations of the MSA over the AWGNC; and (b) performance of the code in (a) over the BEC.

over GF(2) with column and row weights 4 and 8, respectively. The Tanner graph $\mathcal{G}^*_{b,sp,qc,s}(330, 330)$ associated with the array $\mathbf{H}^*_{b,sp,qc,s}(330, 330)$ has girth 6 and contains 7260 cycles of length six, 34 650 cycles of length 8, and 499 620 cycles of length 10 (a very large number of short cycles).

Suppose we mask the base matrix $\mathbf{B}^*_{q,sp,s}$ with the masking matrix $\mathbf{Z}$ given by (7.5), the same masking matrix used in Example 7.1. Masking results in a masked matrix $\mathbf{B}^*_{q,sp,s,mask}$. The CPM-dispersion of $\mathbf{B}^*_{q,sp,s,mask}$ gives a 4×8 masked array $\mathbf{H}^*_{q,sp,qc,s,mask}(330, 330)$ of binary CPMs and ZMs of size 330×330, a 1320×2640 binary matrix with column and row weights 3 and 6, respectively. The null space of the masked array $\mathbf{H}^*_{q,sp,qc,s,mask}(330, 330)$ gives a (3, 6)-regular (2640, 1320) CPM-QC-SP-LDPC code $C^*_{b,sp,qc,mask}$ which has the same length and rate as that of the code $C_{b,sp,qc,mask}$ constructed in Example 7.1. The Tanner graph of the code $C^*_{b,sp,qc,mask}$ has girth 10 and contains 13 200 cycles of length 10. We see that masking increases the girth of the Tanner graph $\mathcal{G}^*_{b,sp,qc,s}(330, 330)$ associated with the unmasked array $\mathbf{H}^*_{b,sp,qc,s}(330, 330)$ from 6 to 10 for the Tanner graph $\mathcal{G}^*_{b,sp,qc,s,mask}(330, 330)$ associated with the masked array $\mathbf{H}^*_{b,sp,qc,s,mask}(330, 330)$. Furthermore, it reduces the number of short cycles greatly, from a total of 541 530 of cycles of lengths 6, 8, and 10 to 13 200 cycles of length 10: a large reduction in short cycles.

Comparing the code $C^*_{b,sp,qc,mask}$ with the code $C_{b,sp,qc,mask}$ constructed in Example 7.1, we find that the Tanner graph $\mathcal{G}^*_{b,sp,qc,s,mask}(330, 330)$ of the code $C^*_{b,sp,qc,mask}$ has a larger girth (girth 10) than the Tanner graph $\mathcal{G}_{b,sp,qc,s,mask}(330, 330)$ of the code $C_{b,sp,qc,mask}$ (girth 8). However, $\mathcal{G}^*_{b,sp,qc,s,mask}(330, 330)$ has a slightly larger number of cycles of length 10 (13 200) than the sum of the numbers of cycles of lengths 8 and 10 (9 570) in $\mathcal{G}_{b,sp,qc,s,mask}(330, 330)$.

In this example, we showed again that proper masking of a base matrix not only enlarges the girth of the Tanner graph of an LDPC code, but also greatly reduces the total number of short cycles. We also showed that the masking matrix $\mathbf{Z}$ given by (7.5) (or the decomposition base matrix given by (4.7) for the algebraic construction of the (2640, 1320) QC-PTG-LDPC code given in Example 4.2) is indeed a good masking matrix.

The BER and BLER performances of the code $C^*_{b,sp,qc,mask}$ over the AWGNC decoded with 5, 10, and 50 iterations of the MSA are shown in Fig. 7.2(a). At a BER of 10^{-10}, the gap between 5 and 10 iterations is about 1.6 dB and the gap between 10 and 50 iterations is within 0.8 dB. From Fig. 7.1(a) and Fig. 7.2(a), we see that the code $C_{b,sp,qc,mask}$ constructed in Example 7.1 and the code $C^*_{b,sp,qc,mask}$ constructed in this example perform almost the same, even though $C^*_{b,sp,qc,mask}$ has a larger girth than $C_{b,sp,qc,mask}$. This is because the total number of short cycles of the Tanner graph of $C^*_{b,sp,qc,mask}$ is slightly larger than that of $C_{b,sp,qc,mask}$. This implies that the cycle distribution of the Tanner graph of an

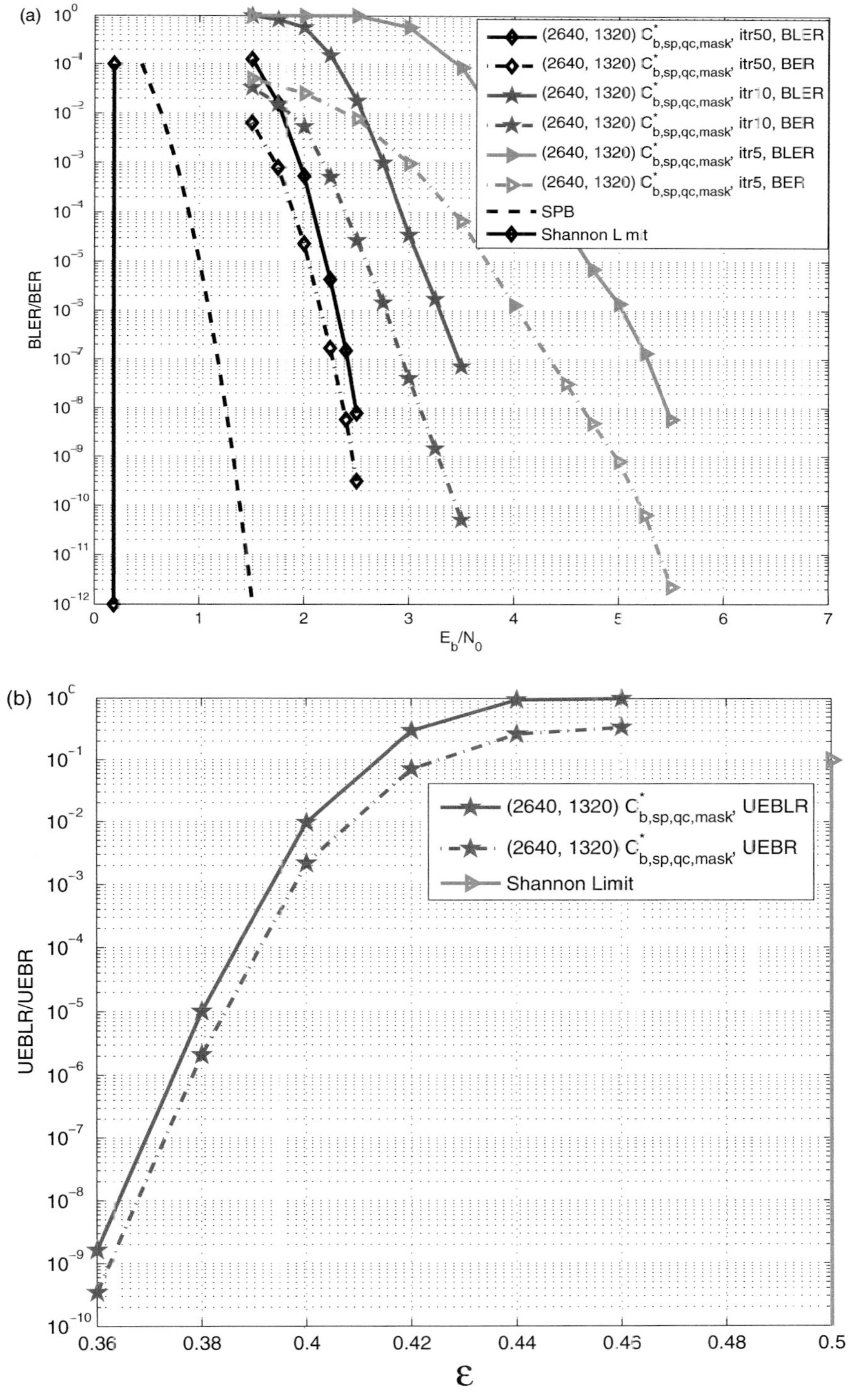

Figure 7.2. (a) The BER and BLER performances of the (2640, 1320) code $C^*_{b,sp,qc,mask}$ given in Example 7.2 decoded with 5, 10, and 50 iterations of the MSA over the AWGNC; and (b) performance of the code in (a) over the BEC.

LDPC code is also a factor affecting the error performance of the code, in addition to the girth.

The UEBR and UEBLR performances of the code $C^*_{b,sp,qc,mask}$ over the BEC are shown in Fig. 7.2(b). At a UEBR of 10^{-10}, the code performs about 0.14 away from the Shannon limit.

7.5 Design of Masking Matrices

In the previous section, we demonstrated that masking is a very effective technique to enlarge the girth and reduce the number of short cycles of the Tanner graph of an algebraically constructed LDPC code. A general guideline to design a good masking matrix $\mathbf{Z}$ is that every 3×3 submatrix of $\mathbf{Z}$ contains at least one 0-entry. This *one-zero constraint* on a 3×3 submatrix of $\mathbf{Z}$ makes it highly possible that a masked 2×2 SM-constrained base matrix $\mathbf{B}_{q,sp,mask}$ satisfies the 3×3 SM-constraint (Theorem 7.2), which is the necessary and sufficient condition for the Tanner graph of the CPM-QC-SP-LDPC code constructed based on $\mathbf{B}_{q,sp,mask}$ to have girth at least 8 and a smaller number of short cycles. In the following, we will present several specific designs of masking matrices.

As shown in Examples 7.1 and 7.2, the 4×8 masking matrix $\mathbf{Z}$ given by (7.5) is very effective for masking 4×8 base matrices in the form of (7.1) to produce CPM-QC-SP-LDPC codes of rate 1/2 whose Tanner graphs have girth 8 or 10 and small numbers of short cycles. We could use this masking matrix and/or its row-permuted versions as building blocks to construct masking matrices of larger sizes. (In the following, we denote the $m \times n$ masking matrix $\mathbf{Z}$ by $\mathbf{Z}(m,n)$ for a clear description, where the pair (m,n) in $\mathbf{Z}(m,n)$ represents the size of the matrix $\mathbf{Z}$.)

The masking matrix $\mathbf{Z}(4,8)$ given by (7.5) has a symmetrical structural property. We partition this masking matrix into two 4×4 submatrices, denoted by $\mathbf{Z}_l(4,4)$ and $\mathbf{Z}_r(4,4)$. The submatrix $\mathbf{Z}_l(4,4)$ consists of the left 4 columns of $\mathbf{Z}(4,8)$ and the submatrix $\mathbf{Z}_r(4,4)$ consists of the right 4 columns of $\mathbf{Z}(4,8)$. $\mathbf{Z}_l(4,4)$ and $\mathbf{Z}_r(4,4)$ are called *left-half* and *right-half* submatrices of $\mathbf{Z}(4,8)$, respectively. From (7.5), we see that $\mathbf{Z}_r(4,4)$ can be obtained by cyclically shifting the rows of $\mathbf{Z}_l(4,4)$ downward two positions. Both the submatrices $\mathbf{Z}_l(4,4)$ and $\mathbf{Z}_r(4,4)$ have constant column weight 3 and two row weights, 2 and 4. Based on the pair, $\mathbf{Z}_l(4,4)$ and $\mathbf{Z}_r(4,4)$, we can construct masking matrices with alternating $\mathbf{Z}_l(4,4)$s and $\mathbf{Z}_r(4,4)$s in the form of:

$$[\mathbf{Z}_l(4,4)\ \mathbf{Z}_r(4,4)\ \mathbf{Z}_l(4,4)\ \mathbf{Z}_r(4,4)\ \mathbf{Z}_l(4,4)\ \ldots].$$

For an even integer k with $k = 2\ell$, we form the following masking matrix:

$$\mathbf{Z}_1(4,4k) = [\mathbf{Z}_l(4,4)\ \mathbf{Z}_r(4,4)\ \mathbf{Z}_l(4,4)\ \mathbf{Z}_r(4,4)\ \ldots\ \mathbf{Z}_l(4,4)\ \mathbf{Z}_r(4,4)], \quad (7.7)$$

that consists of ℓ pairs of $\mathbf{Z}_l(4,4)$ and $\mathbf{Z}_r(4,4)$. In this case, $\mathbf{Z}_1(4,4k)$ has column and row weights 3 and 6ℓ, respectively. For an odd integer $k = 2\ell - 1$ with $\ell > 1$, we form the following masking matrix:

$$\mathbf{Z}_1(4,4k) = [\mathbf{Z}_l(4,4)\ \mathbf{Z}_r(4,4)\ \mathbf{Z}_l(4,4)\ \mathbf{Z}_r(4,4)\ \ldots\ \mathbf{Z}_l(4,4)], \quad (7.8)$$

that consists of $\ell - 1$ pairs of $\mathbf{Z}_l(4,4)$ and $\mathbf{Z}_r(4,4)$ and ends with the left-half submatrix $\mathbf{Z}_l(4,4)$ of $\mathbf{Z}(4,8)$. In this case, $\mathbf{Z}_1(4,4k)$ has column weight 3 and two different row weights, $6(\ell-1)+2$ and $6(\ell-1)+4$. Note that for $k \geq 3$ ($\ell \geq 2$), there are 3×3 submatrices in $\mathbf{Z}_1(4,4k)$ which contain all 1-entries. We call $\mathbf{Z}_1(4,4k)$ the *type-1 masking matrix* constructed using the masking matrix $\mathbf{Z}(4,8)$ given by (7.5) as the building block.

With $k = 2,3,4,\ldots$, we can form a sequence of masking matrices of the forms (7.7) and (7.8) for constructing CPM-QC-SP-LDPC codes with rates 1/2, 2/3, 3/4, ...

For a clear description that is compatible with that of the masking matrices, in the following, we denote $\mathbf{B}_{q,sp}$ by $\mathbf{B}_{q,sp}(m,n)$, where the pair (m,n) in $\mathbf{B}_{q,sp}(m,n)$ represents the size of the base matrix $\mathbf{B}_{q,sp}$.

Example 7.3. For $k = 2,3,\ldots,10$, we use the field GF(2^7) to construct nine $4 \times 4k$ base matrices $\mathbf{B}_{q,sp,s}(4,4k)$s in the form of (7.1) and then mask them with the nine masking matrices $\mathbf{Z}_1(4,4k)$s in the form of (7.7) for even k and (7.8) for odd k. The null spaces of the binary CPM-dispersions of the masked base matrices $\mathbf{B}_{q,sp,s,mask}(4,4k)$s give nine CPM-QC-SP-LDPC codes of lengths $508k$ and rates $(k-1)/k$, denoted by $C_{k-1,mask}$, for $k = 2,3,\ldots,10$. The BER performances of these codes decoded with 50 iterations of the MSA are shown in Fig. 7.3(a). From this figure, we see that the codes perform well.

Consider the (5080, 4572) CPM-QC-SP-LDPC code $C_{9,mask}$ with rate 9/10 ($k = 10$). The unmasked base matrix $\mathbf{B}_{q,sp,s}(4,40)$ for this code is constructed using the two subsets $\boldsymbol{S}_0 = \{\alpha^0, \alpha^1, \alpha^2, \alpha^3\}$ and $\boldsymbol{S}_1 = \{\alpha^5, \alpha^6, \alpha^7, \alpha^8, \alpha^9, \alpha^{10}, \alpha^{11}, \alpha^{12}, \alpha^{14}, \alpha^{15}, \alpha^{16}, \alpha^{17}, \alpha^{18}, \alpha^{19}, \alpha^{20}, \alpha^{21}, \alpha^{23}, \alpha^{24}, \alpha^{25}, \alpha^{26}, \alpha^{27}, \alpha^{28}, \alpha^{29}, \alpha^{30}, \alpha^{33}, \alpha^{34}, \alpha^{35}, \alpha^{36}, \alpha^{37}, \alpha^{38}, \alpha^{39}, \alpha^{40}, \alpha^{41}, \alpha^{42}, \alpha^{43}, \alpha^{44}, \alpha^{45}, \alpha^{46}, \alpha^{47}, \alpha^{48}\}$ of GF(2^7) and setting $\eta = 1$. The null space of the binary CPM-dispersion of $\mathbf{B}_{q,sp,s}(4,40)$ is a (5080, 4589) CPM-QC-SP-LDPC code C_9 whose Tanner graph has girth 6 and contains 245 110 cycles of length six, 24 630 507 cycles of length 8, and 2 165 484 620 cycles of length 10. The (5080, 4572) CPM-QC-SP-LDPC code $C_{9,mask}$ is given by the null space of the CPM-dispersion of the masked base matrix $\mathbf{B}_{q,sp,s,mask}(4,40) = \mathbf{Z}_1(4,40) \odot \mathbf{B}_{q,sp,s}(4,40)$. The Tanner graph of this masked code has girth 6 and contains 30480 cycles of length six, 1 485 773 cycles of length 8, and 65 230 502 cycles of length 10. We see that masking drastically reduces the number of short cycles in the Tanner graph of the unmasked code C_9.

The BER and BLER performances of $C_{9,mask}$ decoded with 50 iterations of the MSA are shown in Fig. 7.3(b). Also included in this figure are the BER and BLER performances of a (5080, 4572) LDPC ode C_{peg} constructed using the PEG algorithm whose parity-check matrix has column weight 3. The Tanner graph of the PEG code C_{peg} has girth 6 and contains 32 723 cycles of length six, 1 567 708 cycles of length 8, and 66 756 528 cycles of length 10. From Fig. 7.3(b), we see that the two codes perform the same above a BER of 10^{-7}.

Another way of constructing masking matrices based on the masking matrix $\mathbf{Z}(4,8)$ given by (7.5) is to repeat the first pair of columns and the third pair of columns in $\mathbf{Z}(4,8)$ k times to obtain a $4 \times 4k$ masking matrix, $\mathbf{Z}_2(4,4k)$, with column weight 3 and row weight $3k$. For example, setting $k = 2$ gives the matrix $\mathbf{Z}(4,8)$ given by (7.5), while setting $k = 3$ gives the masking matrix:

$$\mathbf{Z}_2(4,12) = \begin{bmatrix} 1 & 0 & 1 & 0 & 1 & 0 & 1 & 1 & 1 & 1 & 1 & 1 \\ 0 & 1 & 0 & 1 & 0 & 1 & 1 & 1 & 1 & 1 & 1 & 1 \\ 1 & 1 & 1 & 1 & 1 & 1 & 1 & 0 & 1 & 0 & 1 & 0 \\ 1 & 1 & 1 & 1 & 1 & 1 & 0 & 1 & 0 & 1 & 0 & 1 \end{bmatrix}. \quad (7.9)$$

The masking matrix $\mathbf{Z}_2(4,4k)$ is called the *type-2 masking matrix* constructed using the masking matrix $\mathbf{Z}(4,8)$ given by (7.5) as the building block. $\mathbf{Z}_2(4,4k)$ is simply a uniform expansion of $\mathbf{Z}(4,8)$.

Example 7.4. Let us use GF(331) for the code construction field and let α be a primitive element of GF(331). Suppose we use the subsets $\boldsymbol{S}_0 = \{\alpha^0, \alpha^1, \alpha^2, \alpha^3\}$ and $\boldsymbol{S}_1 = \{\alpha^{14}, \alpha^{15}, \alpha^{16}, \ldots, \alpha^{24}, \alpha^{25}\}$ of GF(331) and set $\eta = -1$ to construct a 4×12 base matrix $\mathbf{B}_{q,sp,s}(4,12)$ over GF(331) in the form of (7.1) which has column and row weights 4 and 12, respectively. The code given by the binary CPM-dispersion of $\mathbf{B}_{q,sp,s}(4,12)$ is a (4, 12)-regular (3960, 2643) CPM-QC-SP-LDPC code $C_{b,sp,qc}$ with rate 0.6674 whose Tanner graph has girth 6 and contains 13 530 cycles of length six, 165 660 cycles of length 8, and 3 835 590 cycles of length 10.

Suppose we mask the base matrix $\mathbf{B}_{q,sp,s}(4,12)$ with $\mathbf{Z}_2(4,12)$ given by (7.9). The masked base matrix $\mathbf{B}_{q,sp,s,mask}(4,12) = \mathbf{Z}_2(4,12) \odot \mathbf{B}_{q,sp,s}(4,12)$ has column and row weights 3 and 9, respectively. The null space of the binary CPM-dispersion of $\mathbf{B}_{q,sp,s,mask}(4,12)$ is a (3, 9)-regular (3960, 2640) CPM-QC-SP-LDPC code $C_{b,sp,qc,mask}$ whose Tanner graph has girth 8 and contains 3960 cycles of length 8 and 117 480 cycles of length 10. We see that

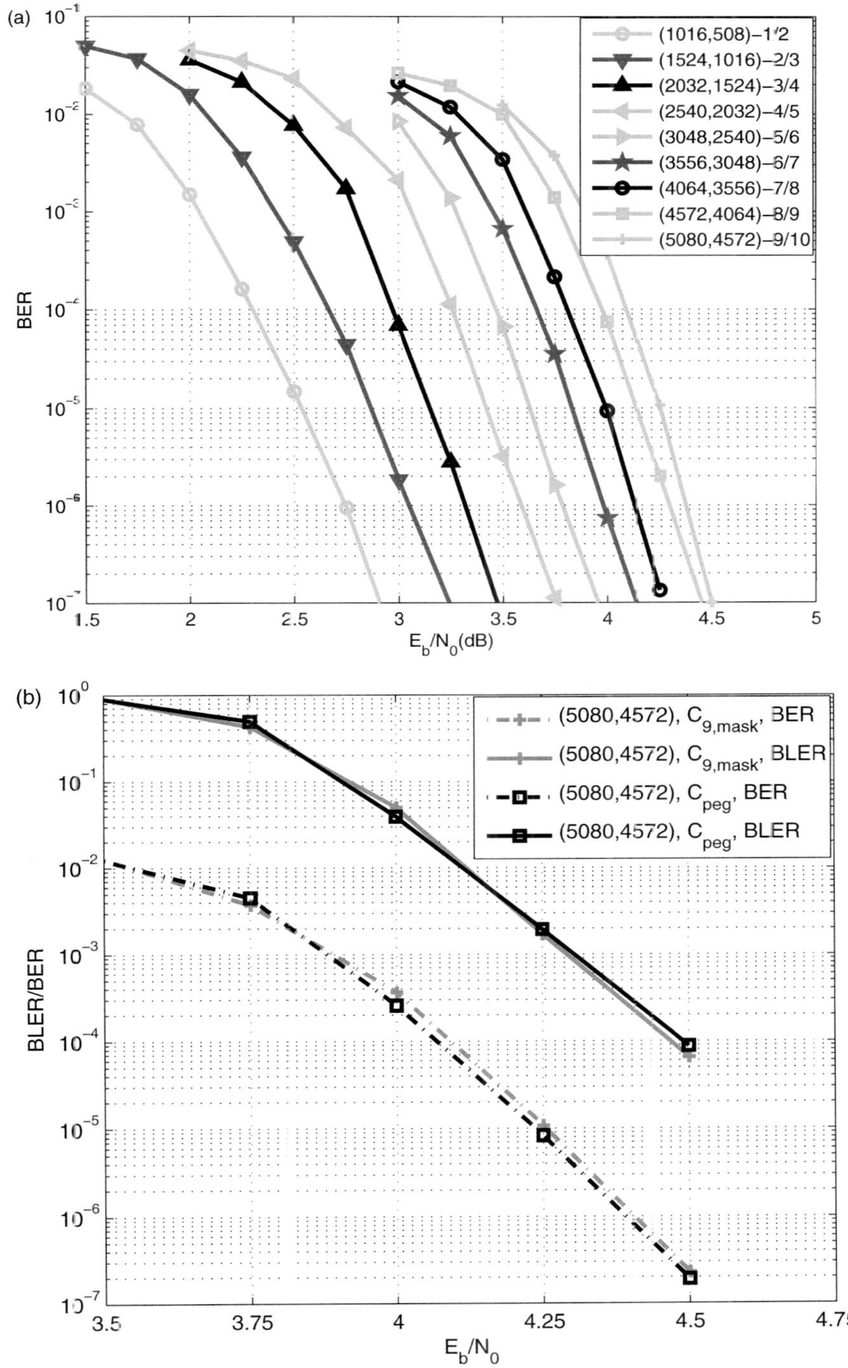

Figure 7.3. (a) The BER performances a sequence of CPM-QC-SP-LDPC codes with rates 1/2, 2/3, 3/4, …, 8/9, 9/10 constructed in Example 7.3 over the AWGNC; and (b) the BER and BLER performances of the (5040, 4572) CPM-QC-SP-LDPC code $C_{9,mask}$ and the (5040, 4572) PEG code C_{peg} in Example 7.3 over the AWGNC.

masking not only increases the girth from 6 to 8 but also significantly reduces the number of short cycles.

The BER and BLER performances of the two codes $C_{b,sp,qc}$ and $C_{b,sp,qc,mask}$ decoded with 50 iterations of the MSA are shown in Fig. 7.4. We see that the masked code $C_{b,sp,qc,mask}$ outperforms the unmasked code $C_{b,sp,qc}$. Therefore, $\mathbf{Z}_2(4, 12)$ is a good masking matrix.

For comparison, also included in Fig. 7.4 are the BER and BLER performances of a (3960, 2640) PEG code C_{peg} constructed using the PEG algorithm, whose parity-check matrix has column weight 3 and average row weight 9. The Tanner graph of the PEG code C_{peg} has girth 8 and contains 712 cycles of length 8 and 118 636 cycles of length 10. From Fig. 7.4, we see that the two codes perform the same above a BER of 10^{-8}.

Note that in the construction of a type-2 masking matrix, we can repeat the first pair of columns of $\mathbf{Z}(4, 8)$ given by (7.5) k_1 times and the third pair of columns of $\mathbf{Z}(4, 8)$ k_2 times. This results in a non-uniform expansion of $\mathbf{Z}(4, 8)$. The masking matrix formed in this way is a $4 \times (2k_1 + 2k_2)$ matrix $\mathbf{Z}_2(4, 2k_1 + 2k_2)$ with column weight 3 and two row weights, $k_1 + 2k_2$ and $2k_1 + k_2$. Non-uniform expansion of $\mathbf{Z}(4, 8)$ results in a larger class of masking matrices.

Next, we present the third type of masking matrices constructed based on a small matrix. Let $\mathbf{A}(4, 2k)$ be a $4 \times 2k$ matrix obtained by repeating the first two columns of $\mathbf{Z}(4, 8)$ given by (7.5) k times. For $\ell \geq 0$, form a $(4 + \ell) \times 2k$ matrix $\mathbf{A}(4 + \ell, 2k)$ by adding ℓ all-zero rows at the bottom of $\mathbf{A}(4, 2k)$. For $f \geq 0$, let $\mathbf{A}^{(f)}(4 + \ell, 2k)$ be a $(4 + \ell) \times 2k$ matrix obtained by cyclically down-shifting the rows of $\mathbf{A}(4+\ell, 2k)$ f positions. For $t \geq 2$, we construct the following $(4+\ell) \times 2kt$ matrix:

$$\mathbf{Z}_3(4 + \ell, 2kt) =$$

$$[\mathbf{A}^{(0)}(4 + \ell, 2k)\ \mathbf{A}^{(f)}(4 + \ell, 2k)\ \mathbf{A}^{(2f)}(4 + \ell, 2k)\ \ldots\ \mathbf{A}^{((t-1)f)}(4 + \ell, 2k)], \tag{7.10}$$

which has column weight 3 and multiple row weights. For various choices of k, ℓ, f, and t, we obtain a sequence of matrices $\mathbf{Z}_3(4 + \ell, 2kt)$ of the form (7.10) that can be used as masking matrices in the construction of a sequence of CPM-QC-SP-LDPC codes of various lengths and rates. The matrix $\mathbf{Z}_3(4 + \ell, 2kt)$ is called a *type-3 masking matrix*. Using performance simulations, we find this type of masking matrix is also quite effective.

Example 7.5. Again, we use the field GF(331) for the code construction. Choose $k = 4$, $\ell = 2$, $f = 4$, and $t = 2$. First, we construct a 6×16 base matrix $\mathbf{B}_{q,sp,s}(6, 16)$ over GF(331) in the form of (7.1) using the following two subsets of GF(331):

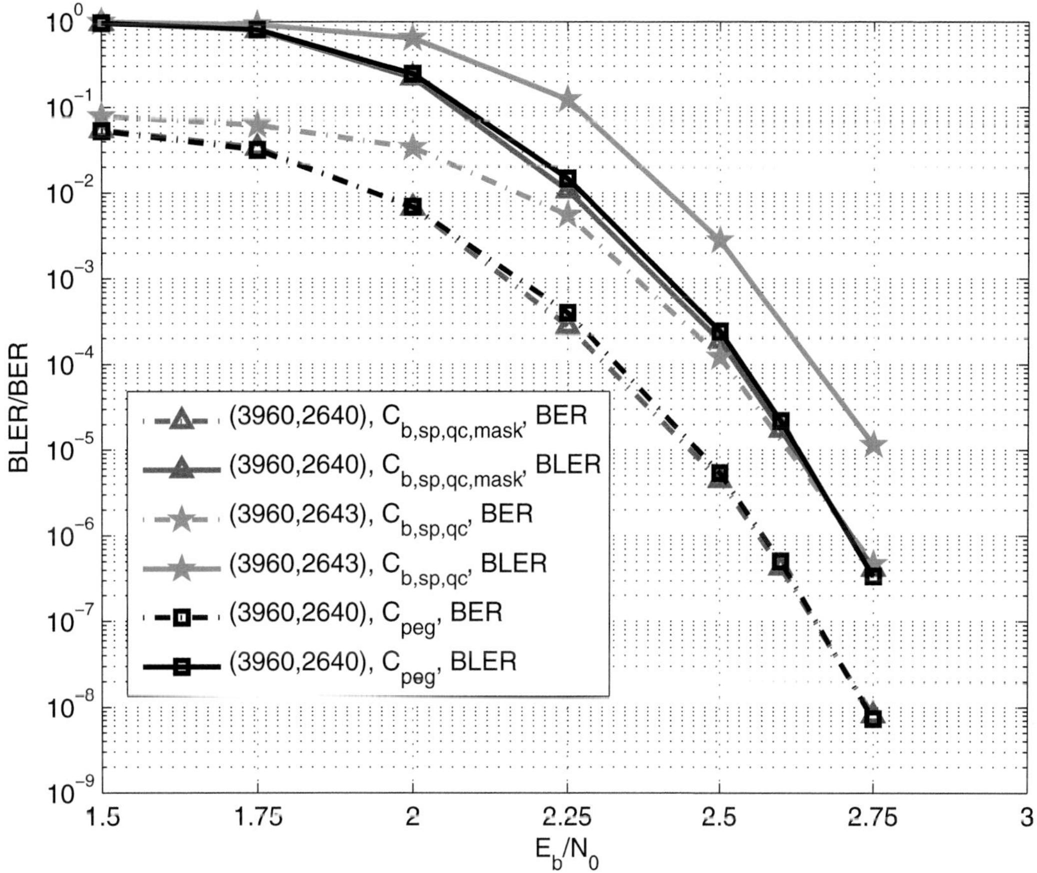

Figure 7.4. The BER and BLER performances of the unmasked (3960, 2643) and the masked (3960, 2640) CPM-QC-SP-LDPC codes $C_{b,sp,qc}$ and $C_{b,sp,qc,mask}$ given in Example 7.4 and a comparable (3960, 2640) PEG code C_{peg} over the AWGNC.

$\mathbf{S}_0 = \{\alpha^0, \alpha^1, \alpha^2, \alpha^3, \alpha^4, \alpha^5\}$ and $\mathbf{S}_1 = \{\alpha^{15}, \alpha^{16}, \alpha^{17}, \ldots, \alpha^{29}, \alpha^{30}\}$. Set $\eta = 1$. The binary CPM-dispersion of $\mathbf{B}_{q,sp,s}(6,16)$ gives a (6, 16)-regular (5280, 3305) CPM-QC-SP-LDPC code $C_{b,sp,qc}$ whose Tanner graph has girth 6 and contains 133 980 cycles of length 6, 4 307 325 cycles of length 8, and 241 141 560 cycles of length 10. With $k=4$, $\ell=2$, $f=4$, and $t=2$, we construct the following 6×16 masking matrix in the form of (7.10):

$$\mathbf{Z}_3(6,16) = \begin{bmatrix} 1 & 0 & 1 & 0 & 1 & 0 & 1 & 0 & 1 & 1 & 1 & 1 & 1 & 1 & 1 & 1 \\ 0 & 1 & 0 & 1 & 0 & 1 & 0 & 1 & 1 & 1 & 1 & 1 & 1 & 1 & 1 & 1 \\ 1 & 1 & 1 & 1 & 1 & 1 & 1 & 1 & 0 & 0 & 0 & 0 & 0 & 0 & 0 & 0 \\ 1 & 1 & 1 & 1 & 1 & 1 & 1 & 1 & 0 & 0 & 0 & 0 & 0 & 0 & 0 & 0 \\ 0 & 0 & 0 & 0 & 0 & 0 & 0 & 0 & 1 & 0 & 1 & 0 & 1 & 0 & 1 & 0 \\ 0 & 0 & 0 & 0 & 0 & 0 & 0 & 0 & 0 & 1 & 0 & 1 & 0 & 1 & 0 & 1 \end{bmatrix}. \tag{7.11}$$

Masking $\mathbf{B}_{q,sp,s}(6,16)$ with $\mathbf{Z}_3(6,16)$, we obtain a 6×16 masked base matrix $\mathbf{B}_{q,sp,s,mask}(6,16)$ with column weight 3 and three row weights 4, 8, and 12. $\mathbf{B}_{q,sp,s,mask}(6,16)$ satisfies the 3×3 SM-constraint. The binary CPM-dispersion of $\mathbf{B}_{q,sp,s,mask}(6,16)$ is a 6×16 array $\mathbf{H}_{b,sp,qc,s,mask}(330,330)$ of CPMs and ZMs of size 330×330, which is a 2980×5280 matrix over GF(2) with column weight 3 and three row weights 4, 8, and 12. The null space of $\mathbf{H}_{b,sp,qc,s,mask}(330,330)$ gives a (5280, 3302) CPM-QC-SP-LDPC code $C_{b,sp,qc,mask}$ with rate slightly higher than 5/8. The Tanner graph of the code $C_{b,sp,qc,mask}$ has girth 8 and contains 4280 cycles of length 8 and 114 840 cycles of length 10. We see that masking not only increases the girth of the Tanner graph of the unmasked code C but also reduces the number of short cycles.

The BER and BLER performances of the masked and the unmasked codes $C_{b,sp,qc,mask}$ and $C_{b,sp,qc}$ decoded with 50 iterations of the MSA are shown in Fig. 7.5. We see that the masked code $C_{b,sp,qc,mask}$ outperforms the unmasked code $C_{b,sp,qc}$ and it reaches down to a BER of 10^{-9} without a visible error-floor. The BER and BLER performances of a comparable (5280, 3300) PEG code C_{peg} constructed using the PEG algorithm are also included in Fig. 7.5 for comparison. The parity-check matrix of the PEG code C_{peg} has constant column weight 3. The Tanner graph of the PEG code C_{peg} has girth 10 and contains 65 745 cycles of length 10. From Fig. 7.5, we see that the constructed code $C_{b,sp,qc,mask}$ performs almost the same as the PEG code C_{peg}.

Once a masking matrix $\mathbf{Z}$ is designed for a given code rate and length, we need to find an RC-constrained base matrix $\mathbf{B}_{q,sp}$ such that the masked base matrix $\mathbf{B}_{q,sp,mask} = \mathbf{Z} \odot \mathbf{B}_{q,sp}$, in conjunction with the CPM-dispersion, produces a CPM-QC-SP-LDPC code whose Tanner graph has a large girth g, say 8 or larger,

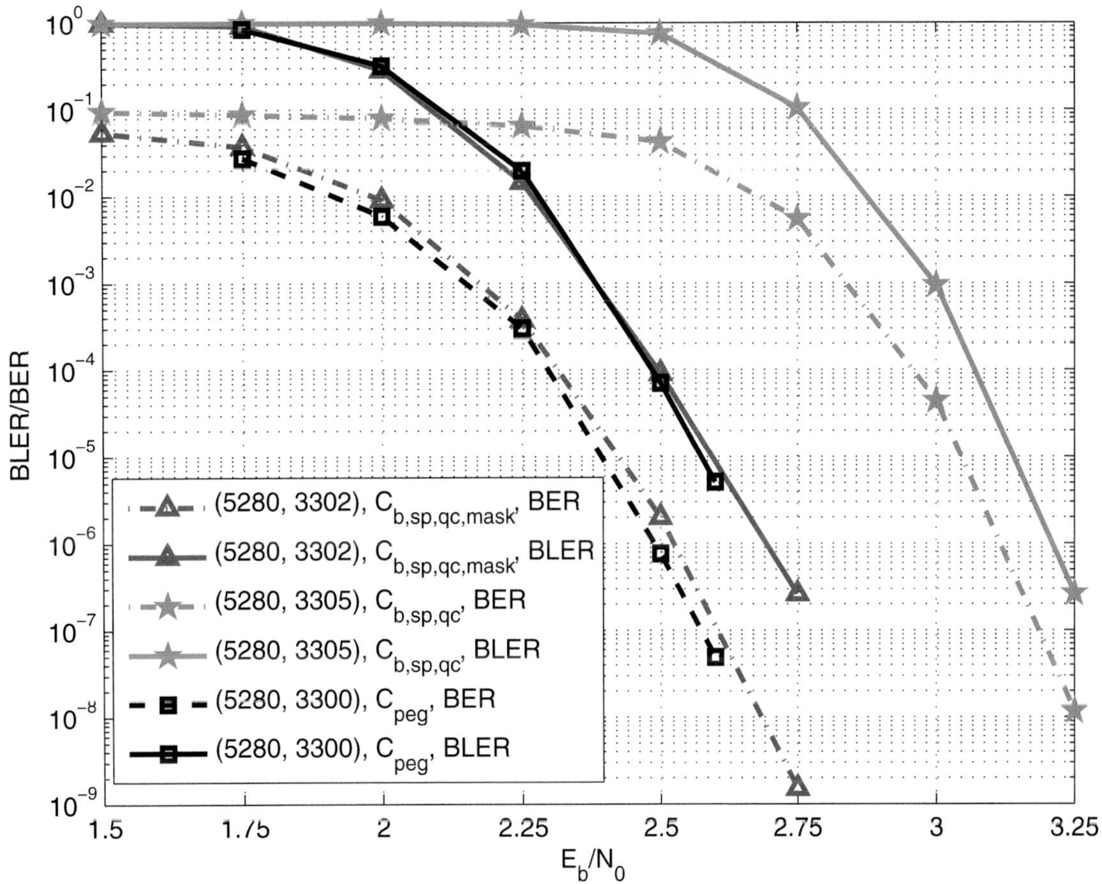

Figure 7.5. The BER and BLER performances of the unmasked (5280, 3305) and the masked (5280, 3302) CPM-QC-SP-LDPC codes $C_{b,sp,qc}$ and $C_{b,sp,qc,mask}$ given in Example 7.5 and a comparable (5280, 3300) PEG code C_{peg} over the AWGNC.

and a small total number of cycles of lengths g and $g+2$ (called short cycles). In this case, we say that the masking matrix $\mathbf{Z}$ and the base matrix $\mathbf{B}_{q,sp}$ are compatible (or a good match to each other).

One approach to find a base matrix that is a good match to a designed masking matrix $\mathbf{Z}$ is that we first construct a large 2×2 SM-constrained mother matrix $\mathbf{B}_{mother}$ over a chosen field using a specific construction method, say the construction method based on two arbitrary subsets of a finite field presented in Section 7.3. Then, we develop a searching algorithm to find a base matrix $B_{q,sp}$, a submatrix of $\mathbf{B}_{mother}$, such that $\mathbf{B}_{q,sp,mask} = \mathbf{Z} \odot \mathbf{B}_{q,sp}$ satisfies the 3×3 SM-constraint or produces a CPM-QC-SP-LDPC code whose Tanner graph has the largest girth (at least 8) and/or the smallest number of short cycles.

In [68], an algorithm was devised to find a 4×8 base matrix from a mother base matrix which is a good match to a designed 4×8 masking matrix $\mathbf{Z}(4,8)$ with column and row weights 3 and 6, respectively. This algorithm was developed to find masked base matrices for constructing $(3,6)$-regular QC-LDPC codes with rate $1/2$. It can be generalized to find masked base matrices for constructing regular QC-LDPC codes with various rates. One such a generalized algorithm is presented in Appendix B.

So far, the design of masking matrices given above has been focused on constructing regular (or near regular) CPM-QC-SP LDPC codes using the CPM-D-SP-construction method. To construct an irregular CPM-QC-SP-LDPC code with optimized VN and CN degree distributions, we need to construct a masking matrix with column and row weight distributions in agreement with (or close to) the optimized VN and CN degree distributions. Such a masking matrix can be constructed by using the PEG algorithm.

7.6 Construction of CPM-QC-SP-LDPC Codes for Correcting Bursts of Erasures by Masking

The masking technique presented in the last two sections can be used to construct CPM-QC-SP-LDPC codes for correcting bursts of erasures, i.e., erasures clustered in bursts. Let GF(q) be the field for the code construction. Let l and t be two positive integers such that $lt \leq q$. We construct a $t \times lt$ 2×2 SM-constrained base matrix $\mathbf{B}_{q,sp,s}(t,lt)$ in the form of (7.1) using two subsets S_0 and S_1 of GF(q) with $|S_0| = t$ and $|S_1| = lt$. Next, we divide $\mathbf{B}_{q,sp,s}(t,lt)$ into l submatrices of size $t \times t$ as follows:

$$\mathbf{B}_{q,sp,s}(t,lt) = [\mathbf{B}_0(t,t) \;\; \mathbf{B}_1(t,t) \;\; \ldots \;\; \mathbf{B}_{l-1}(t,t)]. \tag{7.12}$$

For $0 \leq j < l$, each submatrix $\mathbf{B}_j(t,t)$ consists of t consecutive columns of $\mathbf{B}_{q,sp,s}(t,lt)$.

Let τ be a positive integer with $\tau \leq t-2$. Now, we construct a binary t-tuple $\mathbf{z} = (z_0, z_1, \ldots, z_{t-1})$ in which the first τ components are zeros and the last $t-\tau$ components are ones, i.e.,

$$\mathbf{z} = (\underbrace{0\,0\,\ldots\,0}_{\tau}\ \underbrace{1\,1\,\ldots\,1}_{t-\tau}).$$

Now, we form a $t \times t$ circulant $\mathbf{Z}(t,t)$ with $\mathbf{z}$ as the generator. Each row (or column) of $\mathbf{Z}(t,t)$ contains a *span* of τ 0-entries confined by two 1-entries, including the *end-around case*. This span of τ 0-entries is called a *zero-span* and the parameter τ is called the length of the zero-span.

Next, we mask each $t \times t$ submatrix $\mathbf{B}_j(t,t)$ of $\mathbf{B}_{q,sp,s}(t,lt)$ with $\mathbf{Z}(t,t)$. The masking operation results in a $t \times lt$ masked base matrix:

$$\mathbf{B}_{q,sp,s,mask}(t,lt) = [\mathbf{B}_{0,mask}(t,t)\ \ \mathbf{B}_{1,mask}(t,t)\ \ \ldots\ \ \mathbf{B}_{l-1,mask}(t,t)]. \quad (7.13)$$

It follows from the structure of the masking matrix $\mathbf{Z}(t,t)$ that the masked base matrix $\mathbf{B}_{q,sp,s,mask}(t,lt)$ has the following structural properties: (1) Each row contains l zero-spans of length τ, each separated by $t-\tau$ nonzero entries, including the end-around case; and (2) In each column position, there is one row with a nonzero entry in that position followed by τ 0-entries, including the end-around case.

Dispersing each nonzero entry in $\mathbf{B}_{q,sp,s,mask}(t,lt)$ into a CPM of size $(q-1) \times (q-1)$ and each zero entry into a ZM of size $(q-1) \times (q-1)$, we obtain a binary $t \times lt$ array $\mathbf{H}_{b,sp,qc,s,mask}(q-1,q-1)$ of binary CPMs and ZMs of size $(q-1) \times (q-1)$. The array $\mathbf{H}_{b,sp,qc,s,mask}(q-1,q-1)$ is a $t(q-1) \times lt(q-1)$ matrix with constant column weight $t-\tau$ and constant row weight $l(t-\tau)$. $\mathbf{H}_{b,sp,qc,s,mask}(q-1,q-1)$ consists of t row-blocks and lt column-blocks of CPMs and ZMs of size $(q-1) \times (q-1)$. Each row-block consists of l spans of ZMs, each consisting of τ ZMs of size $(q-1) \times (q-1)$ confined by two CPMs, including the end-around case. Hence, each row of $\mathbf{H}_{b,sp,qc,s,mask}(q-1,q-1)$ consists of l zero-spans, each of length at least $\tau(q-1)$. Based on the structural properties of $\mathbf{B}_{q,sp,qc,s,mask}(t,lt)$ and $\mathbf{H}_{b,sp,qc,s,mask}(q-1,q-1)$, we readily see that in any column position of $\mathbf{H}_{b,sp,qc,s,mask}(q-1,q-1)$, there is exactly one row in $\mathbf{H}_{b,sp,qc,s,mask}(q-1,q-1)$ which has a 1-entry in that column position followed by at least $\tau(q-1)$ zeros entries.

Let $C_{b,sp,qc,mask}$ be the CPM-QC-SP-LDPC code given by the null space of $\mathbf{H}_{b,sp,qc,s,mask}(q-1,q-1)$. The above structural property of $\mathbf{H}_{b,sp,qc,s,mask}(q-1,q-1)$ allows the code $C_{b,sp,qc,mask}$ to correct any erasure pattern $\mathbf{e}$ over the BEC with erased code symbols confined to κ consecutive positions with $0 \leq \kappa \leq$

$\tau(q-1)+1$. This erasure pattern is called an *erasure-burst* of length κ. When such an erasure pattern $\mathbf{e}$ occurs, we find a row in $\mathbf{H}_{b,sp,qc,s,mask}(q-1,q-1)$ which checks only one erasure in $\mathbf{e}$. Based on the zero-constraint on the parity-check sum formed by this row, we can uniquely determine the value of the erased symbol being checked. Recovering this erased code symbol, we obtain an erasure-burst $\mathbf{e}_1$ with one less erasure. Then, we find a row in $\mathbf{H}_{b,sp,qc,s,mask}(q-1,q-1)$ that checks only one erasure in $\mathbf{e}_1$. Based on the zero-constraint on the parity-check sum formed by this row, we recover the code symbol being erased. We continue the above *check-and-recover* process until all the erased code symbols in $\mathbf{e}$ are recovered. Any erasure-burst of length $\tau(q-1)+1$ or less can be corrected by the code $C_{b,sp,qc,mask}$. We say that $C_{b,sp,qc,mask}$ has erasure-burst correcting capability $\tau(q-1)+1$.

We must note that an LDPC code that is efficient for correcting bursts of erasure does not necessarily perform well over the AWGNC or the BEC for correcting random erasures and/or errors.

Example 7.6. Let GF(2^8) be the code construction field and α be a primitive element in GF(2^8). Set $t=7$ and $l=6$. Choose two sets $\boldsymbol{S}_0=\{\alpha^0,\alpha^1,\alpha^2,\ldots,\alpha^6\}$ and $\boldsymbol{S}_1=\{\alpha^7,\alpha^8,\ldots,\alpha^{48}\}$. Set $\eta=-1$. Based on these two subsets, we construct a 7×42 base matrix $\mathbf{B}_{q,sp,s}(7,42)$ over GF(2^8) in the form of (7.1) which satisfies the 2×2 SM-constraint. Dividing $\mathbf{B}_{q,sp,s}(7,42)$ into 6 submatrices, each a 7×7 matrix, we have $\mathbf{B}_{q,sp,s}(7,42)=[\mathbf{B}_0(7,7)\ \mathbf{B}_1(7,7)\ \ldots\ \mathbf{B}_5(7,7)]$. Set $\tau=3$. Form the following 7×7 masking matrix:

$$\mathbf{Z}(7,7)=\begin{bmatrix} 0&0&0&1&1&1&1\\ 1&0&0&0&1&1&1\\ 1&1&0&0&0&1&1\\ 1&1&1&0&0&0&1\\ 1&1&1&1&0&0&0\\ 0&1&1&1&1&0&0\\ 0&0&1&1&1&1&0 \end{bmatrix}. \tag{7.14}$$

For $0\le j<6$, masking $\mathbf{B}_j(7,7)$ of $\mathbf{B}_{q,sp,s}(7,42)$ with $\mathbf{Z}(7,7)$ given by (7.14), we obtain a 7×42 masked base matrix $\mathbf{B}_{q,sp,s,mask}(7,42)=[\ \mathbf{B}_{0,mask}(7,7)\ \mathbf{B}_{1,mask}(7,7)\ \ldots\ \mathbf{B}_{5,mask}(7,7)\ ]$ with column weight 4 and row weight 24. Each row of the masked base matrix $\mathbf{B}_{q,sp,s,mask}(7,42)$ has 6 zero-spans, each of length 3. The CPM-dispersion of $\mathbf{B}_{q,sp,s,mask}(7,42)$ gives a 7×42 array $\mathbf{H}_{b,sp,qc,s,mask}(255,255)$ of CPMs and ZMs of size 255×255 over GF(2). It is a $1785\times 10\,710$ matrix with column weight 4 and row weight 24. Each row of $\mathbf{H}_{b,sp,qc,s,mask}(255,255)$ has 6 zero-spans, each of length at least 765, including the end-around zero-span. In each column position of $\mathbf{H}_{b,sp,qc,s,mask}(255,255)$, there is one row with a 1-entry in that column position followed by at least 765 0-entries.

The null space of the masked array $\mathbf{H}_{b,sp,qc,s,mask}(255, 255)$ gives a (4, 24)-regular (10 710, 8926) CPM-QC-SP-LDPC code $C_{b,sp,qc,mask}$ with rate 0.8334. This code is capable of correcting any erasure-burst of length 766 or less.

The BER and BLER performances of $C_{b,sp,qc,mask}$ over the AWGNC are shown in Fig. 7.6(a). The unresolved erasure performance over the BEC is shown in Fig. 7.6(b).

7.7 Discussion and Remarks

In the construction of a QC-SP-LDPC code using CPM-dispersion, a key ingredient is a 2×2 SM-constrained base matrix over a chosen NB field. In Section 7.3, we presented a powerful and flexible algebraic method for constructing 2×2 SM-constrained base matrices over NB fields. This construction method includes as special cases six major algebraic methods for constructing 2×2 SM-constrained base matrices over NB fields developed since 2003. Besides this method and its special cases, there are other powerful methods. One such a method is the construction of 2×2 SM-constrained base matrices based on cyclic subgroups of *prime order* of an NB field proposed in [64]. This method is not a special case of the construction presented in Section 7.3 and will be discussed in Chapter 10. Another method for constructing 2×2 SM-constrained base matrices is based on partial geometries [29]. This geometric construction method gives a large class of 2×2 SM-constrained base matrices and will be discussed in Appendix A.

Another key ingredient for constructing QC-SP-LDPC codes to have good error performance is masking. In Sections 7.4 and 7.5, we demonstrated that masking is a very effective technique to enlarge the girth and reduce the number of short cycles of the Tanner graph of an algebraically constructed LDPC code. A general guideline to design a good masking matrix $\mathbf{Z}$ is that every 3×3 submatrix of $\mathbf{Z}$ contains at least one 0-entry (or most of the 3×3 submatrices of $\mathbf{Z}$ contain at least one 0-entry). This *one-zero constraint* on a 3×3 submatrix of $\mathbf{Z}$ makes it highly possible that a masked 2×2 SM-constrained base matrix $\mathbf{B}_{q,sp,mask}$ satisfies the 3×3 SM-constraint (Theorem 7.2), which is the necessary and sufficient condition for the Tanner graph of the CPM-QC-SP-LDPC code constructed based on the masked base matrix $\mathbf{B}_{q,sp,mask}$ to have girth at least 8 and a smaller number of short cycles.

For a given rate, a given code length, and a given field, in order to construct a CPM-QC-SP-LDPC code with good error performance, the two key ingredients, base matrix and masking matrix, must be compatible, i.e., a good match to each

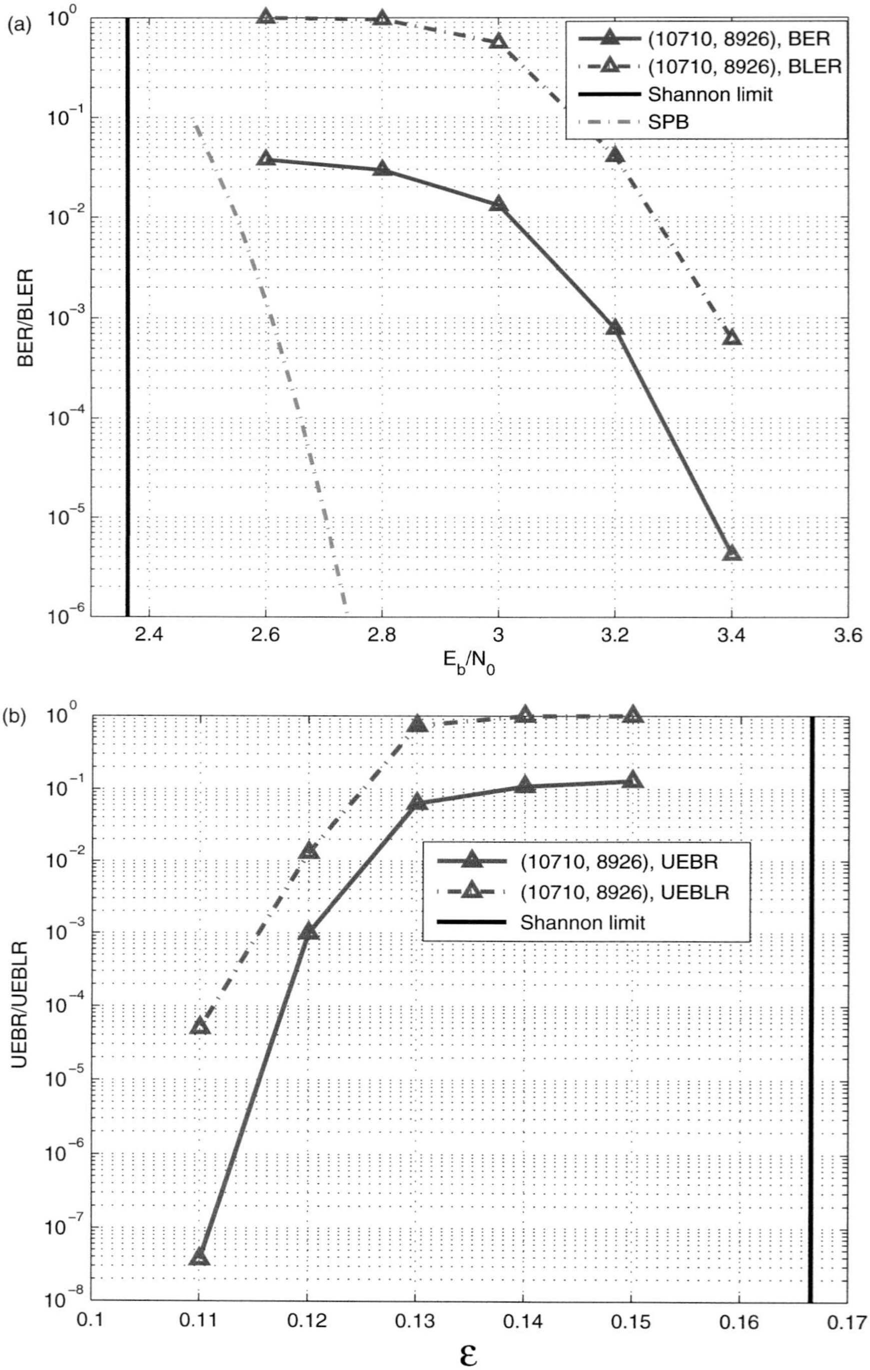

Figure 7.6. (a) The BER and BLER performances of the (4, 24)-regular (10 710, 8926) code $C_{b,sp,qc,mask}$ given in Example 7.6 over the AWGNC; and (b) performance of the code in (a) over the BEC.

other. To achieve a good match, there are two approaches. One is to construct a base matrix based on the chosen field first. Then, we design a masking matrix for the constructed base matrix that results in a code whose Tanner graph has a large girth and/or a small number of short cycles. The other approach is to design a masking matrix first and then find a compatible base matrix. They together produce a code whose Tanner graph has a large girth and/or a small number of short cycles. However, either approach is "suboptimal." A good question is "whether the base and masking matrices can be designed jointly to produce a code whose Tanner graph has good structural properties in terms of girth, cycle distribution, and trapping sets?" This is definitely a good research problem, but also a hard one.

The masking matrices presented in Sections 7.4 and 7.5 are limited to small size or are constructed using small masking matrices as the building blocks. Furthermore, their column weights are limited to 3. To construct masking matrices of large size with various column weights, we could use finite geometries in a way similar to the construction of base matrices for the SP-construction of LDPC codes. An RC-constrained SP-base matrix can be used as a masking matrix. In this case, every 3×3 submatrix of the masking matrix contains at least one 0-entry, i.e., satisfies the one-zero constraint. Masking a 2×2 SM-constrained base matrix with an RC-constrained masking matrix may result in a masked base matrix that satisfies the 3×3 SM-constraint and has other good structural properties. This is also worthy of further investigation.

Another approach to construct a large masking matrix is to start with an $m \times n$ matrix $\mathbf{Z}_0$ with all 1-entries. Then, we perform diagonal replacement of 1-entries by 0-entries. This is done as follows. We start from any position of the top row and move down to the right on a 45° diagonal. When reaching the end of a column, we move to the top of any new column (or the next column). When we reach the end of a row, we move to the leftmost entry of any new row (or the next row). As we make this zig-zag move, we replace each 1-entry on the path by a 0-enty. We continue the zig-zag move and the 1-to-0 replacement until we obtain a matrix $\mathbf{Z}$ for which one or all of the following conditions are met: (1) Any 3×3 submatrix of $\mathbf{Z}$ contains at least one 0-entry; (2) $\mathbf{Z}$ has the desired row and column weight distributions; and (3) $\mathbf{Z}$ satisfies the RC-constraint.

In Section 7.6, we showed that the masking technique is also effective for constructing CPM-QC-SP-LDPC codes for correcting bursts of erasures over the BEC. Therefore, in our belief, design of good masking matrices is a good research problem.

So far, there is no systematic method for constructing good masking matrices, hence the design of good masking matrices is still an art.

8

Doubly QC-LDPC Codes

In this chapter, we present an algebraic construction of CPM-QC-SP-LDPC codes that have both the block- and the section-wise cyclic structures as defined earlier in Section 4.1. Such a CPM-QC-SP-LDPC code is called a *doubly CPM-QC-SP-LDPC code*. The construction of a doubly CPM-QC-SP-LDPC code is based on an SP-base matrix $\mathbf{B}_{q,sp}$ over a finite field that has cyclic structure, called *a cyclic base matrix*. A special case of the base matrix $\mathbf{B}_{q,sp,p}$ given in the form of (7.2) constructed based on two subsets of a finite field $\boldsymbol{S}_0$ and $\boldsymbol{S}_1$ has a cyclic structure. One advantage of a doubly QC-LDPC code is that we can apply either the reduced-complexity iterative decoding scheme proposed in [79] based on its block-cyclic structure or the reduced-complexity iterative decoding scheme presented in [69, 68, 78] based on its section-wise cyclic structure using the same parity-check matrix. Furthermore, well designed doubly QC-LDPC codes perform well over both the AWGNC and the BEC. The base matrix and its partition used in this chapter will be used for constructing spatially and globally coupled LDPC codes in the next two chapters.

8.1 Base Matrices with Cyclic Structure

Let α be a primitive element of GF(q). Set $\boldsymbol{S}_0 = \{\, \alpha^0, \alpha^{-1}, \alpha^{-2}, \ldots, \alpha^{-(q-3)}, \alpha^{-(q-2)} \,\}$, $\boldsymbol{S}_1 = \{\, \alpha^0, \alpha^1, \alpha^2, \ldots, \alpha^{q-3}, \alpha^{q-2} \,\}$, and $\eta = 1$. Using the product-form given by (7.2), we obtain the following 2×2 SM-constrained $(q-1) \times (q-1)$ base matrix over GF(q):

$$\mathbf{B}^*_{q,sp,p} = \begin{bmatrix} \alpha^0 - 1 & \alpha - 1 & \alpha^2 - 1 & \cdots & \alpha^{q-3} - 1 & \alpha^{q-2} - 1 \\ \alpha^{q-2} - 1 & \alpha^0 - 1 & \alpha - 1 & \cdots & \alpha^{q-4} - 1 & \alpha^{q-3} - 1 \\ \vdots & \vdots & \vdots & \ddots & \vdots & \vdots \\ \alpha - 1 & \alpha^2 - 1 & \alpha^3 - 1 & \cdots & \alpha^{q-2} - 1 & \alpha^0 - 1 \end{bmatrix}. \tag{8.1}$$

From (8.1), we readily see that the base matrix $\mathbf{B}^*_{q,sp,p}$ has a cyclic structure. It is a weight-$(q-2)$ circulant over GF(q). This circulant can be obtained by permuting the rows of the circulant given by (7.3). In fact, every row (or column) of $\mathbf{B}^*_{q,sp,p}$ is a minimum weight codeword of the $(q-1,2)$ Reed–Solomon (RS) code over GF(q) with two information symbols and minimum distance $q-2$ [74, 97]. All the entries in a row (or a column) of $\mathbf{B}^*_{q,sp,p}$ are distinct elements in GF(q) and any two rows (or two columns) differ in every location. The $q-1$ entries lying on the main diagonal of $\mathbf{B}^*_{q,sp,p}$ are the zero element of GF(q).

For $1 \le m,n < q$, any $m \times n$ submatrix of $\mathbf{B}^*_{q,sp,p}$ can be used as the base matrix for constructing a CPM-QC-SP-LDPC code. In this chapter, we are interested in submatrices of $\mathbf{B}^*_{q,sp,p}$ from which we can construct doubly QC-LDPC codes.

To simplify the notation and show the connection to RS codes, we use $\mathbf{W}_{RS}$ for the matrix $\mathbf{B}^*_{q,sp,p}$, i.e., $\mathbf{W}_{RS} = \mathbf{B}^*_{q,sp,p}$. Label the rows and columns of $\mathbf{W}_{RS}$ from 0 to $q-2$. Suppose $q-1$ can be factored as the product of two positive integers l and r, i.e., $lr = q-1$. Now, we partition $\mathbf{W}_{RS}$ into r submatrices of size $l \times (q-1)$, denoted by $\mathbf{W}_0, \mathbf{W}_1, \ldots, \mathbf{W}_{r-1}$, where for $0 \le i < r$, the submatrix $\mathbf{W}_i$ consists of the il-th to the $((i+1)l-1)$-th rows of $\mathbf{W}_{RS}$ (l consecutive rows of $\mathbf{W}_{RS}$). Next, we partition the submatrix $\mathbf{W}_0$ into r submatrices of size $l \times l$, denoted by $\mathbf{W}_{0,0}, \mathbf{W}_{0,1}, \ldots, \mathbf{W}_{0,r-1}$, each consisting of l consecutive columns of $\mathbf{W}_0$, so that $\mathbf{W}_0 = [\mathbf{W}_{0,0}\ \mathbf{W}_{0,1}\ \ldots\ \mathbf{W}_{0,r-1}]$. Based on the cyclic structure of $\mathbf{W}_{RS}$, we readily see that, for $0 \le i < r$, the submatrix $\mathbf{W}_i$ can be obtained by cyclically shifting all the rows of $\mathbf{W}_0 = [\mathbf{W}_{0,0}\ \mathbf{W}_{0,1}\ \ldots\ \mathbf{W}_{0,r-1}]$ il positions to the right, i.e., $\mathbf{W}_i = [\mathbf{W}_{0,r-i}\ \ldots\ \mathbf{W}_{0,r-1}\ \mathbf{W}_{0,0}\ \ldots\ \mathbf{W}_{0,r-i-1}]$. Hence, the matrix $\mathbf{W}_{RS}$ can be arranged as an $r \times r$ array of $l \times l$ submatrices in the following block-cyclic structure form:

$$\mathbf{W}_{RS} = \begin{bmatrix} \mathbf{W}_{0,0} & \mathbf{W}_{0,1} & \cdots & \mathbf{W}_{0,r-1} \\ \mathbf{W}_{0,r-1} & \mathbf{W}_{0,0} & \cdots & \mathbf{W}_{0,r-2} \\ \vdots & \vdots & \ddots & \vdots \\ \mathbf{W}_{0,1} & \mathbf{W}_{0,2} & \cdots & \mathbf{W}_{0,0} \end{bmatrix}. \tag{8.2}$$

Each row-block (or each column-block) of $\mathbf{W}_{RS}$ is composed of the same set of constituent matrices $\mathbf{W}_{0,0}, \mathbf{W}_{0,1}, \ldots, \mathbf{W}_{0,r-1}$ in cyclic order. Since $\mathbf{W}_{RS}$ satisfies the 2×2 SM-constraint, each of its constituent matrices $\mathbf{W}_{0,0}, \mathbf{W}_{0,1}, \ldots, \mathbf{W}_{0,r-1}$ also satisfies the 2×2 SM-constraint. Furthermore, all the r constituent submatrices of $\mathbf{W}_{RS}$ satisfy the PW-2×2 SM-constraint. This follows directly from the block-cyclic structure of $\mathbf{W}_{RS}$ shown by (8.2).

For $0 \le j < r$ and $1 \le m,n < l$, we now take an $m \times n$ submatrix $\mathbf{R}_{0,j}$ from $\mathbf{W}_{0,j}$. The submatrices $\mathbf{R}_{0,0}, \mathbf{R}_{0,1}, \ldots, \mathbf{R}_{0,r-1}$ are taken from the constituent matrices $\mathbf{W}_{0,0}, \mathbf{W}_{0,1}, \ldots, \mathbf{W}_{0,r-1}$ under the following *location constraint*: for $j' \neq j$, the locations of the entries of $\mathbf{R}_{0,j}$ in $\mathbf{W}_{0,j}$ are *identical to* the locations of the entries of $\mathbf{R}_{0,j'}$ in $\mathbf{W}_{0,j'}$. We next form the following $r \times r$ array $\mathbf{B}_{q,sp,p}(m,n)$ of $m \times n$ submatrices over GF(q) with a block-cyclic structure:

$$\mathbf{B}_{q,sp,p}(m,n) = \begin{bmatrix} \mathbf{R}_{0,0} & \mathbf{R}_{0,1} & \cdots & \mathbf{R}_{0,r-1} \\ \mathbf{R}_{0,r-1} & \mathbf{R}_{0,0} & \cdots & \mathbf{R}_{0,r-2} \\ \vdots & \vdots & \ddots & \vdots \\ \mathbf{R}_{0,1} & \mathbf{R}_{0,2} & \cdots & \mathbf{R}_{0,0} \end{bmatrix}. \tag{8.3}$$

$\mathbf{B}_{q,sp,p}(m,n)$ is a submatrix of $\mathbf{W}_{RS} = \mathbf{B}^*_{q,sp,p}$ and satisfies the 2×2 SM-constraint. This follows directly from the block-cyclic structure of $\mathbf{W}_{RS}$ and the location constraint on the entries of the submatices $\mathbf{R}_{0,0}, \mathbf{R}_{0,1}, \ldots, \mathbf{R}_{0,r-1}$ taken from $\mathbf{W}_{0,0}, \mathbf{W}_{0,1}, \ldots, \mathbf{W}_{0,r-1}$. Since the submatrices $\mathbf{W}_{0,0}, \mathbf{W}_{0,1}, \ldots, \mathbf{W}_{0,r-1}$ satisfy the PW-2×2 SM-constraint, the submatrices $\mathbf{R}_{0,0}, \mathbf{R}_{0,1}, \ldots, \mathbf{R}_{0,r-1}$ also satisfy the PW-2×2 SM-constraint. (Here, we note that the pair (m,n) in $\mathbf{B}_{q,sp,p}(m,n)$ denotes the size of the submatrices of the base array $\mathbf{B}_{q,sp,p}(m,n)$.)

8.2 CPM-D-SP-Construction of Doubly QC-LDPC Codes

Now, we use the array $\mathbf{B}_{q,sp,p}(m,n)$ as the base matrix for the CPM-D-SP-construction of QC-LDPC codes. Dispersing each nonzero entry of $\mathbf{B}_{q,sp,p}(m,n)$ into a CPM of size $(q-1) \times (q-1)$ and each zero entry into a ZM of size $(q-1) \times (q-1)$, we obtain an $r \times r$ array $\mathbf{H}_{b,sp,d,qc}(q-1,q-1)$ of $m \times n$ subarrays of binary CPMs and/or ZMs of size $(q-1) \times (q-1)$, which is also an $mr \times nr$ array of CPMs and/or ZMs of size $(q-1) \times (q-1)$. The array $\mathbf{H}_{b,sp,d,qc}(q-1,q-1)$ consists of r row-blocks of subarrays, each consisting of r $m \times n$ subarrays of CPMs and/or ZMs of size $(q-1) \times (q-1)$. It follows from the block-cyclic structure of the base array $\mathbf{B}_{q,sp,p}(m,n)$ that $\mathbf{H}_{b,sp,d,qc}(q-1,q-1)$ also has a block-cyclic structure. Each row-block of subarrays of $\mathbf{H}_{b,sp,d,qc}(q-1,q-1)$ is the cyclic-shift of the row-block of subarrays above it one subarray (or $n(q-1)$ positions) to the right and the top row-block of subarrays is the cyclic-shift of the last row-block of subarrays one subarray to the right. Since $\mathbf{H}_{b,sp,d,qc}(q-1,q-1)$ is an $mr \times nr$ array of CPMs and/or ZMs of size $(q-1) \times (q-1)$, it also has a section-wise cyclic structure. Hence, the array $\mathbf{H}_{b,sp,d,qc}(q-1,q-1)$ has both block-cyclic and section-wise cyclic structures.

The null space of $\mathbf{H}_{b,sp,d,qc}(q-1,q-1)$ gives a binary CPM-QC-SP-LDPC $C_{b,sp,d,qc}$ of length $nr(q-1)$ with rate at least $(n-m)/n$. Let $\mathbf{v}$ be a codeword in $C_{b,sp,d,qc}$. Now, suppose we divide a codeword $\mathbf{v}$ in $C_{b,sp,d,qc}$ into r sections, each consisting of $n(q-1)$ bits. If we cyclically shift the codeword $\mathbf{v}$ $n(q-1)$ positions to the right, we obtain another codeword in the code $C_{b,sp,d,qc}$. Hence, the code $C_{b,sp,d,qc}$ has a block-cyclic structure. Suppose we divide a codeword $\mathbf{v}$ into nr sections, each consisting of $q-1$ bits. If we cyclically shift all the nr sections simultaneously one position to the right within the sections, we obtain another codeword in $C_{b,sp,d,qc}$. Hence, the code $C_{b,sp,d,qc}$ has a section-wise cyclic structure. Therefore, the code $C_{b,sp,d,qc}$ is a *doubly QC-LDPC code* which has both block-cyclic and section-wise cyclic structures. The subscripts "*d*" and "*qc*" in $C_{b,sp,d,qc}$ and $\mathbf{H}_{b,sp,d,qc}(q-1,q-1)$ stand for "*doubly QC-structure*."

For $0 \leq j < r$, let CPM($\mathbf{R}_{0,j}$) be the binary CPM-dispersion of $\mathbf{R}_{0,j}$. Then, CPM($\mathbf{R}_{0,j}$) is an $m \times n$ array of CPMs and/or ZMs of size $(q-1) \times (q-1)$, a binary $m(q-1) \times n(q-1)$ matrix. The parity-check matrix $\mathbf{H}_{b,sp,d,qc}$(q-1, $q-1$) of $C_{b,sp,d,qc}$ can be expressed in terms of CPM($\mathbf{R}_{0,0}$), CPM($\mathbf{R}_{0,1}$), ..., CPM($\mathbf{R}_{0,r-1}$) as follows:

$$\mathbf{H}_{b,sp,d,qc}(q-1,q-1) =$$

$$\begin{bmatrix} \text{CPM}(\mathbf{R}_{0,0}) & \text{CPM}(\mathbf{R}_{0,1}) & \cdots & \text{CPM}(\mathbf{R}_{0,r-1}) \\ \text{CPM}(\mathbf{R}_{0,r-1}) & \text{CPM}(\mathbf{R}_{0,0}) & \cdots & \text{CPM}(\mathbf{R}_{0,r-2}) \\ \vdots & \vdots & \ddots & \vdots \\ \text{CPM}(\mathbf{R}_{0,1}) & \text{CPM}(\mathbf{R}_{0,2}) & \cdot & \text{CPM}(\mathbf{R}_{0,0}) \end{bmatrix}. \tag{8.4}$$

From (8.4), we see that, in terms of the SP-construction, the code $C_{b,sp,d,qc}$ is constructed with an $r \times r$ base matrix $\mathbf{B}_{b,sp}$ consisting of all 1-entries and a replacement set $\boldsymbol{R}$ with CPM($\mathbf{R}_{0,0}$), CPM($\mathbf{R}_{0,1}$), ..., CPM($\mathbf{R}_{0,r-1}$) as member matrices. Since the matrices $\mathbf{R}_{0,0}, \mathbf{R}_{0,1}, \ldots, \mathbf{R}_{0,r-1}$ satisfy both the 2×2 and the PW-2×2 SM-constraints, their CPM-dispersions CPM($\mathbf{R}_{0,0}$), CPM($\mathbf{R}_{0,1}$), ..., CPM($\mathbf{R}_{0,r-1}$) satisfy both the RC- and the PW-RC-constraints. The replacement of the 1-entries in the base matrix $\mathbf{B}_{b,sp}$ by the member matrices in $\boldsymbol{R} = \{\text{CPM}(\mathbf{R}_{0,0}), \text{CPM}(\mathbf{R}_{0,1}), \ldots, \text{CPM}(\mathbf{R}_{0,r-1})\}$ is carried out in *a cyclic manner* as shown in (8.4), which satisfies the replacement constraint.

Now suppose we view the binary matrix $\mathbf{H}_{b,sp,d,qc}(q-1,q-1)$ as an $mr \times nr$ array of binary CPMs and/or ZMs of size $(q-1) \times (q-1)$. Using the inverses of the row and column permutations, π_{row} and π_{col}, defined by (4.3) and (4.5), respectively, we can put $\mathbf{H}_{b,sp,d,qc}(q-1,q-1)$ into a $(q-1) \times (q-1)$ array $\mathbf{H}_{b,sp,cyc}(mr,nr)$ of binary $mr \times nr$ matrices in the form of (4.1) with a block-cyclic structure as follows:

$$\mathbf{H}_{b,sp,cyc}(mr,nr) = \begin{bmatrix} \mathbf{D}_0 & \mathbf{D}_1 & \cdots & \mathbf{D}_{q-2} \\ \mathbf{D}_{q-2} & \mathbf{D}_0 & \cdots & \mathbf{D}_{q-3} \\ \vdots & \vdots & \ddots & \vdots \\ \mathbf{D}_1 & \mathbf{D}_2 & \cdots & \mathbf{D}_0 \end{bmatrix}. \tag{8.5}$$

The array $\mathbf{H}_{b,sp,cyc}(mr,nr)$ also satisfies the RC-constraint. From the PTG point of view, the integer sum of constituent matrices $\mathbf{D}_0, \mathbf{D}_1, \ldots, \mathbf{D}_{q-2}$ in the first row-block of $\mathbf{H}_{b,sp,cyc}(mr,nr)$ gives the base matrix $\mathbf{B}_{ptg}$ of a protograph $\mathcal{G}_{ptg}$ with mr CNs and nr VNs (as shown in Section 4.1). Hence, the doubly CPM-QC-SP-LDPC code $C_{b,sp,d,qc}$ can be regarded as a QC-PTG-LDPC code whose Tanner graph is an expansion of the protograph $\mathcal{G}_{ptg}$ by a factor of $(q-1)$.

A binary QC-LDPC code $C_{b,sp,d,qc}$ with a doubly QC-structure can be decoded with either the reduced-complexity iterative decoding method proposed in [79] based on its block-cyclic structure, or the one proposed in [69, 68, 78] based on its section-wise cyclic structure. Either decoding scheme significantly reduces the decoder complexity and the message storage size.

Example 8.1. Consider the construction field GF(17). Construct a 16×16 matrix $\mathbf{W}_{RS}$ over GF(17) in the form given by (8.1). Choose $r = 2$ and $l = 8$ such that $16 = rl$. Then, following (8.2), the matrix $\mathbf{W}_{RS}$ can be viewed as a 2×2 array of 8×8 submatrices with 2 constituent matrices, $\mathbf{W}_{0,0}$ and $\mathbf{W}_{0,1}$. Take two 2×8 submatrices $\mathbf{R}_{0,0}$ and $\mathbf{R}_{0,1}$ from $\mathbf{W}_{0,0}$ and $\mathbf{W}_{0,1}$, respectively. The submatrix $\mathbf{R}_{0,0}$ consists of the first two rows of $\mathbf{W}_{0,0}$ and $\mathbf{R}_{0,1}$ consists of the first two rows of $\mathbf{W}_{0,1}$. Based on these two submatrices $\mathbf{R}_{0,0}$ and $\mathbf{R}_{0,1}$, we can form a 2×2 array $\mathbf{B}_{q,sp,p}(2,8)$ of 2×8 submatrices in the form given by (8.3). If we cyclically shift the first two rows of $\mathbf{B}_{q,sp,p}(2,8)$ simultaneously 8 positions to the right, we obtain the other two rows. Dispersing each nonzero entry in $\mathbf{B}_{q,sp,p}(2,8)$ into a 16×16 binary CPM and each 0-entry into a 16×16 ZM, we obtain a 2×2 array $\mathbf{H}_{b,sp,d,qc}(16,16)$ of 2×8 subarrays of 16×16 CPMs and ZMs. $\mathbf{H}_{b,sp,d,qc}(16,16)$ is a 64×256 binary matrix which can also be viewed as a 4×16 array of 16×16 CPMs and ZMs. Among the 256 columns in $\mathbf{H}_{b,sp,d,qc}(16,16)$, there are 64 columns with weight 3 and 192 columns with weight 4. The null space of $\mathbf{H}_{b,sp,d,qc}(16,16)$ gives an irregular (256, 192) doubly CPM-QC-SP-LDPC code $C_{b,sp,d,qc}$ with rate 0.75. The Tanner graph of the code $C_{b,sp,d,qc}$ has girth 6 and contains 10 048 cycles of length 6. The error performances of the code decoded with 50 iterations of the MSA over the AWGNC and the BEC are shown in Fig. 8.1(a) and Fig. 8.1(b), respectively. At a BLER of 10^{-8}, from Fig. 8.1(a), we see that the code performs 1.8 dB away from the SPB. Fig. 8.1(b) shows that the code performs 0.17 away from the Shannon limit over the BEC at a UEBR of 10^{-7}.

In the following, we present a long high-rate doubly CPM-QC-SP-LDPC code and demonstrate its error performance.

Example 8.2. Suppose we choose the prime field GF(139) for constructing a doubly CPM-QC-SP-LDPC code. Based on the nonzero elements of this field, we can construct a 138×138 matrix $\mathbf{W}_{RS} = \mathbf{B}^*_{q,sp,p}$ in the form of (8.1). Based

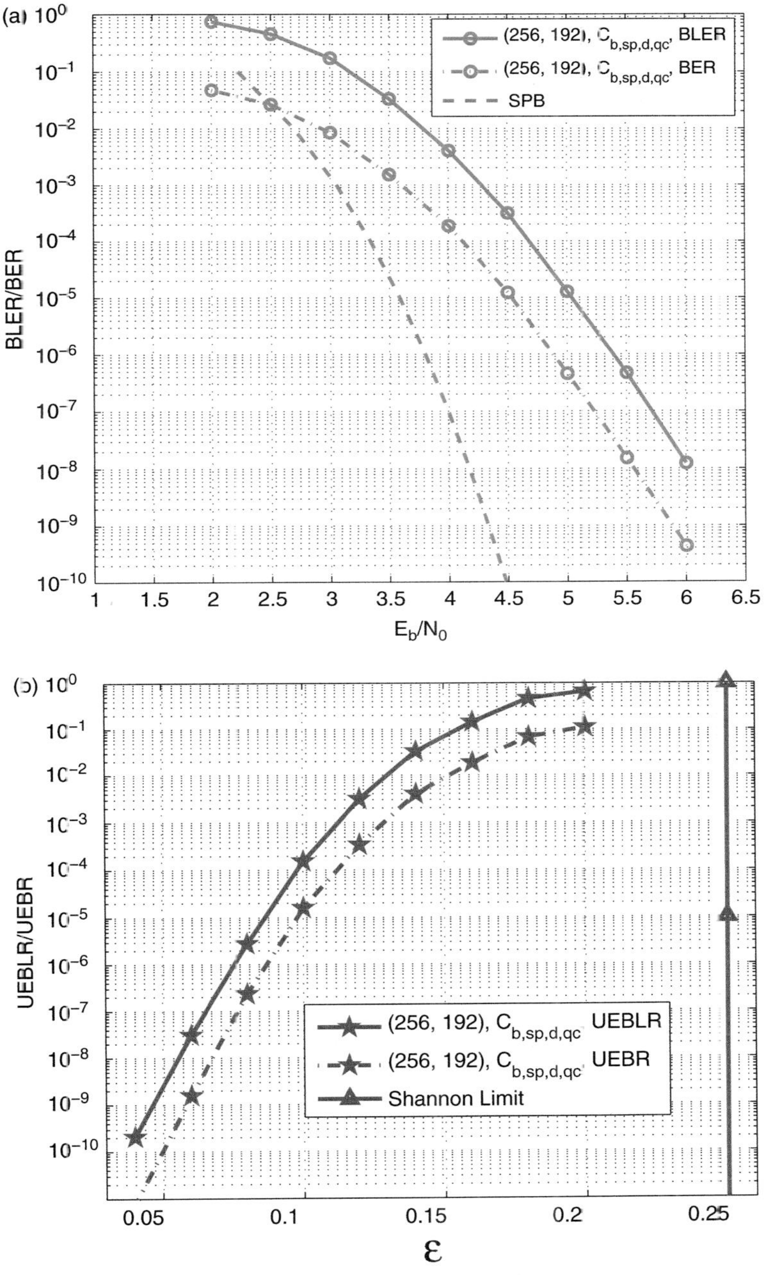

Figure 8.1. (a) The BER and BLER performances of the (256, 192) doubly CPM-QC-SP-LDPC code $C_{b,sp,d,qc}$ given in Example 8.1 over the AWGNC; and (b) performance of the code in (a) over the BEC.

on $\mathbf{W}_{RS}$, we can then construct a family of doubly CPM-QC-SP-LDPC codes of various lengths and rates using the construction method given above.

First we factor 138 as the product of two integers 6 and 23, i.e., $138 = 6 \times 23$, and let $r = 6$ and $l = 23$. The matrix $\mathbf{W}_{RS}$ can then be viewed as a 6×6 array of 23×23 submatrices over GF(139). Next, we choose $m = 1$ and $n = 20$. For $0 \le j < 6$, take a 1×20 submatrix $\mathbf{R}_{0,j}$ from $\mathbf{W}_{0,j}$ in the manner described above. (In this example, we avoid choosing the zero entries of $\mathbf{W}_{RS}$.) Based on these 6 submatrices $\mathbf{R}_{0,0}$, $\mathbf{R}_{0,1}$, $\mathbf{R}_{0,2}$, $\mathbf{R}_{0,3}$, $\mathbf{R}_{0,4}$, and $\mathbf{R}_{0,5}$, we then construct a 6×6 array $\mathbf{B}_{q,sp,p}(1, 20)$ of 1×20 submatrices over GF(139) in the form given by (8.3). $\mathbf{B}_{q,sp,p}(1, 20)$ is a 6×120 matrix over GF(139) with column and row weights 6 and 120, respectively.

The CPM-dispersion of the base matrix $\mathbf{B}_{q,sp,p}(1, 20)$ gives a 6×120 array $\mathbf{H}_{b,sp,d,qc}(138, 138)$ of CPMs of size 138×138. The array $\mathbf{H}_{b,sp,d,qc}(138, 138)$ is an 828×16560 matrix with column and row weights 6 and 120, respectively. Its null space gives a (6, 120)-regular (16 560, 15 737) doubly CPM-QC-SP-LDPC code $C_{b,sp,d,qc}$ with rate 0.9503. Its Tanner graph has girth 6 and contains 34 158 312 cycles of length 6 – a very large number of short cycles. However, this large number of short cycles does not degrade the performance of the code because the Tanner graph of the code also has a very high degree of VN-connectivity. Each VN is connected to 714 other VNs by paths of length 2. As a result, in a few iterations, each VN-MPU collects enough extrinsic information to update its LLR to a level sufficient to make a correct hard decision with high probability.

The BER and BLER performances of the code $C_{b,sp,d,qc}$ decoded with 5, 10, and 50 iterations of the MSA are shown in Fig. 8.2(a). At a BLER of 10^{-7}, the code decoded with 50 iterations performs 0.9 dB away from the SPB. At a BER of 10^{-10}, the code decoded with 50 iterations performs about 1.3 dB away from the Shannon limit with no visible error-floor. From this figure, we also see that the decoding of the code $C_{b,sp,d,qc}$ converges very fast. The performance gap between 5 and 10 iterations is about 0.2 dB and the gap between 10 and 50 iterations is less than 0.1 dB.

The UEBR and UEBLR performances of the code $C_{b,sp,d,qc}$ over the BEC are shown in Fig. 8.2(b). At a UEBR of 10^{-9}, the code performs about 0.026 away from the Shannon limit (0.05).

8.3 Masking and Variations

In the construction of a doubly CPM-QC-SP-LDPC code using the method presented in the last section, masking can be performed on the r constituent matrices $\mathbf{R}_{0,0}, \mathbf{R}_{0,1}, \ldots, \mathbf{R}_{0,r-1}$ of the base matrix $\mathbf{B}_{q,sp,p}(m, n)$. The masked constituent

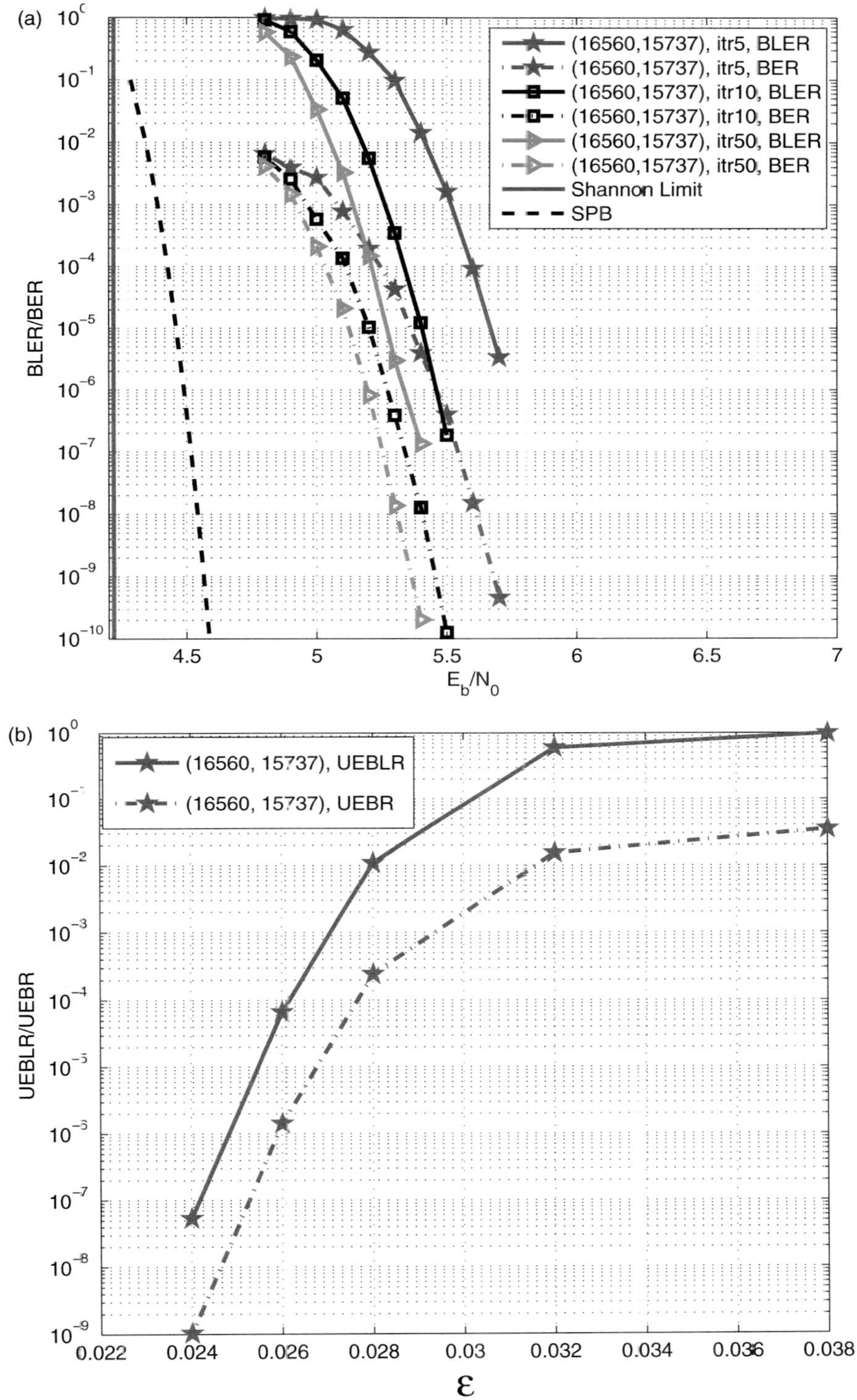

Figure 8.2. (a) The BER and BLER performances of the (16 560, 15 373) doubly CPM-QC-SP-LDPC code $C_{b,sp,d,qc}$ decoded with 50, 10, and 5 iterations of the MSA given in Example 8.2 over the AWGN channel; and (b) performance of the code in (a) over the BEC.

matrices $\mathbf{R}_{0,0,mask}, \mathbf{R}_{0,1,mask}, \ldots, \mathbf{R}_{0,r-1,mask}$ are used to form a cyclic masked base matrix $\mathbf{B}_{q,sp,p,mask}(m,n)$ using $[\mathbf{R}_{0,0,mask}, \mathbf{R}_{0,1,mask}, \ldots, \mathbf{R}_{0,r-1,mask}]$ as the generator row-block. Let $\mathbf{H}_{b,sp,d,qc,mask}(q-1,q-1)$ be the CPM-dispersion of $\mathbf{B}_{q,sp,p,mask}(m,n)$. Then, the null space of $\mathbf{H}_{b,sp,d,qc,mask}(q-1,q-1)$ gives a new doubly CPM-QC-SP-LDPC code $C_{b,sp,d,qc,mask}$ whose Tanner graph will have a smaller number of short cycles or a girth larger (or both) than the unmasked code $C_{b,sp,d,qc}$.

Masking can also be performed on the base array $\mathbf{B}_{q,sp,p}(m,n)$. In this case, some constituent matrices in $\mathbf{B}_{q,sp,p}(m,n)$ are replaced by $m \times n$ ZMs. If we want to maintain the bock-cyclic structure, the masking matrix $\mathbf{Z}_c$ should be an $r \times r$ circulant. Masking the $r \times r$ base array $\mathbf{B}_{q,sp,p}(m,n)$ with $\mathbf{Z}_c$, we obtain a masked block-cyclic base array $\mathbf{B}_{q,sp,p,mask,c}(m,n)$. Then, the null space of the CPM-dispersion of $\mathbf{B}_{q,sp,p,mask,c}(m,n)$ gives a doubly CPM-QC-SP-LDPC code $C_{b,sp,d,qc,mask,c}$. The subscript "c" stands for "*cyclic masking*."

Of course, masking can be performed on both the constituent matrices $\mathbf{R}_{0,0}$, $\mathbf{R}_{0,1}$, ..., $\mathbf{R}_{0,r-1}$ of the base array $\mathbf{B}_{q,sp,p}(m,n)$ and the array itself. This double masking results in a doubly masked base matrix $\mathbf{B}_{q,sp,p,mask,dc}(m,n)$. If the array masking is cyclic, then the null space of the CPM-dispersion of the masked base matrix $\mathbf{B}_{q,sp,p,mask,dc}(m,n)$ gives a doubly CPM-QC-SP-LDPC code. The subscript "dc" stands for "*doubly cyclic masking*."

Since the base array $\mathbf{B}_{q,sp,p}(m,n)$ satisfies the 2×2 SM-constraint, any subarray of $\mathbf{B}_{q,sp,p}(m,n)$ also satisfies the 2×2 SM-constraint and hence its CPM-dispersion gives a CPM-QC-SP-LDPC code whose Tanner graph has girth at least 6. In this case, the code may not have a doubly QC-structure. For example, if we take the first s row-blocks of $\mathbf{B}_{q,sp,p}(m,n)$, $1 \le s < r$, we obtain an $s \times r$ array $\mathbf{B}_{q,sp,p}(s,r,m,n)$ of $m \times n$ matrices over GF(q) which does not have a block-cyclic structure (the cyclic-shift of the last row-block does not give the top row-block). The null space of the CPM-dispersion of $\mathbf{B}_{q,sp,p}(s,r,m,n)$ gives a CPM-QC-SP-LDPC code of length $nr(q-1)$ and rate at least $(nr-ms)/nr$, but not a doubly QC-LDPC code.

If s divides r, then $r = st$ with $1 < t < r$. Suppose we take the 0-th, t-th, $2t$-th, ..., $(s-1)t$-th row-blocks of $\mathbf{B}_{q,sp,p}(m,n)$ to form the following $s \times r$ subarray $\mathbf{B}_{q,sp,p}(s,r,m,n)$:

$$\mathbf{B}_{q,sp,p}(s,r,m,n) = \begin{bmatrix} \mathbf{R}_{0,0} & \mathbf{R}_{0,1} & \cdots & \mathbf{R}_{0,r-1} \\ \mathbf{R}_{0,r-t} & \mathbf{R}_{0,r-t+1} & \cdots & \mathbf{R}_{0,r-t-1} \\ \mathbf{R}_{0,r-2t} & \mathbf{R}_{0,r-2t+1} & \cdots & \mathbf{R}_{0,r-2t-1} \\ \vdots & \vdots & \ddots & \vdots \\ \mathbf{R}_{0,t} & \mathbf{R}_{0,t+1} & \cdots & \mathbf{R}_{0,t-1} \end{bmatrix}. \tag{8.6}$$

From (8.6), we see that each row-block of $\mathbf{B}_{q,sp,p}(s,r,m,n)$ is the cyclic-shift of the row-block above it t constituent matrices (or nt positions) to the right and the top row-block is the cyclic-shift of the last row-block t constituent matrices to the right. So, $\mathbf{B}_{q,sp,p}(s,r,m,n)$ has a block-cyclic structure. In this case, the null space of the CPM-dispersion of $\mathbf{B}_{q,sp,p}(s,r,m,n)$ gives a doubly CPM-QC-SP-LDPC code of length $nr(q-1)$.

Masking can be performed on the array $\mathbf{B}_{q,sp,p}(s,r,m,n)$ (masking its constituent matrices) to obtain a masked array $\mathbf{B}_{q,sp,p,mask}(s,r,m,n)$. The null space of the CPM-dispersion of $\mathbf{B}_{q,sp,p,mask}(s,r,m,n)$ gives a new doubly CPM-QC-SP-LDPC code of length $nr(q-1)$ with rate at least $(nr-ms)/nr$.

Example 8.3. Consider the 16×16 matrix $\mathbf{W}_{RS}$ over GF(17) in the form given by (8.1) constructed in Example 8.1. Suppose, we choose $r = l = 4$ and view the matrix $\mathbf{W}_{RS}$ as a 4×4 array of 4×4 submatrices with 4 constituent matrices, $\mathbf{W}^*_{0,0}$, $\mathbf{W}^*_{0,1}$, $\mathbf{W}^*_{0,2}$, and $\mathbf{W}^*_{0,3}$. Take 2×4 submatrices from these 4 constituent matrices $\mathbf{W}^*_{0,0}$, $\mathbf{W}^*_{0,1}$, $\mathbf{W}^*_{0,2}$, and $\mathbf{W}^*_{0,3}$ in the way we described in this section (here we take the first two rows). Then, we obtain a 4×4 array $\mathbf{B}^*_{q,sp,p}(2,4)$ of 2×4 submatices in the form given by (8.3). It is easy to see that the array $\mathbf{B}^*_{q,sp,p}(2,4)$ has a block-cyclic structure. Consider the following masking matrix:

$$\mathbf{Z} = \begin{bmatrix} 0 & 1 & 1 & 0 \\ 0 & 0 & 1 & 1 \\ 1 & 0 & 0 & 1 \\ 1 & 1 & 0 & 0 \end{bmatrix}.$$

The matrix $\mathbf{Z}$ has a cyclic structure. Replacing each 1-entry in $\mathbf{Z}$ by a 2×4 matrix with all 1-entries and each 0-entry by a 2×4 matrix with all 0-entries, we obtain a new matrix given as below:

$$\mathbf{Z}^* = \begin{bmatrix} 0&0&0&0&1&1&1&1&1&1&1&1&0&0&0&0 \\ 0&0&0&0&1&1&1&1&1&1&1&1&0&0&0&0 \\ 0&0&0&0&0&0&0&0&1&1&1&1&1&1&1&1 \\ 0&0&0&0&0&0&0&0&1&1&1&1&1&1&1&1 \\ 1&1&1&1&0&0&0&0&0&0&0&0&1&1&1&1 \\ 1&1&1&1&0&0&0&0&0&0&0&0&1&1&1&1 \\ 1&1&1&1&1&1&1&1&0&0&0&0&0&0&0&0 \\ 1&1&1&1&1&1&1&1&0&0&0&0&0&0&0&0 \end{bmatrix}. \quad (8.7)$$

We can see that the matrix $\mathbf{Z}^*$ also has a block-cyclic structure. If we cyclically shift the first two rows of $\mathbf{Z}^*$ 4 positions to the right simultaneously, we obtain the next two rows of $\mathbf{Z}^*$. And the first two rows of $\mathbf{Z}^*$ is the cyclic shift of the last two rows 4 positions to the right. It is easy to check that $\mathbf{Z}^*$ has the same block-cyclic structure as the matrix $\mathbf{B}^*_{q,sp,p}(2,4)$. Using $\mathbf{Z}^*$ as a masking matrix to mask the base matrix $\mathbf{B}^*_{q,sp,p}(2,4)$, we obtain a masked 4×4 array $\mathbf{B}^*_{q,sp,p,mask}(2,4)$ of 2×4 submatrices. Since $\mathbf{B}^*_{q,sp,p}(2,4)$ and $\mathbf{Z}^*$ have the same block-cyclic structure, after masking, the masked array $\mathbf{B}^*_{q,sp,p,mask}(2,4)$ also has the same block-cyclic structure. The masked array $\mathbf{B}^*_{q,sp,p,mask}(2,4)$ satisfies the 2×2 SM-constraint. The binary CPM-dispersion of $\mathbf{B}^*_{q,sp,p,mask}(2,4)$ results in an 8×16 array $\mathbf{H}_{b,sp,d,qc,mask}(16,16)$ of CPMs and ZMs of size 16×16. The null space of $\mathbf{H}_{b,sp,d,qc,mask}(16,16)$ gives a (4, 8)-regular (256, 133) doubly CPM-QC-SP-LDPC code $C_{b,sp,d,qc,mask}$ with rate 0.5195. The Tanner graph of the code $C_{b,sp,d,qc,mask}$ has girth 6 and contains 1600 cycles of length 6 and 26 144 cycles of length 8. The performances of the code $C_{b,sp,d,qc,mask}$ over the AWGNC and the BEC are shown in Fig. 8.3(a) and Fig. 8.3(b), respectively. We see that the constructed code performs well over both the AWGNC and the BEC.

8.4 SP-Construction of CPM-QC-SP-LDPC Codes

As shown earlier, in Section 8.2, the matrices CPM($\mathbf{R}_{0,0}$), CPM($\mathbf{R}_{0,1}$), ..., CPM($\mathbf{R}_{0,r-1}$) (the CPM-dispersions of $\mathbf{R}_{0,0}, \mathbf{R}_{0,1}, \ldots, \mathbf{R}_{0,r-1}$) satisfy both the RC- and the PW-RC-constraints. Hence, they can be used as member matrices of a replacement set $\boldsymbol{R}$ for the SP-construction of QC-SP-LDPC codes. By properly choosing a base matrix and using these matrices as the replacement matrices, the SP-construction will result in a CPM-QC-SP-LDPC code whose Tanner graph has girth at least 6.

If the base matrix has a cyclic structure (such as the circulants constructed based on Euclidean geometries presented in Section 6.1), the replacement of its 1-entries with CPM($\mathbf{R}_{0,0}$), CPM($\mathbf{R}_{0,1}$), ..., CPM($\mathbf{R}_{0,r-1}$) can be carried out in a cyclic manner, which will result in a parity-check array with both the block-cyclic and the section-wise cyclic structures. Furthermore, each of the matrices CPM($\mathbf{R}_{0,0}$), CPM($\mathbf{R}_{0,1}$), ..., CPM($\mathbf{R}_{0,r-1}$), or a subset, can be used as an SP-base matrix for the construction of QC-SP-LDPC codes or a decomposition base matrix for the algebraic construction of QC-PTG-LDPC codes.

8.5 Discussion and Remarks

As pointed out in Section 8.2, the doubly CPM-QC-SP-LDPC code $C_{b,sp,d,qc}$ can be decoded with either the reduced-complexity iterative decoding scheme proposed in [79] based on its block-cyclic structure or the reduced-complexity iterative decoding scheme presented in [69, 68, 78] based on its section-wise cyclic

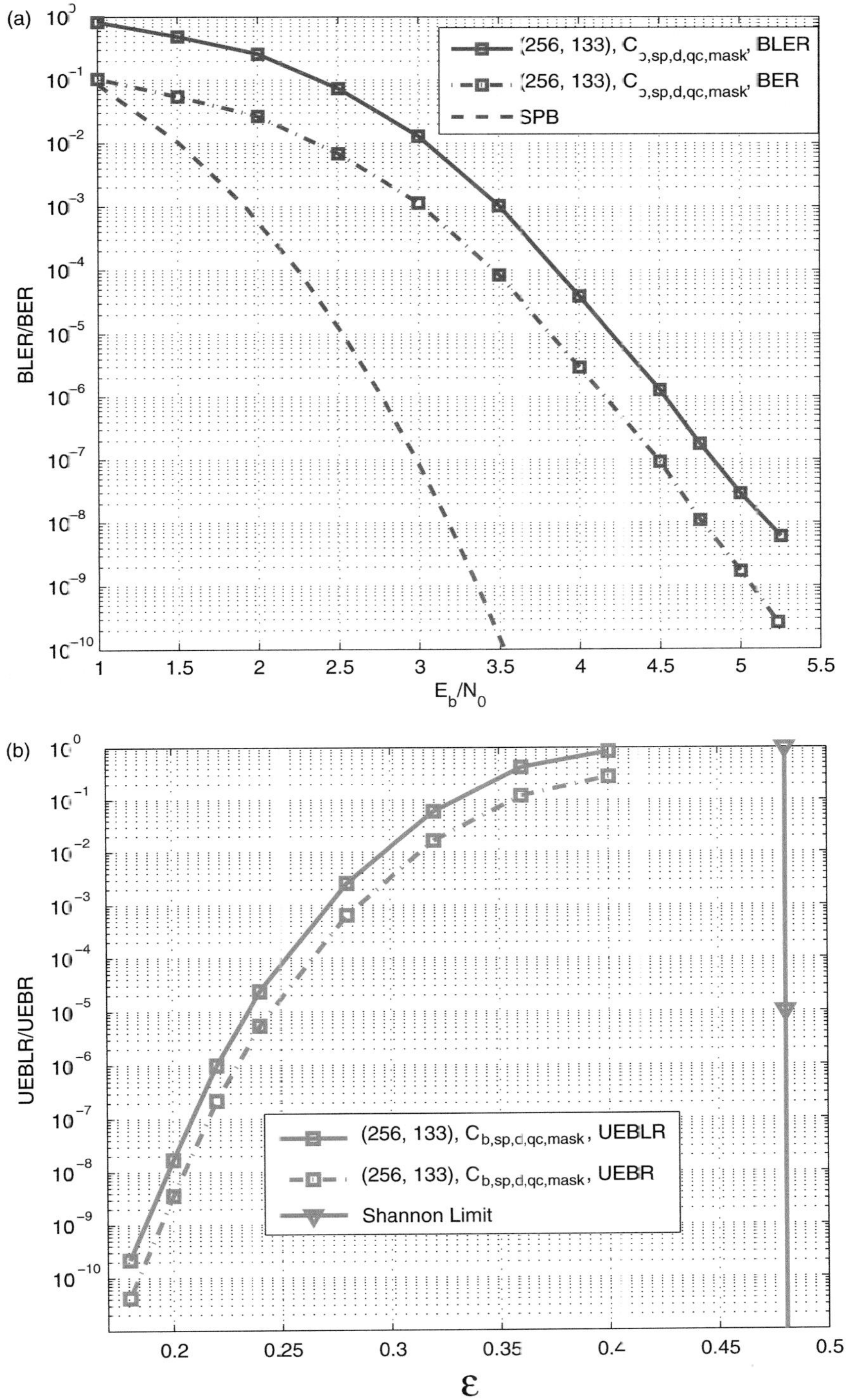

Figure 8.3. (a) The BER and BLER performances of the (256, 133) doubly CPM-QC-SP-LDPC code $C_{b,sp,d,qc,mask}$ given in Example 8.3 over the AWGNC; and (b) performance of the code in (a) over the BEC.

structure. Using the method proposed in [79], we can use the $m(q-1) \times nr(q-1)$ matrix $\mathbf{H}_{dec,0} = [\mathrm{CPM}(\mathbf{R}_{0,0})\ \mathrm{CPM}(\mathbf{R}_{0,1})\ \ldots\ \mathrm{CPM}(\mathbf{R}_{0,r-1})]$ (the top row-block of $\mathbf{H}_{b,sp,d,qc}(q-1,q-1)$ given by (8.4)) as the decoding matrix. Using the method proposed in [69, 68, 78], we can use the $mr \times nr(q-1)$ matrix $\mathbf{H}_{dec,1}$ formed by taking the first rows of the mr row-blocks of CPMs and ZMs of the $mr \times nr$ array $\mathbf{H}_{b,sp,d,qc}(q-1,q-1)$ given by (8.4) as the decoding matrix.

Suppose we implement two decoders to decode the code $C_{b,sp,d,qc}$ jointly in a turbo decoding manner, one based on the decoding matrix $\mathbf{H}_{dec,0}$ and the other based on the decoding matrix $\mathbf{H}_{dec,1}$. Whether this turbo decoding will enhance the performance of the code $C_{b,sp,d,qc}$, in terms waterfall error probability, error-floor, and rate of decoding convergence, is an interesting problem to investigate.

Another interesting problem for investigation is: besides the base matrix given by (8.1), whether there are other base matrices over NB fields that satisfy the 2×2 SM-constraint and have a cyclic structure.

9

SP-Construction of Spatially Coupled QC-LDPC Codes

In this chapter, we present an algebraic construction of a special type of LDPC code whose Tanner graph has a very specific structure. For an LDPC code of this type, its Tanner graph is *locally connected*. Every VN is only connected to the CNs that are confined to a (small) *span* of ρ_{col} *consecutive locations* and every CN is only connected to VNs that are confined to a (small) *span* of ρ_{row} *consecutive locations*. We call such constraints on the connections between VNs and CNs of a Tanner graph $(\rho_{\text{col}}, \rho_{\text{row}})$-*span-constraints*. With this span-constraint, the Tanner graph of such an LDPC code is actually a *chain* of small Tanner graphs in which each graph is connected to its adjacent graphs on either side of it, except the first and the last ones. An LDPC code of this type is called a *span-constrained* LDPC code. The SC-LDPC code investigated in [59, 60, 22, 86] is a type of span-constrained LDPC code.

An SC-LDPC code is an LDPC convolutional (LDPC-C) code viewed from a graphical point of view (or a spatial coupling point of view) [49, 104, 66, 67, 92]. An LDPC-C code [49, 104] is specified by a bi-infinite parity-check matrix whose nonzero entries are confined to a *diagonal band* of a certain width ρ_{row} and a certain depth ρ_{col}. The nonzero entries in every row are confined to a *span* of ρ_{row} consecutive locations and the nonzero entries in every column are confined to a span of ρ_{col} consecutive locations. With these constraints on the locations of the nonzero entries of the parity-check matrix of an LDPC-C code, every VN in its Tanner graph is *only connected* to the CNs that are confined to a span of ρ_{col} consecutive locations and every CN is *only connected* to VNs that are confined to a span of ρ_{row} consecutive locations. These constraints on the locations of the nonzero entries of the parity-check matrix of an LDPC-C code lead to the graphical $(\rho_{\text{col}}, \rho_{\text{row}})$-span-constraint as mentioned above. Hence, an LDPC-C code is a span-constrained LDPC code, an SC-LDPC code.

Over the past 15 years, a great deal of research effort has been expended on the study of LDPC-C codes and impressive theoretical results have been developed [66, 67, 59]. However, the employment of LDPC-C codes in practical applications has lagged behind their theoretical development. Recently, LDPC-C codes have been re-examined from a graphical point of view. As a consequence, they have been renamed spatially coupled (SC) LDPC codes in a series of papers [59, 60, 22, 86], which has generated additional research interest for their asymptotically good ensemble performance and structural properties. Reference [22] gives a very good tutorial presentation of the structure of LDPC-C and SC-LDPC codes and also presents some open research problems. Their advantages and disadvantages compared with typical block LDPC codes will be discussed at the end of this chapter.

In this chapter, we present SP-based methods (or CPM-dispersion methods) for constructing algebraic QC-SC-LDPC codes and their terminated codes. Two types of QC-SC-LDPC codes are presented. In our presentation, we consider only the construction of binary QC-SC-LDPC codes.

9.1 Base Matrices and Their Structural Properties

A base matrix for a QC-SC-LDPC code can be constructed from the 2×2 SM-constrained $r \times r$ array $\mathbf{B}_{q,sp,p}(m,n)$ of $m \times n$ submatrices over GF(q) with a block-cyclic structure given by (8.3). To give a better picture of the construction, we permute the row-blocks of $\mathbf{B}_{q,sp,p}(m,n)$ to put it in the following block-cyclic structure form:

$$\mathbf{B}'_{q,sp,p}(m,n) = \begin{bmatrix} \mathbf{R}_{0,0} & \mathbf{R}_{0,1} & \cdots & \mathbf{R}_{0,r-2} & \mathbf{R}_{0,r-1} \\ \mathbf{R}_{0,1} & \mathbf{R}_{0,2} & \cdots & \mathbf{R}_{0,r-1} & \mathbf{R}_{0,0} \\ \vdots & \vdots & \ddots & \vdots & \vdots \\ \mathbf{R}_{0,r-1} & \mathbf{R}_{0,0} & \cdots & \mathbf{R}_{0,r-3} & \mathbf{R}_{0,r-2} \end{bmatrix}. \tag{9.1}$$

From (9.1), we see that, for $0 \le i < r$, the i-th column-block of $\mathbf{B}'_{q,sp,p}(m,n)$ is identical to the i-th row-block. Actually, the permuted array $\mathbf{B}'_{q,sp,p}(m,n)$ can be obtained by cyclically shifting the top row-block of $\mathbf{B}_{q,sp,p}(m,n)$ given by (8.3) to the left $r-1$ times, one submatrix at a time. Since permuting the row-blocks of $\mathbf{B}_{q,sp,p}(m,n)$ does not affect its 2×2 SM-constrained property, the permuted array $\mathbf{B}'_{q,sp,p}(m,n)$ still satisfies the 2×2 SM-constraint and all the constituent matrices $\mathbf{R}_{0,0}, \mathbf{R}_{0,1}, \ldots, \mathbf{R}_{0,r-1}$ of the top row-block of the array $\mathbf{B}'_{q,sp,p}(m,n)$ still satisfy both the 2×2 and the PW-2×2 SM-constraints.

Let e and τ be two positive integers such that $e \ge 2$ and $\tau e < r$. Using the constituent matrices $\mathbf{R}_{0,0}, \mathbf{R}_{0,1}, \ldots, \mathbf{R}_{0,r-1}$ of the left most column-block (or the top row-block) of the array $\mathbf{B}'_{q,sp,p}(m,n)$, we form the following semi-infinite array over GF(q):

$$\Lambda_{q,sp,sc}(\tau,m,n) = \left(\begin{array}{cccc|cccc|c}
\mathbf{R}_{0,0} & & & & & & & & \\
\mathbf{R}_{0,1} & \mathbf{R}_{0,e} & & & & & & & \\
\vdots & \mathbf{R}_{0,e+1} & \ddots & & & & & & \\
\vdots & \vdots & \ddots & \mathbf{R}_{0,(\tau-1)e} & & & & & \\
\vdots & \vdots & \ddots & \mathbf{R}_{0,(\tau-1)e+1} & \mathbf{R}_{0,0} & & & & \\
\mathbf{R}_{0,e-1} & \vdots & \ddots & \vdots & \mathbf{R}_{0,1} & \mathbf{R}_{0,e} & & & \\
 & \mathbf{R}_{0,2e-1} & \ddots & \vdots & \vdots & \mathbf{R}_{0,e+1} & \ddots & & \\
 & & \ddots & \vdots & \vdots & \vdots & \ddots & \mathbf{R}_{0,(\tau-1)e} & \\
 & & & \mathbf{R}_{0,\tau e-1} & \vdots & \vdots & \ddots & \mathbf{R}_{0,(\tau-1)e+1} & \ddots \\
 & & & & \mathbf{R}_{0,e-1} & \vdots & \ddots & & \ddots \\
 & & & & & \mathbf{R}_{0,2e-1} & \ddots & & \ddots \\
 & & & & & & \ddots & & \ddots \\
 & & & & & & & \mathbf{R}_{0,\tau e-1} & \ddots \\
 & & & & & & & & \ddots
\end{array}\right). \tag{9.2}$$

Note that all the nonzero matrices in the semi-infinite array $\Lambda_{q,sp,sc}(\tau,m,n)$ given above lie on a diagonal band of width en and depth em. This semi-infinite array $\Lambda_{q,sp,sc}(\tau,m,n)$ has the following structural properties: (1) Every column-block consists of e distinct nonzero $m \times n$ matrices taken from the PW-2×2 SM-constrained set $\{\mathbf{R}_{0,0}, \mathbf{R}_{0,1}, \ldots, \mathbf{R}_{0,r-1}\}$; (2) The e nonzero matrices in every column-block are confined to a span of em consecutive locations, called a *nonzero column-block-span*; (3) For $0 \leq j < \tau$, the e matrices in the nonzero column-block-span with the matrix $\mathbf{R}_{0,je}$ as the first matrix are the top e matrices of the je-th column-block of the $r \times r$ cyclic array $\mathbf{B}'_{q,sp,p}(m,n)$ given by (9.1); (4) All the τe nonzero matrices in the nonzero column-block-spans of τ consecutive column-blocks are *distinct*; (5) Every nonzero column-block-span repeats itself every τ consecutive nonzero column-block-spans but is shifted downward τ positions; (6) Every $(e + \tau - 1) \times \tau$ subarray that consists of the nonzero column-block-spans of any τ *consecutive* column-blocks of $\Lambda_{q,sp,sc}(\tau,m,n)$, called a τ-*span subarray*, is a subarray of the $r \times r$ cyclic array $\mathbf{B}'_{q,sp,p}(m,n)$ given by (9.1) after replacing some submatrices in $\mathbf{B}'_{q,sp,p}(m,n)$ by ZMs; and (7) Every τ-span subarray consists of the same set of τ nonzero column-block-spans, but not necessarily in the same order. A τ-span subarray of $\Lambda_{q,sp,sc}(\tau,m,n)$ is an $m(e+\tau-1) \times n\tau$ matrix over GF(q).

Structural properties 3, 6, and 7 of the semi-infinite array $\Lambda_{q,sp,sc}(\tau,m,n)$ simply follow from the block-cyclic structure of $\mathbf{B}'_{q,sp,p}(m,n)$. Since every τ-span subarray of $\Lambda_{q,sp,sc}(\tau,m,n)$ is a subarray of $\mathbf{B}'_{q,sp,p}(m,n)$ and since the array $\mathbf{B}'_{q,sp,p}(m,n)$ satisfies the 2×2 SM-constraint, every τ-span subarray of $\Lambda_{q,sp,sc}(\tau,m,n)$ satisfies the 2×2 SM-constraint. It follows that the semi-infinite array $\Lambda_{q,sp,sc}(\tau,m,n)$ satisfies the 2×2 SM-constraint.

For $j \geq 0$, the $j\tau$-th to the $((j+1)\tau-1)$-th column-blocks of $\Lambda_{q,sp,sc}(\tau,m,n)$ form a *period* of $\Lambda_{q,sp,sc}(\tau,m,n)$. Two periods of $\Lambda_{q,sp,sc}(\tau,m,n)$ do not overlap and they cover the same set of nonzero column-block-spans in the same order. Hence, $\Lambda_{q,sp,sc}(\tau,m,n)$ consists of infinite repetitions of the first period formed by the 0-th to the $(\tau-1)$-th column-blocks of $\Lambda_{q,sp,sc}(\tau,m,n)$. The semi-infinite array $\Lambda_{q,sp,sc}(\tau,m,n)$ is said to have a period τ. The τ nonzero column-block-spans of a period of $\Lambda_{q,sp,sc}(\tau,m,n)$ forms a τ-span subarray of $\Lambda_{q,sp,sc}(\tau,m,n)$.

Let $\mathbf{B}_{q,sp,sc}(\tau,m,n)$ be the first τ-span subarray of the semi-infinite array $\Lambda_{q,sp,sc}(\tau,m,n)$ (i.e., the first τ nonzero column-block-spans of $\Lambda_{q,sp,sc}(\tau,m,n)$) and let $\mathcal{G}_{q,sp,sc}(\tau,m,n)$ be the Tanner graph associated with $\mathbf{B}_{q,sp,sc}(\tau,m,n)$. Then, from the structure of $\Lambda_{q,sp,sc}(\tau,m,n)$ displayed in (9.2), we see that the Tanner graph associated with $\Lambda_{q,sp,sc}(\tau,m,n)$, denoted by $\mathcal{G}_{q,sp,sc}(\Lambda)$, consists of an infinite number of *ordered* copies of $\mathcal{G}_{q,sp,sc}(\tau,m,n)$, each connected to the two adjacent ones on either side of it by edges, except for the first one. The set of connecting edges between any two consecutive periods (or two adjacent copies of $\mathcal{G}_{q,sp,sc}(\tau,m,n)$) is the same. Therefore, the Tanner graph $\mathcal{G}_{q,sp,sc}(\Lambda)$ associated with $\Lambda_{q,sp,sc}(\tau,m,n)$ has an infinite *chain-like structure*.

The semi-infinite array $\Lambda_{q,sp,sc}(\tau,m,n)$ of $m \times n$ submatrices over GF(q) is the base array for constructing a QC-SC-LDPC code.

9.2 Type-1 QC-SC-LDPC Codes

Let $\mathbf{H}_{b,sp,sc}(\tau,m,n)$ be the binary CPM-dispersion of $\Lambda_{q,sp,sc}(\tau,m,n)$. Then, $\mathbf{H}_{b,sp,sc}(\tau,m,n)$ is a semi-infinite array of binary CPMs and ZMs of size $(q-1) \times (q-1)$ with period τ. It follows from the 2×2 SM-constrained property of the array $\Lambda_{q,sp,sc}(\tau,m,n)$ that the array $\mathbf{H}_{b,sp,sc}(\tau,m,n)$ satisfies the

RC-constraint. The null space of the array $\mathbf{H}_{b,sp,sc}(\tau, m, n)$ gives a *periodically time-varying* CPM-QC-SC-LDPC code, denoted by $C_{b,sp,sc}(\tau)$, with period τ. The Tanner graph, denoted by $\mathcal{G}_{b,sp,sc}(\Lambda)$, of the CPM-QC-SC-LDPC code $C_{b,sp,sc}(\tau)$, contains no cycle of length 4 and hence has girth at least 6. It is a periodically time-varying LDPC-C code with constraint length $en(q-1)$ (in symbols). The CPM-QC-SC-LDPC code $C_{b,sp,sc}(\tau)$ constructed based on the array $\Lambda_{q,sp,sc}(\tau, m, n)$ is referred to as a *type-1 CPM-QC-SC-LDPC code*.

For $0 \le j < r$, let CPM($\mathbf{R}_{0,j}$) be the binary CPM-dispersion of the $m \times n$ matrix $\mathbf{R}_{0,j}$ over GF(q). From the SP-construction point of view, the parity-check matrix $\mathbf{H}_{b,sp,sc}(\tau, m, n)$ of the time-varying CPM-QC-SC-LDPC code $C_{b,sp,sc}(\tau)$ is constructed by using the set

$$\boldsymbol{R} = \{\text{CPM}(\mathbf{R}_{0,0}), \text{CPM}(\mathbf{R}_{0,1}), \ldots, \text{CPM}(\mathbf{R}_{0,r-1})\} \tag{9.3}$$

of $m(q-1) \times n(q-1)$ matrices ($m \times n$ array of CPMs of size $(q-1) \times (q-1)$) as the replacement set $\boldsymbol{R}$ and the binary matrix $\mathbf{B}_{sp}$ given by (9.4) as the SP-base matrix. Hence, the CPM-QC-SC-LDPC code $C_{b,sp,sc}(\tau)$ is also a CPM-QC-SP-LDPC code.

From the PTG-construction point of view, the Tanner graph $\mathcal{G}_{q,sp,sc}(\tau, m, n)$ of the first τ-span subarray $\mathbf{B}_{q,sp,sc}(\tau, m, n)$ of the array $\Lambda_{q,sp,sc}(\tau, m, n)$ may be regarded as the protograph for the PTG-construction of the CPM-QC-SC-LDPC code $C_{b,sp,sc}(\tau)$. The edges of $\mathcal{G}_{q,sp,sc}(\tau, m, n)$ are labeled with nonzero elements in GF(q). In the PTG-based construction of the parity-check matrix $\mathbf{H}_{b,sp,sc}(\tau, m, n)$ of the code $C_{b,sp,sc}(\tau)$, there are two expansions of the protograph $\mathcal{G}_{q,sp,sc}(\tau, m, n)$, first expanding $\mathcal{G}_{q,sp,sc}(\tau, m, n)$ into $\mathcal{G}_{q,sp,sc}(\Lambda)$ by taking an infinite number of copies of $\mathcal{G}_{q,sp,sc}(\tau, m, n)$ and connecting them into a chain and then expanding $\mathcal{G}_{q,sp,sc}(\Lambda)$ by binary CPM-dispersing the q-ary label of each edge in $\mathcal{G}_{q,sp,sc}(\Lambda)$ to construct the Tanner graph $\mathcal{G}_{b,sp,sc}(\Lambda)$ of the code $C_{b,sp,sc}(\tau)$ as described above. The adjacency matrix of $\mathcal{G}_{b,sp,sc}(\Lambda)$ gives the parity-check matrix $\mathbf{H}_{b,sp,sc}(\tau, m, n)$ of the code $C_{b,sp,sc}(\tau)$. Therefore, the CPM-QC-SC-LDPC code $C_{b,sp,sc}(\tau)$ can be viewed as a QC-PTG-LDPC code.

Since $\mathbf{H}_{b,sp,sc}(\tau, m, n)$ is the CPM-dispersion of $\Lambda_{q,sp,sc}(\tau, m, n)$, the Tanner graph of the time-varying CPM-QC-SC-LDPC code $C_{b,sp,sc}(\tau)$ is a chain of identical subgraphs, each being an expansion of the graph $\mathcal{G}_{q,sp,sc}(\tau, m, n)$ by a factor of $q-1$. Each of these subgraphs is the Tanner graph of the CPM-dispersion of $\mathbf{B}_{q,sp,sc}(\tau, m, n)$, the first τ-span subarray of the semi-infinite array $\Lambda_{q,sp,sc}(\tau, m, n)$.

Since a type-1 CPM-QC-SC-LDPC code is an LDPC-C code, the methods devised for decoding an LDPC-C code [74, 47] can be applied to decode the

CPM-QC-SC-LDPC code $C_{b,sp,sc}(\tau)$. The rate of the code $C_{b,sp,sc}(\tau)$ is at least $(n-m)/n$.

$$
\mathbf{B}_{sp} = \left(\begin{array}{cccc|cccc|c}
1 & & & & & & & & \\
1 & 1 & & & & & & & \\
\vdots & 1 & \ddots & & & & & & \\
\vdots & \vdots & \ddots & 1 & & & & & \\
\vdots & \vdots & \ddots & 1 & 1 & & & & \\
1 & \vdots & \ddots & \vdots & 1 & 1 & & & \\
 & 1 & \ddots & \vdots & \vdots & 1 & \ddots & & \\
 & & \ddots & \vdots & \vdots & \vdots & \ddots & 1 & \\
 & & & 1 & \vdots & \vdots & \ddots & 1 & \ddots \\
 & & & & 1 & \vdots & \ddots & \vdots & \ddots \\
 & & & & & 1 & \ddots & \vdots & \ddots \\
 & & & & & & \ddots & \vdots & \ddots \\
 & & & & & & & 1 & \ddots \\
 & & & & & & & & \ddots
\end{array}\right). \tag{9.4}
$$

Example 9.1. In this example, we use the field GF(127) for constructing a CPM-QC-SC-LDPC code. First, we construct a 126×126 cyclic matrix $\mathbf{B}_{q,sp,p}$ over GF(127) in the form given by (8.2) which satisfies the 2×2 SM-constraint. Factor $127 - 1 = 126$ as the product of 63 and 2, and set $r = 63$ and $l = 2$. Choose $m = 1$ and $n = 2$. Then, following the construction process in Section 9.1, we form a 63×63 cyclic array $\mathbf{B}'_{q,sp,p}(1,2)$ of 1×2 matrices in the form of (9.1). The top row-block of this array consists of 63 constituent matrices $\mathbf{R}_{0,0}, \mathbf{R}_{0,1}, \ldots, \mathbf{R}_{0,62}$ of size 1×2.

Choose $e = 4$ (nonzero column-block-span) and $\tau = 4$ (period). Using the constituent matrices $\mathbf{R}_{0,0}, \mathbf{R}_{0,1}, \ldots, \mathbf{R}_{0,15}$, we form the following semi-infinite base array:

$$
\Lambda_{q,sp,sc}(4,1,2) =
$$

$$
\left(\begin{array}{cccc|cccc|c}
\mathbf{R}_{0,0} & & & & & & & & \\
\mathbf{R}_{0,1} & \mathbf{R}_{0,4} & & & & & & & \\
\mathbf{R}_{0,2} & \mathbf{R}_{0,5} & \mathbf{R}_{0,8} & & & & & & \\
\mathbf{R}_{0,3} & \mathbf{R}_{0,6} & \mathbf{R}_{0,9} & \mathbf{R}_{0,12} & & & & & \\
 & \mathbf{R}_{0,7} & \mathbf{R}_{0,10} & \mathbf{R}_{0,13} & \mathbf{R}_{0,0} & & & & \\
 & & \mathbf{R}_{0,11} & \mathbf{R}_{0,14} & \mathbf{R}_{0,1} & \mathbf{R}_{0,4} & & & \\
 & & & \mathbf{R}_{0,15} & \mathbf{R}_{0,2} & \mathbf{R}_{0,5} & \mathbf{R}_{0,8} & & \\
 & & & & \mathbf{R}_{0,3} & \mathbf{R}_{0,6} & \mathbf{R}_{0,9} & \mathbf{R}_{0,12} & \\
 & & & & & \mathbf{R}_{0,7} & \mathbf{R}_{0,10} & \mathbf{R}_{0,13} & \ddots \\
 & & & & & & \mathbf{R}_{0,11} & \mathbf{R}_{0,14} & \ddots \\
 & & & & & & & \mathbf{R}_{0,15} & \ddots \\
 & & & & & & & & \ddots
\end{array}\right). \tag{9.5}
$$

The CPM-dispersion of the above matrix $\Lambda_{q,sp,sc}(4,1,2)$ gives a semi-infinite array $\mathbf{H}_{b,sp,sc}(4,1,2)$ of binary CPMs and ZMs of size 126×126, which is a (504, 1008)-span-constrained matrix. The null space of $\mathbf{H}_{b,sp,sc}(4,1,2)$ gives a type-1 periodically time-varying CPM-QC-SC-LDPC code $C_{b,sp,sc}(4)$ with a period of 4 whose Tanner graph has girth 6. The code can achieve a rate of 0.5 with an unterminated decoding.

Suppose we decode this code with 1000 iterations of the SPA with a decoding block of length $L = 88$ (i.e., over the span of $M = 22$ periods). In this case, the code length is 22 176 and the rate is 0.4831. The Tanner graph of the terminated code $C^t_{b,sp,sc}(4)$ has girth 6 and contains 13 608 cycles of length 6 and 203 238 cycles of length 8. (The subscript "t" in $C^t_{b,sp,sc}(4)$ stands for "*termination*.") The bit error probability performance of the terminated code over the AWGNC is shown in Fig. 9.1(a). At a BER of 10^{-5}, the code performs about 1.5 dB away from the Shannon limit. Also included in this figure is the BER performance of a QC-SC-LDPC code C_{cjl} of length 25 000 bits and rate 0.488 reported in [14]. The subscript "cjl" stands for the first letters of the last names of the authors, Chandrasetty, Johnson, and Lechner. The code C_{cjl}, designed with a good decoding threshold, is also decoded with 1000 iterations of the SPA. From the figure, we see that the algebraic CPM-QC-SC-LDPC code $C^t_{b,sp,sc}(4)$ constructed in this example outperforms the PTG-based QC-SC-LDPC code C_{cjl} given in [14] by about 0.1 dB.

The UEBR and UEBLR performances of the code $C^t_{b,sp,sc}(4)$ over the BEC are shown in Fig. 9.1(b).

The base matrix $\mathbf{B}_{q,sp,sc}(4,1,2)$ for constructing the semi-infinite CPM-QC-SC-LDPC code $C_{b,sp,sc}(4)$ (or the terminated code $C^t_{b,sp,sc}(4)$) is

$$\mathbf{B}_{q,sp,sc}(4,1,2) = \begin{bmatrix} \mathbf{R}_{0,0} & & & \\ \mathbf{R}_{0,1} & \mathbf{R}_{0,4} & & \\ \mathbf{R}_{0,2} & \mathbf{R}_{0,5} & \mathbf{R}_{0,8} & \\ \mathbf{R}_{0,3} & \mathbf{R}_{0,6} & \mathbf{R}_{0,9} & \mathbf{R}_{0,12} \\ & \mathbf{R}_{0,7} & \mathbf{R}_{0,10} & \mathbf{R}_{0,13} \\ & & \mathbf{R}_{0,11} & \mathbf{R}_{0,14} \\ & & & \mathbf{R}_{0,15} \end{bmatrix},$$

which is a 7×8 matrix over GF(127). The Tanner graph $\mathcal{G}_{q,sp,sc}(4,1,2)$ associated with the base matrix $\mathbf{B}_{q,sp,sc}(4,1,2)$ consists of 7 CNs, 8 VNs, and 32 edges labeled with elements from GF(127) as shown in Fig. 9.2. From the graphical point of view, $\mathcal{G}_{q,sp,sc}(4,1,2)$ is the protograph for constructing the CPM-QC-SC-LDPC code $C^t_{b,sp,sc}(4)$.

If we choose various r, l, m, n, e, and τ, we can construct a sequence of CPM-QC-SC-LDPC codes with various rates based on base matrices in the form of (9.1) over an NB field GF(q).

Example 9.2. In this example, we use the same field GF(127) for code construction as in Example 9.1. This time, we factor out 126 as the product of 21 and 6. Set $r = 21$ and $l = 6$ and choose $m = 1$, $n = 4$, $e = 4$, and $\tau = 4$. With these parameters, we can construct a semi-infinite array $\mathbf{H}_{b,sp,sc}(4,1,4)$ of binary CPMs and ZMs of size 126×126. The null space of $\mathbf{H}_{b,sp,sc}(4,1,4)$ gives a type-1 periodically time-varying CPM-QC-SC-LDPC code $C_{b,sp,sc}(4)$ with period 4. The rate of this code is 0.75 (with unterminated decoding).

If we terminate $C_{b,sp,sc}(4)$ by setting $M = 11$ ($L = 44$), we obtain a terminated CPM-QC-SC-LDPC code $C^t_{b,sp,sc}(4)$ with length 22 176 and rate 0.7331. The bit error probability performance of this terminated code over the AWGNC decoded with 1000 iterations of the SPA is shown in Fig. 9.3(a) and its erasure error performance over the BEC is shown in Fig. 9.3(b). At a BER of 10^{-6}, the code performs within 1 dB of the Shannon limit.

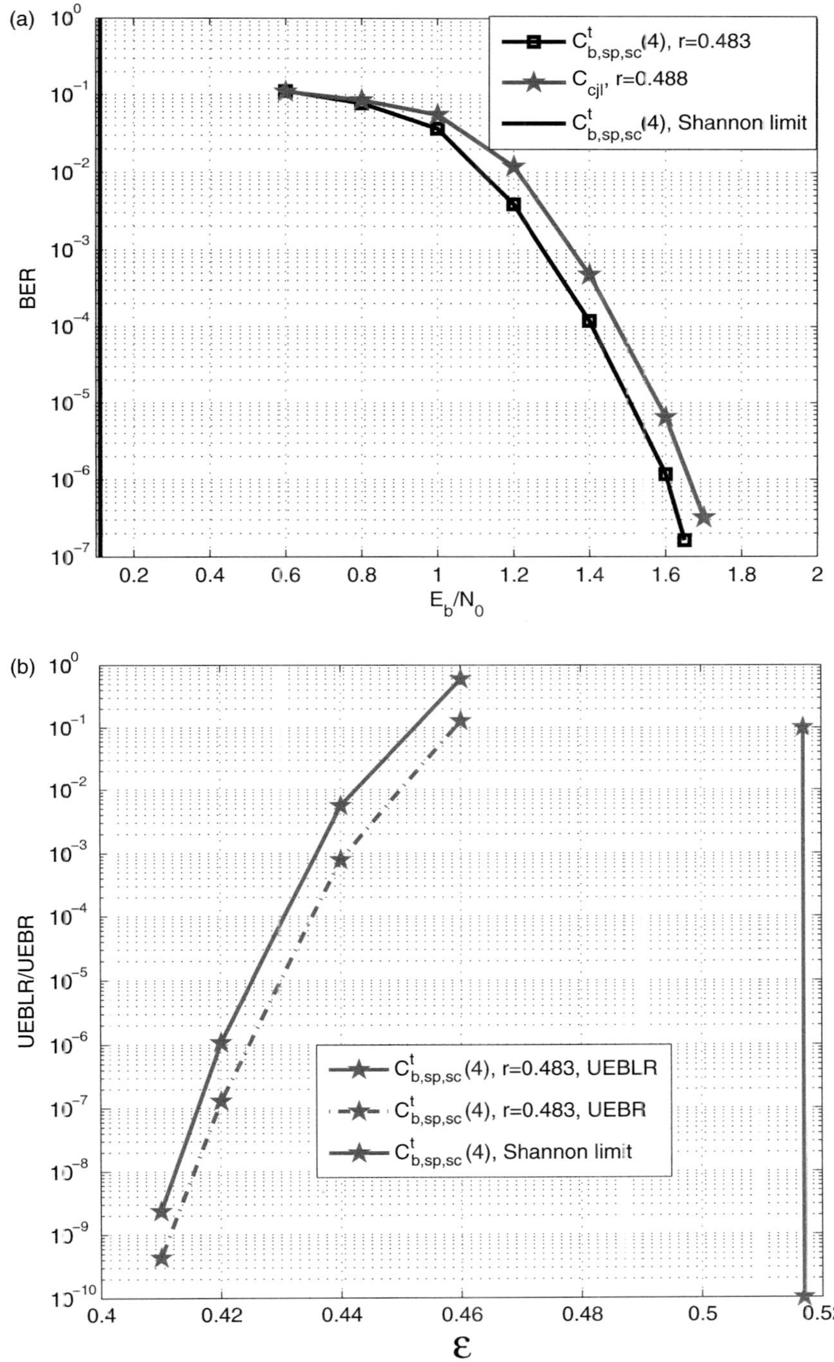

Figure 9.1. (a) The bit error probability performances of the CPM-QC-SC-LDPC code $C^t_{b,sp,sc}(4)$ given in Example 9.1 and the code C_{cjl} given in [14] over the AWGNC; (b) performance of the code $C^t_{b,sp,sc}(4)$ in (a) over the BEC.

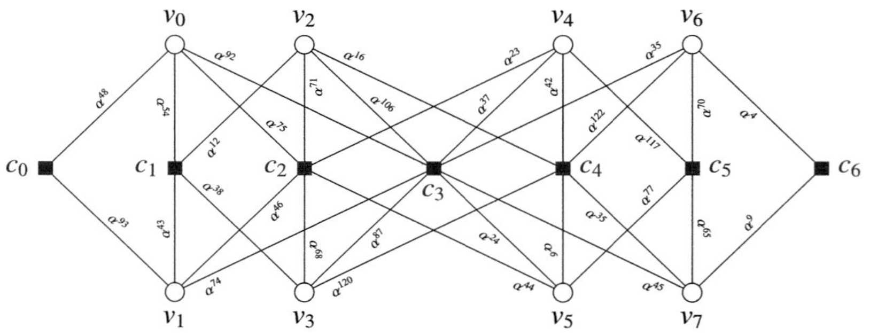

Figure 9.2. The Tanner graph $\mathcal{G}_{q,sp,sc}(4,1,2)$ of the base matrix $\mathbf{B}_{q,sp,sc}(4,1,2)$ given in Example 9.1.

9.3 Type-2 QC-SC-LDPC Codes

Suppose we form a semi-infinite array $\Lambda^{\star}_{q,sp,sc}(\tau,m,n)$ in which the i-th row-block is identical to the i-th column-block of the array $\Lambda_{q,sp,sc}(\tau,m,n)$ given by (9.2), with $i=0,1,\ldots,$ as follows:

$$\Lambda^{\star}_{q,sp,sc}(\tau,m,n) =$$

$$\begin{pmatrix}
\mathbf{R}_{0,0}\mathbf{R}_{0,1} & \cdots & \cdots & \cdots & \mathbf{R}_{0,e-1} & & & & & & \\
 & \mathbf{R}_{0,e}\mathbf{R}_{0,e+1} & \cdots & \cdots & \cdots & \mathbf{R}_{0,2e-1} & & & & & \\
 & \ddots & \ddots & \ddots & \ddots & \ddots & \ddots & & & & \\
 & & \mathbf{R}_{0,(\tau-1)e}\mathbf{R}_{0,(\tau-1)e+1} & \cdots & \cdots & \cdots & \mathbf{R}_{0,\tau e-1} & & & & \\
 & & & \mathbf{R}_{0,0} & \mathbf{R}_{0,1} & \cdots & \cdots & \cdots & \mathbf{R}_{0,e-1} & & \\
 & & & & \mathbf{R}_{0,e} & \mathbf{R}_{0,e+1} & \cdots & \cdots & \cdots & \mathbf{R}_{0,2e-1} & \\
 & & & & & \ddots & \ddots & \ddots & \ddots & \ddots & \ddots \\
 & & & & & & \mathbf{R}_{0,(\tau-1)e}\mathbf{R}_{0,(\tau-1)e+1} & \cdots & \cdots & \cdots & \mathbf{R}_{0,\tau e-1} \\
 & & & & & & & \ddots & \ddots & \ddots & \ddots \ \ddots \ \ddots
\end{pmatrix}. \tag{9.6}$$

Note that all the nonzero matrices in the semi-infinite array $\Lambda^{\star}_{q,sp,sc}(\tau,m,n)$ given above lie in a diagonal band of width en and depth em. The structural properties of $\Lambda^{\star}_{q,sp,sc}(\tau,m,n)$ can be developed from the structural properties of the array $\Lambda_{q,sp,sc}(\tau,m,n)$ just by changing "column" to "row" and "downward" to "rightward." Every nonzero row-block-span repeats itself every τ consecutive nonzero row-block-spans but shifted to the right by τ positions. Every $\tau \times (e+\tau-1)$ subarray of $\Lambda^{\star}_{q,sp,sc}(\tau,m,n)$ that consists of the nonzero row-block-spans of any τ *consecutive* row-blocks of $\Lambda^{\star}_{q,sp,sc}(\tau,m,n)$, called a τ-*span subarray*, is a subarray of the $r \times r$ cyclic array $\mathbf{B}'_{q,sp,p}(m,n)$ given by (9.1) after replacing some entries with ZMs. Furthermore, every such τ-span subarray consists of the

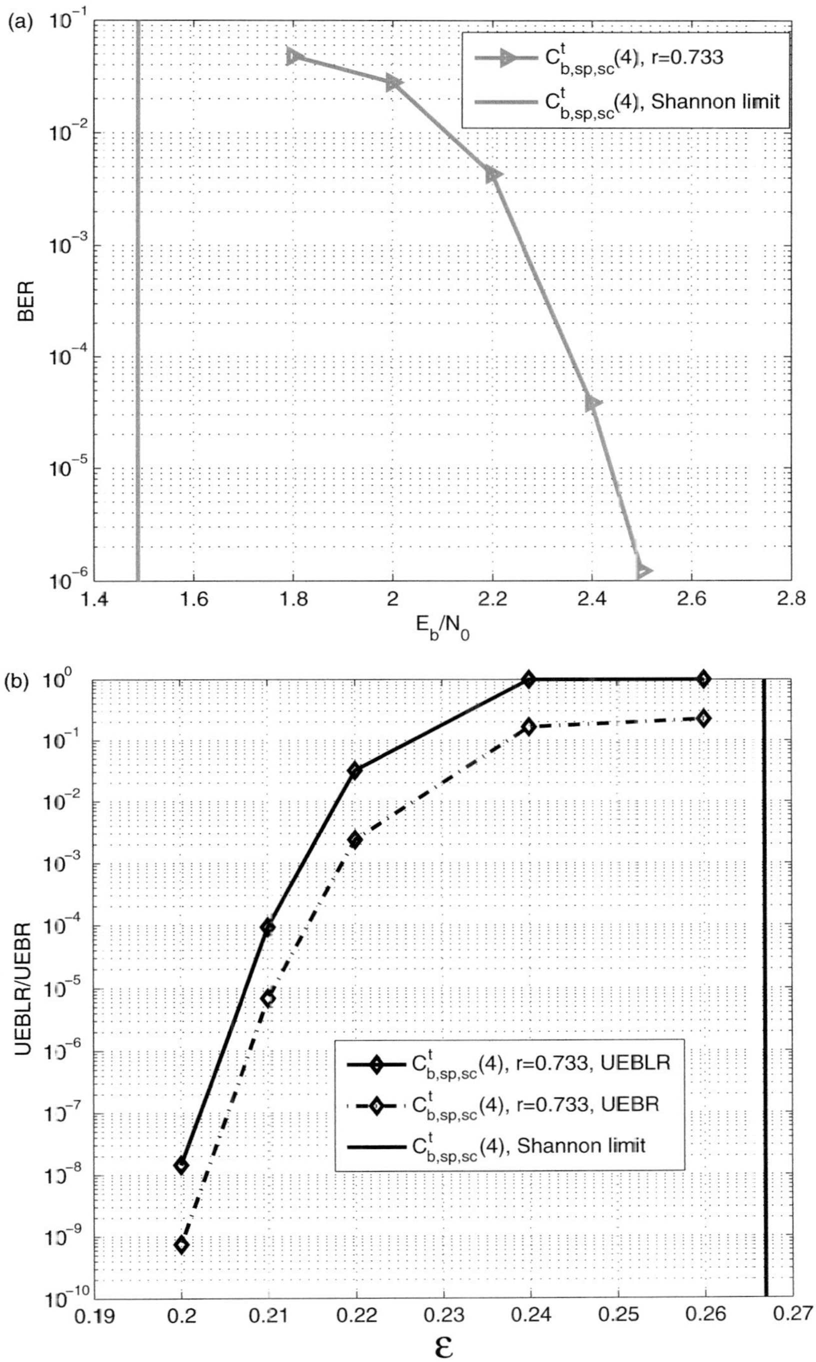

Figure 9.3. (a) The bit error probability performance of the CPM-QC-SC-LDPC code $C^t_{b,sp,sc}(4)$ given in Example 9.2 over the AWGNC; and (b) performance of the code in (a) over the BEC.

same set of τ nonzero row-block-spans, but not necessarily in the same order. Note that a τ-span subarray of $\Lambda^{\star}_{q,sp,sc}(\tau, m, n)$ is an $m\tau \times n(e + \tau - 1)$ matrix over GF(q).

The structure of the Tanner graph $\mathcal{G}_{q,sp,sc}(\Lambda^{\star})$ associated with the array $\Lambda^{\star}_{q,sp,sc}(\tau, m, n)$ can be analyzed in a similar way as the Tanner graph $\mathcal{G}_{q,sp,sc}(\Lambda)$ of the array $\Lambda_{q,sp,sc}(\tau, m, n)$ in the construction of type-1 CPM-QC-SC-LDPC codes. It is an infinite chain of an infinite number of copies of the Tanner graph $\mathcal{G}^{\star}_{q,sp,sc}(\tau, m, n)$ of the first τ-span subarray of $\Lambda^{\star}_{q,sp,sc}(\tau, m, n)$.

Dispersing each nonzero entry in $\Lambda^{\star}_{q,sp,sc}(\tau, m, n)$ into a CPM of size $(q-1) \times (q-1)$ and each zero entry to a $(q-1) \times (q-1)$ ZM, we obtain a semi-infinite array $\mathbf{H}^{\star}_{b,sp,sc}(\tau, m, n)$ of CPMs and ZMs of size $(q-1) \times (q-1)$. The null space of $\mathbf{H}^{\star}_{b,sp,sc}(\tau, m, n)$ gives a *type-2 periodically time-varying CPM-QC-SC-LDPC code* $C^{\star}_{b,sp,sc}(\tau)$.

From the graph-theoretic point of view, the graph $\mathcal{G}^{\star}_{q,sp,sc}(\tau, m, n)$ is regarded as the protograph for constructing the CPM-QC-SC-LDPC code $C^{\star}_{b,sp,sc}(\tau)$. From the SP point of view, the base matrix for the SP-construction of $C^{\star}_{b,sp,sc}(\tau)$ is:

$$\mathbf{B}^{\star}_{sp} = \begin{bmatrix} 1 & 1 & \cdots & \cdots & \cdots & 1 & & & & & & & \\ & 1 & 1 & \cdots & \cdots & \cdots & 1 & & & & & & \\ & & & & \ddots & \ddots & \ddots & \ddots & \ddots & \ddots & & & \\ & & & & & 1 & 1 & \cdots & \cdots & \cdots & 1 & & \\ & & & & & & 1 & 1 & \cdots & \cdots & \cdots & 1 & \\ & & & & & & & \ddots & \ddots & \ddots & \ddots & \ddots & \ddots \end{bmatrix}, \tag{9.7}$$

and the replacement set is $\{\text{CPM}(\mathbf{R}_{0,0}), \text{CPM}(\mathbf{R}_{0,1}), \ldots, \text{CPM}(\mathbf{R}_{0,r-1})\}$.

Example 9.3. In Example 9.1, we construct a 63×63 cyclic array $\mathbf{B}'_{q,sp,p}(1,2)$ of 1×2 matrices in the form of (9.1). The top row of this array consists of 63 constituent 1×2 matrices $\mathbf{R}_{0,0}, \mathbf{R}_{0,1}, \ldots, \mathbf{R}_{0,62}$.

Choose $e = 4$ (row-block-span) and $\tau = 4$ (period). Using the constituent matrices $\mathbf{R}_{0,0}, \mathbf{R}_{0,1}, \ldots, \mathbf{R}_{0,15}$, we form the following semi-infinite array:

$$\Lambda^{\star}_{q,sp,sc}(4,1,2) = \left(\begin{array}{cccccccccccc} \mathbf{R}_{0,0} & \mathbf{R}_{0,1} & \mathbf{R}_{0,2} & \mathbf{R}_{0,3} & & & & & & & & \\ & \mathbf{R}_{0,4} & \mathbf{R}_{0,5} & \mathbf{R}_{0,6} & \mathbf{R}_{0,7} & & & & & & & \\ & & \mathbf{R}_{0,8} & \mathbf{R}_{0,9} & \mathbf{R}_{0,10} & \mathbf{R}_{0,11} & & & & & & \\ & & & \mathbf{R}_{0,12} & \mathbf{R}_{0,13} & \mathbf{R}_{0,14} & \mathbf{R}_{0,15} & & & & & \\ & & & & \mathbf{R}_{0,0} & \mathbf{R}_{0,1} & \mathbf{R}_{0,2} & \mathbf{R}_{0,3} & & & & \\ & & & & & \mathbf{R}_{0,4} & \mathbf{R}_{0,5} & \mathbf{R}_{0,6} & \mathbf{R}_{0,7} & & & \\ & & & & & & \mathbf{R}_{0,8} & \mathbf{R}_{0,9} & \mathbf{R}_{0,10} & \mathbf{R}_{0,11} & & \\ & & & & & & & \mathbf{R}_{0,12} & \mathbf{R}_{0,13} & \mathbf{R}_{0,14} & \mathbf{R}_{0,15} & \\ & & & & & & & & \ddots & \ddots & \ddots & \ddots \end{array}\right). \tag{9.8}$$

The CPM-dispersion of the above matrix $\Lambda^{\star}_{q,sp,sc}(4,1,2)$ gives a semi-infinite array $\mathbf{H}^{\star}_{b,sp,sc}(4,1,2)$ of binary CPMs and ZMs of size 126×126. The null space of $\mathbf{H}^{\star}_{b,sp,sc}(4,1,2)$ gives a type-2 periodically time-varying CPM-QC-SC-LDPC code $C^{\star}_{b,sp,sc}(4)$ with a period of 4 whose Tanner graph has girth 6.

If we terminate $\mathbf{H}^{\star}_{b,sp,sc}(4,1,2)$ by setting $M = 22$ ($L = 88$) and then remove the first two and last two column-blocks, we obtain a terminated and shortened array $\mathbf{H}^{\star,t}_{b,sp,sc}(4,1,2)$. The null space of the new array $\mathbf{H}^{\star,t}_{b,sp,sc}(4,1,2)$ gives a CPM-QC-SC-LDPC code $C^{\star,t}_{b,sp,sc}(4)$ with length 22 050 and rate 0.4972. The BER performances of this terminated code over the AWGNC decoded with 1000 and 50 iterations of the SPA are shown in Fig. 9.4(a) and its erasure error performance over the BEC is shown in Fig. 9.4(b). The gap between 1000 iterations and 50 iterations of the SPA is about 0.2 dB.

The base matrix $\mathbf{B}^{\star}_{q,sp,sc}(4,1,2)$ for constructing the semi-infinite CPM-QC-SC-LDPC code $C^{\star,t}_{b,sp,sc}(4)$ is

$$\mathbf{B}^{\star}_{q,sp,sc}(4,1,2) = \begin{bmatrix} \mathbf{R}_{0,0} & \mathbf{R}_{0,1} & \mathbf{R}_{0,2} & \mathbf{R}_{0,3} & & & \\ & \mathbf{R}_{0,4} & \mathbf{R}_{0,5} & \mathbf{R}_{0,6} & \mathbf{R}_{0,7} & & \\ & & \mathbf{R}_{0,8} & \mathbf{R}_{0,9} & \mathbf{R}_{0,10} & \mathbf{R}_{0,11} & \\ & & & \mathbf{R}_{0,12} & \mathbf{R}_{0,13} & \mathbf{R}_{0,14} & \mathbf{R}_{0,15} \end{bmatrix}.$$

9.4 Terminated and Tailbiting CPM-QC-SC-LDPC Codes

Suppose we terminate the semi-infinite base array $\Lambda_{q,sp,sc}(\tau,m,n)$ by taking its first M periods and removing all the zero row-blocks. This results in a $(\tau M +$

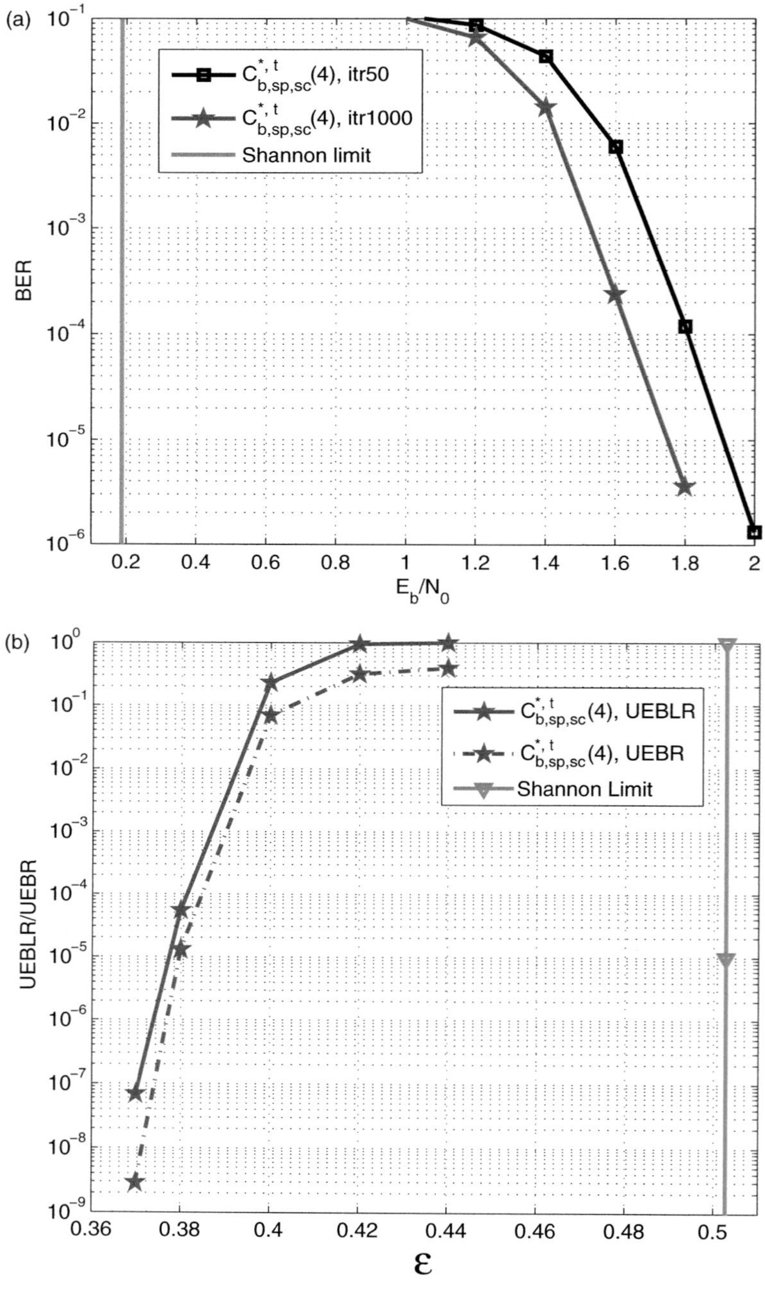

Figure 9.4. (a) The BER performances of the CPM-QC-SC-LDPC code $C^{\star,t}_{b,sp,sc}(4)$ decoded with 1000 and 50 iterations of the SPA given in Example 9.3 over the AWGNC; and (b) performance of the code in (a) over the BEC.

$e-1) \times \tau M$ array of $m \times n$ matrices over GF(q), denoted by $\Lambda^t_{q,sp,sc}(M,\tau,m,n)$. Since $\Lambda^t_{q,sp,sc}(M,\tau,m,n)$ is a subarray of $\Lambda_{q,sp,sc}(\tau,m,n)$ and $\Lambda_{q,sp,sc}(\tau,m,n)$ satisfies the 2×2 SM-constraint, $\Lambda^t_{q,sp,sc}(M,\tau,m,n)$ also satisfies the 2×2 SM-constraint. The Tanner graph of this array consists of M copies of the Tanner graph $\mathcal{G}_{q,sp,sc}(\tau,m,n)$ connected in a finite-length chain. The last copy of $\mathcal{G}_{q,sp,sc}(\tau,m,n)$ and the first copy of $\mathcal{G}_{q,sp,sc}(\tau,m,n)$ may not be directly connected.

The CPM-dispersion of the finite array $\Lambda^t_{q,sp,sc}(M,\tau,m,n)$ gives an $m(M\tau + e-1) \times nM\tau$ array, denoted by $\mathbf{H}^t_{b,sp,sc}(M,\tau,m,n)$, of binary CPMs and ZMs of size $(q-1) \times (q-1)$. The null space of $\mathbf{H}^t_{b,sp,sc}(M,\tau,m,n)$ gives a block CPM-QC-SC-LDPC code $C^t_{b,sp,sc}(\tau)$ of length $nM\tau(q-1)$ whose Tanner graph $\mathcal{G}_{b,sp,sc}(\Lambda^t)$ contains no cycle of length 4 and has a finite-length chain-like structure. We call this code $C^t_{b,sp,sc}(\tau)$ an *L-block-terminated* type-1 CPM-QC-SC-LDPC code where $L = M\tau$. The superscript "t" in $\Lambda^t_{q,sp,sc}(\tau,m,n)$, $\mathbf{H}^t_{b,sp,sc}(\tau,m,n)$, and $C^t_{b,sp,sc}(\tau)$ stands for "*terminated*." An L-block-terminated type-1 CPM-QC-SC-LDPC code can be decoded with any conventional iterative decoding algorithm of a block LDPC code, such as the SPA, the MSA, or their simplified variations. The code $C^t_{b,sp,sc}(\tau)$ has a rate at least $(n-m)/n - (e-1)m/nL$. Therefore, there is a rate loss of $(e-1)m/nL$ compared to the unterminated code. For a large L, the rate loss is very small.

Now suppose we terminate the semi-infinite base matrix $\Lambda_{q,sp,sc}(\tau,m,n)$ by taking the first M τ-span subarrays and wrapping the last $e-1$ row-blocks to the top in the way as shown in (9.9). This results in an $M\tau \times M\tau$ array, denoted by $\Lambda^{tb}_{q,sp,sc}(M,\tau,m,n)$ in the form of (9.9), of $m \times n$ matrices over GF(q). Every row-block of the array $\Lambda^{tb}_{q,sp,sc}(M,\tau,m,n)$ contains e nonzero $m \times n$ matrices and every column-block contains e nonzero $m \times n$ matrices. The Tanner graph $\mathcal{G}^{tb}_{q,sp,sc}(\Lambda^{tb})$ of $\Lambda^{tb}_{q,sp,sc}(M,\tau,m,n)$ consists of M copies of $\mathcal{G}_{q,sp,sc}(\tau,m,n)$ connected as a *ring* in which the last copy is directly connected to the first copy of $\mathcal{G}_{q,sp,sc}(\tau,m,n)$. The wrapping around operation is called *tailbiting*.

The binary CPM-dispersion of the finite array $\Lambda^{tb}_{q,sp,sc}(M,\tau,m,n)$ gives an $mM\tau \times nM\tau$ array $\mathbf{H}^{tb}_{b,sp,sc}(M,\tau,m,n)$ of CPMs and ZMs of size $(q-1) \times (q-1)$. The null space of $\mathbf{H}^{tb}_{b,sp,sc}(M,\tau,m,n)$ gives a block CPM-QC-SC-LDPC code, called a *type-1 tailbiting (TB) CPM-QC-SC-LDPC code*, denoted by $C^{tb}_{b,sp,sc}(\tau)$. The superscript "tb" in $\Lambda^{tb}_{q,sp,sc}(M,\tau,m,n)$, $\mathbf{H}^{tb}_{b,sp,sc}(M,\tau,m,n)$, and $C^{tb}_{b,sp,sc}(\tau)$ stands for "*tailbiting*." The rate of this code is at least $(n-m)/n$.

For a given field GF(q), based on the block-cyclic array $\mathbf{B}'_{q,sp,p}(m,n)$ given by (9.1), we can construct a large family of CPM-QC-SC-LDPC, L-terminated and tailbiting CPM-QC-SC-LDPC codes with various rates and periods using different choices of parameters, m, n, τ, and M.

By terminating (or tailbiting) the semi-infinite array $\Lambda^{\star}_{q,sp,sc}(\tau, m, n)$, we can construct a type-2 terminated (or tailbiting) CPM-QC-SC-LDPC code. Suppose we take the top M τ-span subarrays from $\Lambda^{\star}_{q,sp,sc}(\tau, m, n)$ to form an $M\tau \times (M\tau + e - 1)$ subarray of $\Lambda^{\star}_{q,sp,sc}(\tau, m, n)$, denoted by $\Lambda^{\star,t}_{q,sp,sc}(M, \tau, m, n)$. The binary CPM-dispersion of $\Lambda^{\star,t}_{q,sp,sc}(M, \tau, m, n)$ gives an $mM\tau \times n(M\tau + e - 1)$ array of CPMs and ZMs of size $(q-1) \times (q-1)$, denoted by $\mathbf{H}^{\star,t}_{b,sp,sc}(M, \tau, m, n)$. The null space of $\mathbf{H}^{\star,t}_{b,sp,sc}(M, \tau, m, n)$ gives a terminated type-2 CPM-QC-SC-LDPC code $C^{\star,t}_{b,sc,sp}(\tau)$ of length $n(M\tau + e - 1)(q - 1)$ with rate at least $(M\tau(n - m) + (e - 1)n)/n(M\tau + e - 1)$. If we wrap the last $e - 1$ column-blocks of $\Lambda^{\star,t}_{q,sp,sc}(M, \tau, m, n)$ around to the left, we obtain an $M\tau \times M\tau$ tailbiting array $\Lambda^{\star,tb}_{q,sp,sc}(M, \tau, m, n)$ of $m \times n$ matrices over GF(q). The binary CPM-dispersion of $\Lambda^{\star,tb}_{q,sp,sc}(M, \tau, m, n)$ gives an $mM\tau \times nM\tau$ array $\mathbf{H}^{\star,tb}_{b,sp,sc}(M, \tau, m, n)$ of CPMs and ZMs of size $(q-1) \times (q-1)$. The null space of $\mathbf{H}^{\star,tb}_{b,sp,sc}(M, \tau, m, n)$ gives a type-2 tailbiting CPM-QC-SC-LDPC code $C^{\star,tb}_{b,sc,sp}(\tau)$ of length $nM\tau(q - 1)$.

Example 9.4. Consider the cyclic base matrix $\mathbf{B}_{q,sp,p}$ over GF(127) constructed in Example 9.1. We factor 126 as the product of 9 and 14 and set $r = 9$ and $l = 14$. Then, we partition $\mathbf{B}_{q,sp,p}$ into a 9×9 array of 14×14 matrices over GF(127) in the form given by (8.2). Next, we choose $m = 2$ and $n = 8$ and construct a 9×9 cyclic array $\mathbf{B}'_{q,sp,p}(2, 8)$ of 2×8 matrices in the form of (9.1). The top row of $\mathbf{B}'_{q,sp,p}(2, 8)$ consists of 9 constituent 2×8 matrices $\mathbf{R}_{0,0}$, $\mathbf{R}_{0,1}$, ..., $\mathbf{R}_{0,8}$. In the code construction, we choose $e = 2$ and $\tau = 1$ and use $\mathbf{R}_{0,0}$ and $\mathbf{R}_{0,1}$ to construct the base matrix $\Lambda_{q,sp,sc}(1, 2, 8)$.

Set $M = 4$. Taking the first four 1-span subarrays from $\Lambda_{q,sp,sc}(1, 2, 8)$ and wrapping the last row-block around to the top as described above, we obtain a 4×4 array $\Lambda^{tb}_{q,sp,sc}(4, 1, 2, 8)$ of 2×8 submatrices over GF(127) given in the form of (9.9). The binary CPM-dispersion of the base matrix $\Lambda^{tb}_{q,sp,sc}(4, 1, 2, 8)$ gives an 8×32 array $\mathbf{H}^{bt}_{b,sp,sc}(4, 1, 2, 8)$ of CPMs and ZMs of size 126×126, which is a 1008×4032 binary matrix with column and row weights 4 and 8, respectively. The null space of this array gives a (4, 16)-regular (4032, 3029) type-1 tailbiting CPM-QC-SC-LDPC code $C^{tb}_{b,sp,sc,0}(1)$ with rate 0.7512.

The BER and BLER performances of this code decoded with 50 iterations of the MSA over the AWGNC are shown in Fig. 9.5(a). From this figure, we can see that the code performs about 1.6 dB away from the Shannon limit at a BER of 10^{-8}. At a BLER of 10^{-6}, the code performs 0.95 dB from the SPB.

If we choose $m = 2$, $n = 10$, $e = 2$, $\tau = 1$, and $M = 4$, we can construct an 8×40 array $\mathbf{H}^{tb}_{b,sp,sc}(4, 1, 2, 10)$ of CPMs and ZMs of size 126×126, which is a 1008×5040 binary matrix with column and row weights 4 and 10, respectively. The null space of $\mathbf{H}^{tb}_{b,sp,sc}(4, 1, 2, 10)$ gives a (4, 20)-regular (5040, 4037) type-1

$$
\Lambda_{q,sp,sc}(M,\tau,m,n)=\left(\begin{array}{ccccccccccccc}
\mathbf{R}_{0,0} & & & & & & & & & \vdots & \vdots & \ddots & \mathbf{R}_{0,(\tau-1)e+1} \\
\mathbf{R}_{0,1} & \mathbf{R}_{0,e} & & & & & & & & \mathbf{R}_{0,e-1} & \vdots & \ddots & \vdots \\
\vdots & \mathbf{R}_{0,e+1} & \ddots & & & & & & & & \mathbf{R}_{0,2e-1} & \ddots & \vdots \\
\vdots & \vdots & \ddots & \mathbf{R}_{0,(\tau-1)e} & & & & & & & & \ddots & \vdots \\
\vdots & \vdots & \ddots & \mathbf{R}_{0,(\tau-1)e+1} & \mathbf{R}_{0,0} & & & & & & & & \mathbf{R}_{0,\tau e-1} \\
\mathbf{R}_{0,e-1} & \vdots & \ddots & \vdots & \mathbf{R}_{0,1} & \mathbf{R}_{0,e} & & & & & & & \\
 & \mathbf{R}_{0,2e-1} & \ddots & \vdots & \vdots & \mathbf{R}_{0,e+1} & \ddots & & & & & & \\
 & & \ddots & \vdots & \vdots & \vdots & \ddots & \mathbf{R}_{0,(\tau-1)e} & & & & & \\
 & & & \mathbf{R}_{0,\tau e-1} & \vdots & \vdots & \ddots & \mathbf{R}_{0,(\tau-1)e+1} & \ddots & & & & \\
 & & & & \mathbf{R}_{0,e-1} & \vdots & \ddots & \vdots & \ddots & \mathbf{R}_{0,0} & & & \\
 & & & & & \mathbf{R}_{0,2e-1} & \ddots & \vdots & \ddots & \mathbf{R}_{0,1} & \mathbf{R}_{0,e} & & \\
 & & & & & & \ddots & \vdots & \ddots & \vdots & \mathbf{R}_{0,e+1} & \ddots & \\
 & & & & & & & \mathbf{R}_{0,\tau e-1} & \ddots & \vdots & \vdots & \ddots & \mathbf{R}_{0,(\tau-1)e}
\end{array}\right). \tag{9.9}
$$

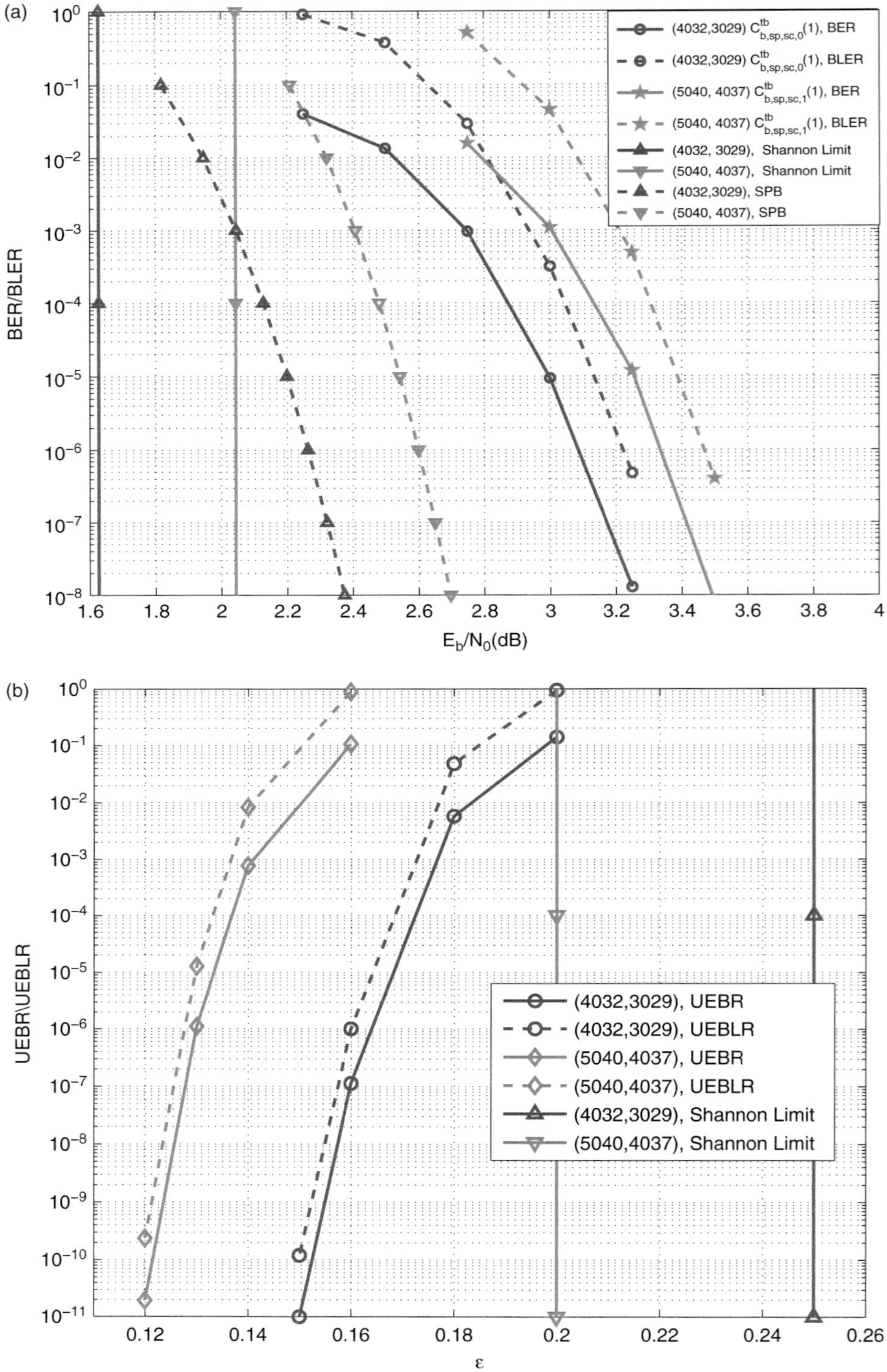

Figure 9.5. (a) The BER and BLER performances of the (4032, 3029) and the (5040, 4037) tailbiting type-1 CPM-QC-SC-LDPC codes $C^{tb}_{b,sp,sc,0}(1)$ and $C^{tb}_{b,sp,sc,1}(1)$ given in Example 9.4 over the AWGNC; and (b) performances of the codes in (a) over the BEC.

tailbiting CPM-QC-SC-LDPC code $C^{tb}_{b,sp,sc,1}(1)$ with rate 0.8010. Its BER and BLER performances over the AWGNC decoded with 50 iterations of the MSA are also shown in Fig. 9.5(a). From Fig. 9.5(a), we see that, at a BER of 10^{-8}, this code performs 1.4 dB away from the Shannon limit. At a BLER of 10^{-6}, the code performs 0.9 dB from the SPB.

The performances of the two codes $C^{tb}_{b,sp,sc,0}(1)$ and $C^{tb}_{b,sp,sc,1}(1)$ over the BEC are shown in Fig. 9.5(b). From this figure, we see that the code $C^{tb}_{b,sp,sc,0}(1)$ performs about 0.09 away from the Shannon limit at a UEBR of 10^{-7} and the code $C^{tb}_{b,sp,sc,1}(1)$ performs about 0.07 away from the Shannon limit at a UEBR of 10^{-6}.

Note that the two codes $C^{tb}_{b,sp,sc,0}(1)$ and $C^{tb}_{b,sp,sc,1}(1)$ constructed in this example are two doubly CPM-QC-SP-LDPC codes , as we can see by examining the structure of their base arrays.

9.5 A More General Construction of Type-1 CPM-QC-SC-LDPC Codes

In Sections 9.1 and 9.2, we presented a specific construction of type-1 CPM-QC-SC-LDPC codes using the $r \times r$ cyclic array $\mathbf{B}'_{q,sp,p}(m,n)$ given by (9.1) as the base matrix. Based on the base matrix $\mathbf{B}'_{q,sp,p}(m,n)$, we first constructed an $(e+\tau-1)\times\tau$ array $\mathbf{B}_{q,sp,sc}(\tau,m,n)$ of $m\times n$ matrices over GF(q) with $\tau e < r$, and then we formed a semi-infinite base array $\Lambda_{q,sp,sc}(\tau,m,n)$ that consists of infinite copies of $\mathbf{B}_{q,sp,sc}(\tau,m,n)$ lying on a diagonal band with width en and depth em shown in (9.2). The null space of the binary CPM-dispersion $\mathbf{H}_{b,sp,sc}(\tau,m,n)$ of $\Lambda_{q,sp,sc}(\tau,m,n)$ gives a type-1 periodically time-varying CPM-QC-SC-LDPC code $C_{b,sp,sc}(\tau)$ with period τ. In the construction of $\mathbf{B}_{q,sp,sc}(\tau,m,n)$, we restricted all the constituent matrices to be distinct. In this section, we generalize the construction of $\mathbf{B}_{q,sp,sc}(\tau,m,n)$ by removing this restriction. Removing this restriction gives a larger class of type-1 CPM-QC-SC-LDPC codes.

The first step of the generalized construction is to cut the $r\times r$ cyclic base matrix $\mathbf{B}'_{q,sp,p}(m,n)$ along its main diagonal and form two *triangular subarrays* of $m\times n$ matrices over GF(q) as follows:

$$\mathbf{T}_{lower}(r,r,m,n)=\begin{bmatrix} \mathbf{R}_{0,0} & \mathbf{O} & \mathbf{O} & \cdots & \mathbf{O} & \mathbf{O} \\ \mathbf{R}_{0,1} & \mathbf{R}_{0,2} & \mathbf{O} & \cdots & \mathbf{O} & \mathbf{O} \\ \mathbf{R}_{0,2} & \mathbf{R}_{0,3} & \mathbf{R}_{0,4} & \cdots & \mathbf{O} & \mathbf{O} \\ \vdots & \vdots & \vdots & \ddots & \vdots & \vdots \\ \mathbf{R}_{0,r-2} & \mathbf{R}_{0,r-1} & \mathbf{R}_{0,0} & \cdots & \mathbf{R}_{0,r-4} & \mathbf{O} \\ \mathbf{R}_{0,r-1} & \mathbf{R}_{0,0} & \mathbf{R}_{0,1} & \cdots & \mathbf{R}_{0,r-3} & \mathbf{R}_{0,r-2} \end{bmatrix}, \quad (9.10)$$

and

$$\mathbf{T}_{upper}(r,r,m,n) = \begin{bmatrix} \mathbf{O} & \mathbf{R}_{0,1} & \mathbf{R}_{0,2} & \cdots & \mathbf{R}_{0,r-2} & \mathbf{R}_{0,r-1} \\ \mathbf{O} & \mathbf{O} & \mathbf{R}_{0,3} & \cdots & \mathbf{R}_{0,r-1} & \mathbf{R}_{0,0} \\ \vdots & \vdots & \vdots & \ddots & \vdots & \vdots \\ \mathbf{O} & \mathbf{O} & \mathbf{O} & \cdots & \mathbf{O} & \mathbf{R}_{0,r-3} \\ \mathbf{O} & \mathbf{O} & \mathbf{O} & \cdots & \mathbf{O} & \mathbf{O} \end{bmatrix}, \quad (9.11)$$

where $\mathbf{O}$ is an $m \times n$ ZM.

The subarrays $\mathbf{T}_{lower}(r,r,m,n)$ and $\mathbf{T}_{upper}(r,r,m,n)$ are called the *lower* and *upper triangular* subarrays of $\mathbf{B}'_{q,sp,p}(m,n)$. They are disjoint and form a partition of $\mathbf{B}'_{q,sp,p}(m,n)$. Note that each nonzero constituent matrix in either $\mathbf{T}_{lower}(r,r,m,n)$ or $\mathbf{T}_{upper}(r,r,m,n)$ is a member matrix in the set $\boldsymbol{R} = \{\mathbf{R}_{0,0}, \mathbf{R}_{0,1}, \ldots, \mathbf{R}_{0,r-1}\}$ and each member matrix in $\boldsymbol{R}$ appears r times in the union of $\mathbf{T}_{lower}(r,r,m,n)$ and $\mathbf{T}_{upper}(r,r,m,n)$. Since $\mathbf{B}'_{q,sp,p}(m,n)$ satisfies the 2×2 SM-constraint and $\mathbf{T}_{lower}(r,r,m,n)$ and $\mathbf{T}_{upper}(r,r,m,n)$ are two disjoint subarrays of $\mathbf{B}'_{q,sp,p}(m,n)$, $\mathbf{T}_{lower}(r,r,m,n)$ and $\mathbf{T}_{upper}(r,r,m,n)$ satisfy both the 2×2 and the PW-2×2 SM-constraints.

Now, using $\mathbf{T}_{lower}(r,r,m,n)$ and $\mathbf{T}_{upper}(r,r,m,n)$ as building blocks, we form a semi-infinite array as follows:

$$\Delta_{q,sp,sc}(r,r,m,n) =$$

$$\begin{bmatrix} \mathbf{T}_{lower}(r,r,m,n) & & & \\ \mathbf{T}_{upper}(r,r,m,n) & \mathbf{T}_{lower}(r,r,m,n) & & \\ & \mathbf{T}_{upper}(r,r,m,n) & \mathbf{T}_{lower}(r,r,m,n) & \\ & & \mathbf{T}_{upper}(r,r,m,n) & \mathbf{T}_{lower}(r,r,m,n) \\ & & & \ddots \end{bmatrix}. \quad (9.12)$$

Consider the following $2r \times r$ subarray of $\Delta_{q,sp,sc}(r,r,m,n)$:

$$\mathbf{B}_{q,sp,sc,v}(r,r,m,n) = \begin{bmatrix} \mathbf{T}_{lower}(r,r,m,n) \\ \mathbf{T}_{upper}(r,r,m,n) \end{bmatrix}, \quad (9.13)$$

that consists of r column-blocks, each containing all the r member matrices of the set $\{\mathbf{R}_{0,0}, \mathbf{R}_{0,1}, \ldots, \mathbf{R}_{0,r-1}\}$ and r ZMs of size $m \times n$. The nonzero constituent matrices in each column-block of $\mathbf{B}_{q,sp,sc,v}(r,r,m,n)$ are confined to a span of r consecutive locations (called a nonzero column-span) and are ordered *cyclically* within the span. For $1 \le j < r$, the j-th nonzero column-span is obtained from the 0-th nonzero column-span by cyclically shifting upward j positions and wrapping the top j nonzero submatrices around to the bottom of the nonzero column-span. Since $\mathbf{T}_{lower}(r,r,m,n)$ and $\mathbf{T}_{upper}(r,r,m,n)$ satisfy the PW-2 $\times$ 2 SM-constraint, the r-span subarray $\mathbf{B}_{q,sp,sc,v}(r,r,m,n)$ of $\Delta_{q,sp,sc}(r,r,m,n)$ satisfies the 2×2 SM-constraint. $\mathbf{B}_{q,sp,sc,v}(r,r,m,n)$ is called a *vertical r-span subarray* of $\Delta_{q,sp,sc}(r,r,m,n)$. The subscript "$v$" stands for "*vertical*."

Next, we consider the structure of the following $r \times 2r$ subarray of $\Delta_{q,sp,sc}(r,r,m,n)$:

$$\mathbf{B}_{q,sp,sc,h}(r,r,m,n) = \left[\begin{array}{cc} \mathbf{T}_{upper}(r,r,m,n) & \mathbf{T}_{lower}(r,r,m,n) \end{array} \right], \tag{9.14}$$

that consists of r row-blocks, each containing all the member matrices of the set $\{\mathbf{R}_{0,0}, \mathbf{R}_{0,1}, \ldots, \mathbf{R}_{0,r-1}\}$ and r ZMs of size $m \times n$. The nonzero constituent matrices of each row-block of $\mathbf{B}_{q,sp,sc,h}(r,r,m,n)$ are confined to a span of r consecutive locations and ordered cyclically within the nonzero row-span. For $1 \le i < r$, the i-th nonzero row-span is obtained from the 0-th nonzero row-span by cyclically shifting i positions to the left and wrapping the first i nonzero submatrices around to the right of the nonzero row-span. Each member matrix of the set $\{\mathbf{R}_{0,0}, \mathbf{R}_{0,1}, \ldots, \mathbf{R}_{0,r-1}\}$ appears r times in $\mathbf{B}_{q,sp,sc,h}(r,r,m,n)$. The array $\mathbf{B}_{q,sp,sc,h}(r,r,m,n)$ is called a *horizontal r-span subarray* of $\Delta_{q,sp,sc}(r,r,m,n)$, where the subscript "$h$" in $\mathbf{B}_{q,sp,sc,h}(r,r,m,n)$ stands for "*horizontal*." Again, since $\mathbf{T}_{lower}(r,r,m,n)$ and $\mathbf{T}_{upper}(r,r,m,n)$ satisfy the PW-2×2 SM-constraint, the subarray $\mathbf{B}_{q,sp,sc,h}(r,r,m,n)$ satisfies the 2×2 SM-constraint.

From the structure of $\Delta_{q,sp,sc}(r,r,m,n)$ in the form given by (9.12), we see that $\Delta_{q,sp,sc}(r,r,m,n)$ consists of infinite copies of the vertical r-span subarray $\mathbf{B}_{q,sp,sc,v}(r,r,m,n)$. Two adjacent copies of $\mathbf{B}_{q,sp,sc,v}(r,r,m,n)$ are connected by a horizontal r-span subarray $\mathbf{B}_{q,sp,sc,h}(r,r,m,n)$. It follows from the structure of $\Delta_{q,sp,sc}(r,r,m,n)$ and the PW-2×2 SM-constrained structure of $\mathbf{T}_{lower}(r,r,m,n)$ and $\mathbf{T}_{upper}(r,r,m,n)$ that the semi-infinite array $\Delta_{q,sp,sc}(r,r,m,n)$ satisfies the 2×2 SM-constraint.

Dispersing each nonzero entry of $\Delta_{q,sp,sc}(r,r,m,n)$ into a binary CPM of size $(q-1) \times (q-1)$ and each 0-entry into a ZM of size $(q-1) \times (q-1)$, we obtain a semi-infinite array $\mathbf{H}_{b,sp,sc}(r,r,m,n)$ of CPMs and ZMs of size $(q-1) \times (q-1)$. The null space of $\mathbf{H}_{b,sp,sc}(r,r,m,n)$ gives a type-1 periodically time-varying CPM-QC-SC-LDPC code $C_{b,sp,sc}(r)$ with period r and rate close to $(n-m)/n$ (assuming $m < n$) whose Tanner graph has girth at least 6.

Let c be a positive integer such that $0 \le c < r$. Suppose we remove the top c row-blocks from both $\mathbf{T}_{lower}(r,r,m,n)$ and $\mathbf{T}_{upper}(r,r,m,n)$. We obtain two $(r-c) \times r$ arrays of $m \times n$ matrices over GF(q), denoted by $\mathbf{T}_{lower,row}(r-c,r,m,n)$ and $\mathbf{T}_{upper,row}(r-c,r,m,n)$, respectively. Next, we form the following semi-infinite array using $\mathbf{T}_{lower,row}(r-c,r,m,n)$ and $\mathbf{T}_{upper,row}(r-c,r,m,n)$ as building blocks:

$$\Delta_{q,sp,sc,row}(r-c,r,m,n) = \begin{bmatrix} \mathbf{T}_{lower,row}(r-c,r,m,n) & & & \\ \mathbf{T}_{upper,row}(r-c,r,m,n) & \mathbf{T}_{lower,row}(r-c,r,m,n) & & \\ & \mathbf{T}_{upper,row}(r-c,r,m,n) & \mathbf{T}_{lower,row}(r-c,r,m,n) & \\ & & & \ddots \end{bmatrix}. \tag{9.15}$$

The subscript "*row*" in $\mathbf{T}_{lower,row}(r-c,r,m,n)$, $\mathbf{T}_{upper,row}(r-c,r,m,n)$, and $\Delta_{q,sp,sc,row}(r-c,r,m,n)$ stands for "*row removal*."

In $\Delta_{q,sp,sc,row}(r-c,r,m,n)$, each vertical subarray:

$$\mathbf{B}_{q,sp,sc,v,row}(r-c,r,m,n) = \begin{bmatrix} \mathbf{T}_{lower,row}(r-c,r,m,n) \\ \mathbf{T}_{upper,row}(r-c,r,m,n) \end{bmatrix}, \tag{9.16}$$

is a $2(r-c) \times r$ array of $m \times n$ matrices over GF(q) and each horizontal subarray:

$$\mathbf{B}_{q,sp,sc,h,row}(r-c,r,m,n) = \begin{bmatrix} \mathbf{T}_{upper,row}(r-c,r,m,n) & \mathbf{T}_{lower,row}(r-c,r,m,n) \end{bmatrix}, \tag{9.17}$$

is an $(r-c) \times 2r$ array of $m \times n$ matrices over GF(q). For $r-c \geq 2$, two adjacent copies of $\mathbf{B}_{q,sp,sc,v,row}(r-c,r,m,n)$ in $\Delta_{q,sp,sc,row}(r-c,r,m,n)$ are connected (or coupled) by the row-blocks of $\mathbf{B}_{q,sp,sc,h,row}(r-c,r,m,n)$. As c increases from 0 to $r-2$, the *degree of coupling* between two adjacent copies of $\mathbf{B}_{q,sp,sc,v,row}(r-c,r,m,n)$ decreases. If $c = r-1$, then two adjacent copies of $\mathbf{B}_{q,sp,sc,v,row}(r-c,r,m,n)$ in $\Delta_{q,sp,sc,row}(r-c,r,m,n)$ are disconnected.

The null space of the binary CPM-dispersion $\mathbf{H}_{b,sp,sc,row}(r-c,r,m,n)$ of $\Delta_{q,sp,sc,row}(r-c,r,m,n)$ also gives a type-1 periodically time-varying CPM-QC-SC-LDPC code $C_{b,sp,sc,row}(r)$ with period r whose Tanner graph has girth at least 6.

In the above construction of type-1 CPM-QC-SC-LDPC codes, we removed row-blocks from the triangular arrays $\mathbf{T}_{lower}(r,r,m,n)$ and $\mathbf{T}_{upper}(r,r,m,n)$. Construction of type-1 CPM-QC-SC-LDPC codes can also be achieved by removing column-blocks from the two triangular arrays.

Let f be a positive integer such that $0 \leq f < r-1$. Suppose we remove the leftmost f column-blocks from both $\mathbf{T}_{lower}(r,r,m,n)$ and $\mathbf{T}_{upper}(r,r,m,n)$. We obtain two $r \times (r-f)$ arrays of $m \times n$ matrices over GF(q), $\mathbf{T}_{lower,col}(r,r-f,m,n)$ and $\mathbf{T}_{upper,col}(r,r-f,m,n)$. Next, we form the following semi-infinite array using $\mathbf{T}_{lower,col}(r,r-f,m,n)$ and $\mathbf{T}_{upper,col}(r,r-f,m,n)$ as building blocks:

$$\Delta_{q,sp,sc,col}(r, r-f, m, n) =$$

$$\begin{bmatrix} \mathbf{T}_{lower,col}(r, r-f, m, n) & & & \\ \mathbf{T}_{upper,col}(r, r-f, m, n) & \mathbf{T}_{lower,col}(r, r-f, m, n) & & \\ & \mathbf{T}_{upper,col}(r, r-f, m, n) & \mathbf{T}_{lower,col}(r, r-f, m, n) & \\ & & & \ddots \end{bmatrix}. \tag{9.18}$$

The subscript "*col*" in $\mathbf{T}_{lower,col}(r, r-f, m, n)$, $\mathbf{T}_{upper,col}(r, r-f, m, n)$, and $\Delta_{q,sp,sc,col}(r, r-f, m, n)$ stands for "*column removal*."

In this case, the null space of the binary CPM-dispersion $\mathbf{H}_{b,sp,sc,col}(r, r-f, m, n)$ of $\Delta_{q,sp,sc,col}(r, r-f, m, n)$ gives a type-1 periodically time-varying CPM-QC-SC-LDPC code $C_{b,sp,sc,col}(r-f)$ with period $r-f$ whose Tanner graph has girth at least 6. Note that the top f rows of $\Delta_{q,sp,sc,col}(r, r-f, m, n)$ are rows of zeros which must be removed in code construction.

For $0 \leq c, f < r-1$, suppose we remove the top c row-blocks and the leftmost f column-blocks from $\mathbf{T}_{lower}(r, r, m, n)$ and $\mathbf{T}_{upper}(r, r, m, n)$ given by (9.10) and (9.11), respectively. We obtain two $(r-c) \times (r-f)$ arrays, $\mathbf{T}_{lower,row,col}(r-c, r-f, m, n)$ and $\mathbf{T}_{upper,row,col}(r-c, r-f, m, n)$ of $m \times n$ matrices over GF(q). Using these two arrays as building blocks, we can construct the following 2×2 SM-constrained semi-infinite array:

$$\Delta_{q,sp,sc,row,col}(r-c, r-f, m, n) =$$

$$\begin{bmatrix} \mathbf{T}_{lower,row,col}(r-c, r-f, m, n) & & & \\ \mathbf{T}_{upper,row,col}(r-c, r-f, m, n) & \mathbf{T}_{lower,row,col}(r-c, r-f, m, n) & & \\ & \mathbf{T}_{upper,row,col}(r-c, r-f, m, n) & \mathbf{T}_{lower,row,col}(r-c, r-f, m, n) & \\ & & & \ddots \end{bmatrix}. \tag{9.19}$$

The null space of the CPM-dispersion $\mathbf{H}_{b,sp,sc,row,col}(r-c, r-f, m, n)$ of $\Delta_{q,sp,sc,row,col}(r-c, r-f, m, n)$ gives another type-1 periodically time-varying CPM-QC-SC-LDPC code $C_{b,sp,sc,row,col}(r-f)$ with period $r-f$ whose Tanner graph has girth at least 6. Note that the top f rows of $\Delta_{q,sp,sc,row,col}(r-c, r-f, m, n)$ are rows of zeros which must be removed in code construction.

For various choices of parameters, r, l, m, n, c, and f, we can construct type-1 CPM-QC-SC-LDPC codes with various rates and periods.

Clearly, tailbiting CPM-QC-SC-LDPC codes can be constructed from any of the above four generalized type-1 CPM-QC-SC-LDPC codes.

9.6 A More General Construction of Type-2 CPM-QC-SC-LDPC Codes

Let e and τ be two positive integers such that $e\tau = r$. Suppose we divide the top (or the 0-th) row-block $\mathbf{R}_0 = [\mathbf{R}_{0,0}\ \mathbf{R}_{0,1}\ \ldots\ \mathbf{R}_{0,r-1}]$ of the $r \times r$ array $\mathbf{B}'_{q,sp,p}(m,n)$ given by (9.1) into τ sections, each consisting of e consecutive constituent matrices of the first row-block $\mathbf{R}_0$ of $\mathbf{B}'_{q,sp,p}(m,n)$. For $0 \le i < \tau$, let $\mathbf{B}_i = [\mathbf{R}_{0,ie}\ \mathbf{R}_{0,ie+1}\ \ldots\ \mathbf{R}_{0,(i+1)e-1}]$ denote the i-th section of $\mathbf{R}_0$. Then, $\mathbf{R}_0 = [\mathbf{B}_0\ \mathbf{B}_1\ \ldots\ \mathbf{B}_{\tau-1}]$. Next, we form a $\tau \times \tau$ array by cyclically shifting $\mathbf{R}_0$ to the left $\tau - 1$ times, one section at a time, as follows:

$$\mathbf{B}_q(\tau,\tau) = \begin{bmatrix} \mathbf{B}_0 & \mathbf{B}_1 & \mathbf{B}_2 & \cdots & \mathbf{B}_{\tau-2} & \mathbf{B}_{\tau-1} \\ \mathbf{B}_1 & \mathbf{B}_2 & \mathbf{B}_3 & \cdots & \mathbf{B}_{\tau-1} & \mathbf{B}_0 \\ \vdots & \vdots & \vdots & \ddots & \vdots & \vdots \\ \mathbf{B}_{\tau-1} & \mathbf{B}_0 & \mathbf{B}_1 & \cdots & \mathbf{B}_{\tau-3} & \mathbf{B}_{\tau-2} \end{bmatrix}. \quad (9.20)$$

Note that the τ row-blocks of $\mathbf{B}_q(\tau,\tau)$ are simply the 0-th, e-th, ..., $(\tau-1)e$-th row-blocks of $\mathbf{B}'_{q,sp,p}(m,n)$. Next, we cut $\mathbf{B}_q(\tau,\tau)$ along the main diagonal into two triangular arrays, denoted by $\mathbf{T}_{q,upper}(\tau,\tau)$ and $\mathbf{T}_{q,lower}(\tau,\tau)$, where $\mathbf{T}_{q,upper}(\tau,\tau)$ contains all the nonzero sections on or above the main diagonal of $\mathbf{B}_q(\tau,\tau)$ and $\mathbf{T}_{q,lower}(\tau,\tau)$ contains only nonzero sections below the main diagonal. Then, we use $\mathbf{T}_{q,upper}(\tau,\tau)$ and $\mathbf{T}_{q,lower}(\tau,\tau)$ to form a $\tau \times 2\tau$ array as follows:

$$\mathbf{B}_{q,sc}(\tau,\tau) = \left[\mathbf{T}_{q,upper}(\tau,\tau)\ \ \mathbf{T}_{q,lower}(\tau,\tau)\right] =$$

$$\left[\begin{array}{cccccccc|ccccc} \mathbf{B}_0 & \mathbf{B}_1 & \mathbf{B}_2 & \cdots & \cdots & \mathbf{B}_{\tau-2} & \mathbf{B}_{\tau-1} & & \mathbf{O} & & & & \\ & \mathbf{B}_2 & \mathbf{B}_3 & \mathbf{B}_4 & \cdots & \mathbf{B}_{\tau-1} & \mathbf{B}_0 & & \mathbf{B}_1 & \mathbf{O} & & & \\ & & \mathbf{B}_4 & \mathbf{B}_5 & \cdots & \mathbf{B}_0 & \mathbf{B}_1 & & \mathbf{B}_2 & \mathbf{B}_3 & \mathbf{O} & & \\ & & & \ddots & \ddots & \ddots & \ddots & & \ddots & \ddots & \ddots & \ddots & \\ & & & & & & \mathbf{B}_{\tau-2} & & \mathbf{B}_{\tau-1} & \mathbf{B}_0 & \cdots & \mathbf{B}_{\tau-3} & \mathbf{O} \end{array}\right], \quad (9.21)$$

where $\mathbf{O}$ is an $m\tau \times ne$ ZM.

Using an argument similar to that given in the last section, the array $\mathbf{B}_{q,sc}(\tau,\tau)$ as an $m\tau \times 2ne\tau$ matrix over GF(q), satisfies the 2×2 SM-constraint. Now, we take an infinite number of copies of $\mathbf{B}_{q,sc}(\tau,\tau)$ to form the following semi-infinite array of $m \times ne$ matrices over GF(q):

$$\Upsilon_{q,sp,sc}(\tau,\tau,m,n) =$$

$$\begin{bmatrix} \mathbf{T}_{q,upper}(\tau,\tau) & \mathbf{T}_{q,lower}(\tau,\tau) & & \\ & \mathbf{T}_{q,upper}(\tau,\tau) & \mathbf{T}_{q,lower}(\tau,\tau) & \\ & & \ddots & \ddots \end{bmatrix}. \quad (9.22)$$

The array $\Upsilon_{q,sp,sc}(\tau,\tau,m,n)$, as a semi-infinite matrix, satisfies the 2×2 SM-constraint. $\mathbf{B}_{q,sc}(\tau,\tau)$ is called the 2τ-span subarray of $\Upsilon_{q,sp,sc}(\tau.\tau,m,n)$.

Let $\mathbf{H}_{b,sp,sc}(\tau,\tau,m,n)$ be the CPM-dispersion of $\Upsilon_{q,sp,sc}(\tau,\tau,m,n)$. The null space of $\mathbf{H}_{b,sp,sc}(\tau,\tau,m,n)$ gives a type-2 periodically time-varying CPM-QC-SC-LDPC code with period τ.

Similar to the constructions of various generalized type-1 CPM-QC-SC-LDPC codes, we can construct various generalized type-2 CPM-QC-SC-LDPC codes by removing corresponding columns and/or rows from $\mathbf{T}_{q,upper}(\tau,\tau)$ and $\mathbf{T}_{q,lower}(\tau,\tau)$. If we terminate $\Upsilon_{q,sp,sc}(\tau,\tau,m,n)$ by taking M row-blocks and wrap the tail $\mathbf{T}_{q,lower}(\tau,\tau)$ of the last τ row-span array around to the left-end, we obtain a tailbiting CPM-QC-SC-LDPC code.

9.7 Discussion and Remarks

In this chapter, several algebraic methods for constructing CPM-QC-SC-LDPC codes were presented. All the constructions presented were based on a special type of 2×2 SM-constrained base matrix with a cyclic structure in the form given by (7.3) (or (8.1)) which is a special case of the construction presented in Section 7.3, using the cyclic group GF(q)\{0} of a finite field GF(q). Every row (or column) of $\mathbf{B}'_{q,sp,p}(m,n)$ is a minimum weight codeword of the cyclic $(q-1,2)$ RS code over GF(q) with minimum distance $q-2$ [97, 64]. Besides this type of cyclic base matrix, there are two other types of 2×2 SM-constrained cyclic base matrices which can also be used to construct CPM-QC-SC-LDPC codes. One type is constructed based on two cyclic subgroups of the cyclic group of GF(q) [100, 114], which is also a special case of the construction presented in Section 7.3. The other construction is based on the 2-dimensional Euclidean geometry EG($2,q$) over GF(q) [46, 27] (see Appendix A). Based on these two types of cyclic base matrices, CPM-QC-SC-LDPC codes can be constructed.

Note that CPM-QC-SC-LDPC codes can also be constructed based on any 2×2 SM-constrained base matrix even though it does not have a cyclic structure. One such a base matrix is a Latin square over a finite field as presented in Section 7.3.

The construction of CPM-QC-SC-LDPC codes using a non-cyclic base matrix may be a little more complicated than using a cyclic base matrix. This may deserve further study. A good coverage of various constructions of 2×2 SM-constrained base matrices can be found in [97, 46].

Another interesting question is how the degree of coupling between two adjacent τ-span subarrays (or two adjacent periods) in the semi-infinite parity-check array of an SC-LDPC code affects the error performance of the code. It definitely affects the rate of decoding convergence. A higher degree of coupling makes the decoding converge faster.

Section 9.4 showed that a tailbiting code can be constructed by terminating the semi-infinite parity-check matrix of an SC-LDPC code and then wrapping the tail of the last column-span array (or last row-span array) around to the top (or to the left). Conversely, an SC-LDPC code can be constructed by unwrapping the parity-check matrix of a tailbiting code and continuing the downward (rightward) shift of the nonzero column-span (row-span) array (or the first period) to form a semi-infinite spatially coupled chain. Two classes of good and structured CPM-QC-LDPC codes that can be used for tailbiting were constructed in [65] and [101] based on finite Euclidean geometries and BIBDs, respectively. Construction of QC-SC-LDPC codes from these two classes of tailbiting codes needs further investigation.

Structured tailbiting codes, such as tailbiting Reed–Muller (RM) codes, have good trellis structure and can be decoded with the MAP decoding algorithm [75]. A good question, then, is whether the tailbiting SC-LDPC codes presented in this chapter are suitable for MAP decoding. This may be an interesting problem to investigate.

SC-LDPC codes have several advantages and some disadvantages compared to LDPC block codes. For asymptotically large termination lengths and base matrix expansion factors (resulting in very long codes), their iterative decoding performance is known to achieve the performance of the optimum MAP decoding. This is not possible with LDPC block codes. Also, in the asymptotic case, if the underlying structure is regular, the minimum distance of SC-LDPC codes grows linearly with block length, which suggests that they are not subject to severe error-floors. So, for very large block lengths, SC-LDPC codes normally have advantages.

In the practical finite-length case emphasized in this chapter, there are trade-offs. On an equal latency basis, SC-LDPC codes have been shown to achieve several tenths of a dB coding gain in the waterfall region of the BER curve compared to LDPC block codes with the same underlying structure. However, block codes typically have a sharper BER curve slope, which means that they will outperform

SC-LDPC codes in the error-floor. There are also differences in the implementation of iterative decoding in the two cases, and a code designer must carefully consider these differences in light of the particular application being considered. Another major disadvantage of SC-LDPC codes is that iterative BP-decoding of these codes takes a large number of iterations to converge.

10

Globally Coupled QC-LDPC Codes

In this chapter, we present another type of LDPC code with structure related to, but different from that of the SC-LDPC codes presented in Chapter 9. For an LDPC code of this type, its Tanner graph is composed of a set of *disjoint* Tanner graphs which are connected together by a group of *overall* CNs, called *global* CNs. The other CNs are called *local CNs*. A code of this type is called a *CN-based* GC-LDPC code.

Two types of CN-based GC-LDPC codes and their constructions will be presented in this chapter. The first type of CN-based GC-LDPC codes is constructed based on the block-cyclic base array $\mathbf{B}_{q,sp,p}(m,n)$ given by (8.3). Examples show that codes of this type perform very well over both the AWGNC and the BEC. The second type of codes are direct products of two LDPC codes, and they not only perform well over both the AWGNC and the BEC, but they are also very effective in correcting bursts of erasures, i.e., erasures clustered in bursts as discussed in Section 7.6. A method for constructing CN-based product GC-LDPC codes with QC-structure is given. Both types of CN-based GC-LDPC codes are special cases of SP-LDPC codes.

Also presented in this chapter is a *reduced-complexity local/global two-phase iterative decoding scheme* for CN-based GC-LDPC codes. This decoding scheme allows correction of both local and global random errors and/or erasures.

10.1 Construction of CN-Based QC-GC-LDPC Codes: Method-1

In forming the $r \times r$ cyclic array $\mathbf{B}_{q,sp,p}(m,n)$ in the form of (8.3) from the RS-based $r \times r$ array $\mathbf{W}_{RS}$ of $l \times l$ submatrices over GF(q) given by (8.2), there are $l-m$ rows in each row-block and $l-n$ columns in each column-block of $\mathbf{W}_{RS}$ which are unused. So, a total of $r(l-m)$ rows of $\mathbf{W}_{RS}$ are not used in forming the array $\mathbf{B}_{q,sp,p}(m,n)$. We denote the set of $r(l-m)$ unused rows of $\mathbf{W}_{RS}$ by Π. For each row $\mathbf{w}$ in Π, we remove the components at the locations that correspond to the columns that are not used in forming the array $\mathbf{B}_{q,sp,p}(m,n)$ from $\mathbf{W}_{RS}$. This results in a shortened row $\mathbf{w}^* = (\mathbf{w}_{0,0}, \mathbf{w}_{0,1}, \ldots, \mathbf{w}_{0,r-1})$ that consists of r sections, each consisting of n components. The locations of the n components of the i-th section $\mathbf{w}_{0,i}$ of $\mathbf{w}^*$ correspond to the locations of the n columns of the submatrix $\mathbf{R}_{0,i}$ of the submatrix $\mathbf{W}_{0,i}$ in the array $\mathbf{W}_{RS}$. Let Π^* denote the set of $r(l-m)$ shortened versions of the rows in Π. The set Π^* of rows and the set of rows in the first row-block of $\mathbf{B}_{q,sp,p}(m,n)$ are *disjoint*.

Let s and t be two positive integers with $1 \le s \le r(l-m)$ and $1 \le t \le r$. Take s rows from Π^* and remove the last $r-t$ sections from each of the s chosen rows. With these s shortened rows, we form an $s \times nt$ matrix $\mathbf{X}_{gc,cn}(s,t)$ over GF(q). Next, we form the following array over GF(q):

$$\mathbf{B}_{gc,cn} = \left[\begin{array}{ccccc} \mathbf{R}_{0,0} & & & & \\ & \mathbf{R}_{0,0} & & & \\ & & \mathbf{R}_{0,0} & & \\ & & & \ddots & \\ & & & & \mathbf{R}_{0,0} \\ \hdashline & & \mathbf{X}_{gc,cn}(s,t) & & \end{array}\right]. \tag{10.1}$$

The upper subarray of $\mathbf{B}_{gc,cn}$ is a $t \times t$ diagonal array with t copies of $\mathbf{R}_{0,0}$ on its main diagonal and the lower part is the $s \times nt$ matrix $\mathbf{X}_{gc,cn}(s,t)$. The matrix $\mathbf{B}_{gc,cn}$ is an $(mt+s) \times nt$ matrix over GF(q) which is a submatrix of the matrix $\mathbf{W}_{RS}$ after replacing some entries in $\mathbf{W}_{RS}$ by zeros. Since $\mathbf{W}_{RS}$ satisfies the 2×2 SM-constraint, $\mathbf{B}_{gc,cn}$ must also satisfy the 2×2 SM-constraint. The subscripts "*gc*" and "*cn*" in $\mathbf{X}_{gc,cn}(s,t)$ and $\mathbf{B}_{gc,cn}$ stand for "*global coupling*" and "*check node*," respectively.

Let $\mathcal{G}_{0,0}$ and $\mathcal{G}_{gc,cn}$ be the Tanner graphs associated with $\mathbf{R}_{0,0}$ and $\mathbf{B}_{gc,cn}$, respectively. It follows from the structure of the matrix $\mathbf{B}_{gc,cn}$ that the Tanner graph $\mathcal{G}_{gc,cn}$ associated with $\mathbf{B}_{gc,cn}$ is composed of t disjoint copies of the Tanner graph $\mathcal{G}_{0,0}$ associated with the matrix $\mathbf{R}_{0,0}$ connected by s global CNs that correspond to the s rows of the matrix $\mathbf{X}_{gc,cn}(s,t)$, as shown in Fig. 10.1. (In this figure, the edge labels are ignored.) The s global CNs provide the only connections between any two disjoint copies of $\mathcal{G}_{0,0}$, i.e., two disjoint copies of $\mathcal{G}_{0,0}$ are only connected to each other through a set of global CNs.

Let CPM($\mathbf{R}_{0,0}$) and CPM($\mathbf{X}_{gc,cn}(s,t)$) denote the binary CPM-dispersions of $\mathbf{R}_{0,0}$ and $\mathbf{X}_{gc,cn}(s,t)$, respectively. Then, the CPM-dispersion of $\mathbf{B}_{gc,cn}$ gives the following $(mt+s) \times nt$ array $\mathbf{H}_{gc,cn,sp,qc}(q-1,q-1)$ of binary CPMs and ZMs

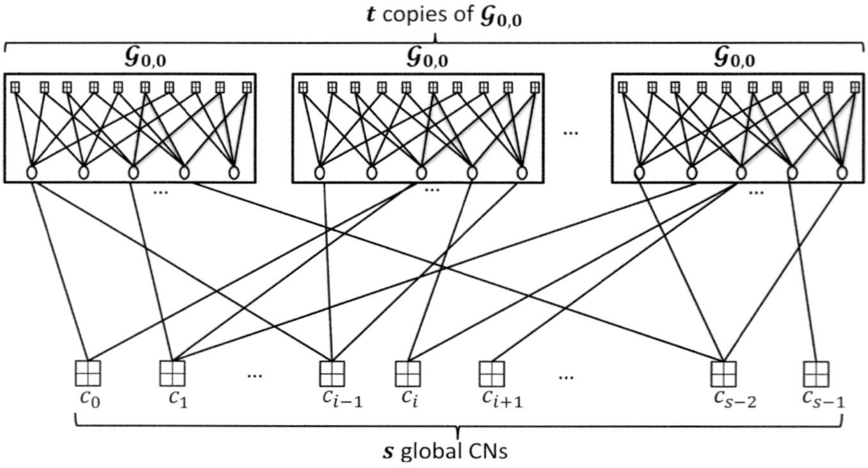

Figure 10.1. The Tanner graph $\mathcal{G}_{gc,cn}$ of the base matrix $\mathbf{B}_{gc,cn}$.

of size $(q-1) \times (q-1)$:

$$\mathbf{H}_{gc,cn,sp,qc}(q-1, q-1) =$$

$$\left[\begin{array}{ccccc} \text{CPM}(\mathbf{R}_{0,0}) & & & & \\ & \text{CPM}(\mathbf{R}_{0,0}) & & & \\ & & \text{CPM}(\mathbf{R}_{0,0}) & & \\ & & & \ddots & \\ & & & & \text{CPM}(\mathbf{R}_{0,0}) \\ \hdashline & & \text{CPM}(\mathbf{X}_{gc,cn}(s,t)) & & \end{array}\right]. \tag{10.2}$$

The array $\mathbf{H}_{gc,cn,sp,qc}(q-1, q-1)$ is a binary $(mt+s)(q-1) \times nt(q-1)$ matrix. Since the base matrix $\mathbf{B}_{gc,cn}$ satisfies the 2×2 SM-constraint, $\mathbf{H}_{gc,cn,sp,qc}(q-1, q-1)$ must satisfy the RC-constraint. The null space of $\mathbf{H}_{gc,cn,sp,qc}(q-1, q-1)$ gives a *CN-based CPM-QC-GC-LDPC code* of length $nt(q-1)$, denoted by $C_{gc,cn,sp,qc}$. The Tanner graph $\mathcal{G}_{gc,cn,sp,qc}(q-1, q-1)$ of the code $C_{gc,cn,sp,qc}$ has girth at least 6 and consists of $s(q-1)$ global CNs.

The $s(q-1) \times nt(q-1)$ matrix $\text{CPM}(\mathbf{X}_{gc,cn}(s,t))$ is an $s \times nt$ array of CPMs and/or ZMs of size $(q-1) \times (q-1)$. Divide this array into t subarrays, $\text{CPM}(\mathbf{X}_{gc,cn,0})$, $\text{CPM}(\mathbf{X}_{gc,cn,1})$, $\ldots$, $\text{CPM}(\mathbf{X}_{gc,cn,t-1})$, each an $s \times n$ array of CPMs and/or ZMs of size $(q-1) \times (q-1)$ and consisting of n consecutive column-blocks of the matrix CPM($\mathbf{X}_{gc,cn}(s,t)$). From the SP point of view, the code $C_{gc,cn,sp,qc}$ is constructed with $\boldsymbol{R} = \{$ CPM($\mathbf{R}_{0,0}$), CPM($\mathbf{X}_{gc,cn,0}$), CPM($\mathbf{X}_{gc,cn,1}$), ..., CPM($\mathbf{X}_{gc,cn,t-1}$) $\}$ as the replacement set and the following $(t+1) \times t$ matrix as the SP-base matrix:

$$\mathbf{B}_{sp} = \left[\begin{array}{ccccc} 1 & & & & \\ & 1 & & & \\ & & 1 & & \\ & & & \ddots & \\ & & & & 1 \\ \hdashline 1 & 1 & 1 & \cdots & 1 \end{array}\right]. \tag{10.3}$$

Hence, the CN-based CPM-QC-GC-LDPC code $C_{gc,cn,sp,qc}$ is a CPM-QC-SP-LDPC code. Note that the member matrices of $\boldsymbol{R}$ have two different sizes, but they all have the same number of columns, $n(q-1)$.

From the PTG point of view, we may regard the Tanner graph $\mathcal{G}_{0,0}$ associated with the $m \times n$ matrix $\mathbf{R}_{0,0}$ as the protograph for constructing the Tanner graph $\mathcal{G}_{gc,cn,sp,qc}(q-1, q-1)$ of $C_{gc,cn,sp,qc}$. In the construction of $\mathcal{G}_{gc,cn,sp,qc}(q-1, q-1)$, we first take $q-1$ copies of $\mathcal{G}_{0,0}$ and then connect them into a bipartite graph $\mathcal{G}_{0,0}(q-1, q-1)$ using the conventional PTG-methods as presented in Section 3.1. Next, we take t copies of $\mathcal{G}_{0,0}(q-1, q-1)$ and connect them by $s(q-1)$ global CNs. The connection is based on the *connection matrix* CPM($\mathbf{X}_{gc,cn}(s,t)$). We see that the construction of $\mathcal{G}_{gc,cn,sp,qc}(q-1, q-1)$ from the protograph $\mathcal{G}_{0,0}$ takes two steps of expansion. Hence, $C_{gc,cn,sp,qc}$ can be viewed as a *generalized* PTG-LDPC code. However, we may also regard the Tanner graph $\mathcal{G}_{gc,cn}$ associated with the base matrix $\mathbf{B}_{gc,cn}$ given by (10.1) as the protograph. In this case, the Tanner graph $\mathcal{G}_{gc,cn,sp,qc}(q-1, q-1)$ of the CN-based CPM-QC-GC-LDPC code $C_{gc,cn,sp,qc}$ is obtained by taking $q-1$ copies of $\mathcal{G}_{gc,cn}$ and connecting them by permuting the edges among the copies.

In forming the array $\mathbf{B}_{gc,cn}$, we can use t different member matrices in the set $\{\mathbf{R}_{0,0}, \mathbf{R}_{0,1}, \ldots, \mathbf{R}_{0,r-1}\}$ as the matrices on the main diagonal of the upper $t \times t$ subarray of $\mathbf{B}_{gc,cn}$ as follows:

$$\mathbf{B}^*_{gc,cn} = \left[\begin{array}{ccccc} \mathbf{R}_{0,0} & & & & \\ & \mathbf{R}_{0,1} & & & \\ & & \mathbf{R}_{0,2} & & \\ & & & \ddots & \\ & & & & \mathbf{R}_{0,t-1} \\ \hdashline \multicolumn{5}{c}{\mathbf{X}_{gc,cn}(s,t)} \end{array}\right]. \tag{10.4}$$

Then, the null space of the binary CPM-dispersion $\mathbf{H}^*_{gc,cn,sp,qc}(q-1, q-1)$ of $\mathbf{B}^*_{gc,cn}$ gives a *time-varying* CN-based CPM-QC-GC-LDPC code.

Since a CN-based CPM-QC-GC-LDPC code is a special type of LDPC block code, all the conventional BP-algorithms for decoding an LDPC block code can be applied for decoding a CN-based CPM-QC-GC-LDPC code. For different choices of the parameters r, l, m, n, s, and t, we can construct a large family of CN-based CPM-QC-GC-LDPC codes from a given field GF(q), with various lengths and rates.

In the next two examples, we construct three high-rate codes based on the same finite field, but using three different sets of the parameters m, n, s and t. We label these three codes with indices 0, 1 and 2, respectively.

Example 10.1. In this and the next examples, we use the prime field GF(127) for the code construction. Based on this field, we can construct a 126×126 cyclic matrix $\mathbf{W}_{RS}$ over GF(127) in the form given by (8.2) which satisfies the 2×2 SM-constraint.

For the construction of the first code, we factor 126 as the product of 6 and 21 and set $r = 6$ and $l = 21$. Then, the matrix $\mathbf{W}_{RS}$ can be viewed as a 6×6 array of 21×21 submatrices, whose first row-block consists of six submatrices $\mathbf{W}_{0,0}, \mathbf{W}_{0,1}, \ldots, \mathbf{W}_{0,5}$ of size 21×21. Taking six 3×21 submatrices, $\mathbf{R}_{0,0}, \mathbf{R}_{0,1}, \ldots, \mathbf{R}_{0,5}$, from these submatrices $\mathbf{W}_{0,0}, \mathbf{W}_{0,1}, \ldots, \mathbf{W}_{0,5}$ (i.e., choose $m = 3, n = 21$) under the location constraint as presented in Section 8.1, we obtain a 6×6 array $\mathbf{B}_{q,sp,p,0}(3, 21)$ of 3×21 submatrices over GF(127) in the form given by (8.3). (Note that the submatrices $\mathbf{R}_{0,0}, \mathbf{R}_{0,1}, \ldots, \mathbf{R}_{0,5}$ are taken *arbitrarily* from $\mathbf{W}_{0,0}, \mathbf{W}_{0,1}, \ldots, \mathbf{W}_{0,5}$ under the location constraint.) The submatrix $\mathbf{R}_{0,0}$ is a 3×21 matrix over GF(127). The set Π^* consists of $r(l-m) = 6 \times (21-3) = 108$ rows of length 126 of the matrix $\mathbf{W}_{RS}$. The rows in Π^* are disjoint from the rows in $[\mathbf{R}_{0,0}\ \mathbf{R}_{0,1}\ \ldots\ \mathbf{R}_{0,5}]$ of $\mathbf{B}_{q,sp,p,0}(3, 21)$.

Next, we choose $t = 6$ and $s = 1$. Then, the base matrix $\mathbf{B}_{gc,cn,0}$ given in the form of (10.1) for the code construction consists of 6 copies of $\mathbf{R}_{0,0}$ lying on its main diagonal and one randomly chosen row (or 1×6 array of 1×21 submatrices) from the set Π^*. $\mathbf{B}_{gc,cn,0}$ is a 19×126 matrix over GF(127) and satisfies the 2×2 SM-constraint. The binary CPM-dispersion of $\mathbf{B}_{gc,cn,0}$ gives a 19×126 array $\mathbf{H}_{gc,cn,sp,qc,0}(126, 126)$ of CPMs and ZMs of size 126×126. The array $\mathbf{H}_{gc,cn,sp,qc,0}(126, 126)$ is a $2394 \times 15\,876$ binary matrix. With the construction of $\mathbf{R}_{0,0}, \mathbf{R}_{0,1}, \ldots, \mathbf{R}_{0,5}$ from $\mathbf{W}_{0,0}, \mathbf{W}_{0,1}, \ldots, \mathbf{W}_{0,5}$ and the randomly chosen row from Π^*, we find that $\mathbf{H}_{gc,cn,sp,qc,0}(126, 126)$ has two column weights 3 and 4, two row weights 21 and 125, and average column and row weights 3.99 and 26.47, respectively. The null space of $\mathbf{H}_{gc,sp,qc,0}(126, 126)$ gives a (15 876, 13 494) CN-based CPM-QC-GC-LDPC code $C_{gc,cn,sp,qc,0}$ with rate 0.85. The Tanner graph $\mathcal{G}_{gc,cn,sp,qc,0}(126, 126)$ of the code $C_{gc,cn,sp,qc,0}$ has girth 6 and contains 204 876 cycles of length 6 and 21 677 544 cycles of length 8. From the above, we see that the parameters chosen to construct the code $C_{gc,cn,sp,qc,0}$ are: $r = 6$, $l = 21$, $m = 3$, $n = 21$, $s = 1$, and $t = 6$.

The BER and BLER performances of the code $C_{gc,cn,sp,qc,0}$ over the AWGNC decoded with 50 iterations of the MSA are shown in Fig. 10.2(a). From this figure, we see that at a BER of 10^{-8} and a BLER of 10^{-6}, the code performs within 1.5 dB and 1.0 dB of the Shannon limit and the SPB, respectively.

The unresolved erasure probability performance of this code over the BEC is shown in Fig. 10.2(b). We see that the code also performs well over the BEC.

Example 10.2. In this example, we use the same field GF(127) as the one used in Example 10.1 for constructing two other high-rate CN-based CPM-QC-GC-LDPC codes. The two sets of parameters used in the code construction are: (1) $r = 3$, $l = 42$, $m = 2$, $n = 42$, $s = 2$, and $t = 3$; and (2) $r = 3$, $l = 42$, $m = 1$, $n = 42$, $s = 3$, and $t = 3$. Based on these two sets of parameters and the construction described in Example 10.1, we construct two CN-based CPM-QC-GC-LDPC codes, denoted by $C_{gc,cn,sp,qc,1}$ and $C_{gc,cn,sp,qc,2}$.

The code $C_{gc,cn,sp,qc,1}$ is a (15 876, 14 871) code with rate 0.9367. Its parity-check matrix $\mathbf{H}_{gc,cn,sp,qc,1}(126, 126)$ has average column and row weights 3.98 and 62.75, respectively. The Tanner graph of $C_{qc,cn,sp,qc,1}$ has girth 6 and contains 1 675 296 cycles of length 6 and 384 179 481 cycles of length 8. The code $C_{gc,cn,sp,qc,2}$ is a (15 876, 15 120) code with rate 0.9524. The average column and row weights of its parity-check matrix are 3.9762 and 83.5, respectively. The Tanner graph of $C_{gc,cn,sp,qc,2}$ has girth 6 and contains 3 783 780 cycles of length 6 and 1 037 792 362 cycles of length 8.

The BER and BLER performances of these two codes $C_{gc,cn,sp,qc,1}$ and $C_{gc,cn,sp,qc,2}$ over the AWGNC decoded with 5, 10, and 50 iterations of the MSA are included in Fig. 10.3(a) and Fig. 10.3(b), respectively. From these two figures, we see that both codes perform well. For example, with 50 iterations, the (15 876, 14 871) CPM-QC-GC-LDPC code $C_{gc,cn,sp,qc,1}$ with rate 0.9367 has no visible error-floor all the way down to a BER of 10^{-10}, where it performs within 1.15 dB of the Shannon limit. Also at a BLER of 10^{-7}, it performs only about 0.75 dB away from the SPB. Even though the Tanner graph of $C_{gc,cn,sp,qc,1}$ has a very large number of cycles of lengths 6 and 8, its MSA decoding converges very fast, as shown in Fig. 10.3(a). At a BER of 10^{-10}, the performance gap between 5 and 10 iterations is about 0.4 dB and the gap between 10 and 50 iterations is within 0.2 dB. This fast rate of decoding convergence is due to the large *connectivity* of each VN. On average, each VN is connected to 246 other VNs with paths of length 2.

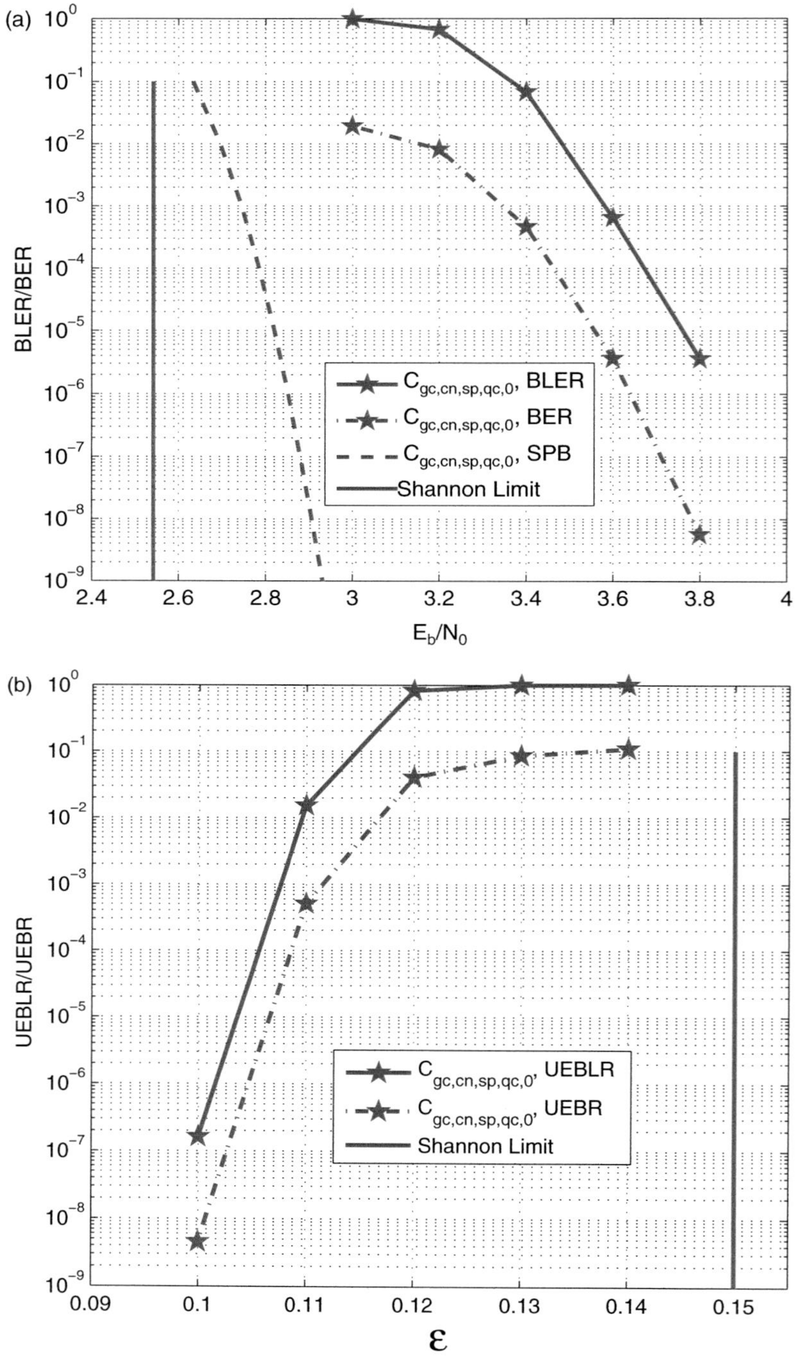

Figure 10.2. (a) The BER and BLER performances of the (15 876, 13 494) CN-based CPM-QC-GC-LDPC codes $C_{gc,cn,sp,qc,0}$ given in Example 10.1 over the AWGNC; and (b) performance of the code in (a) over the BEC.

(a)
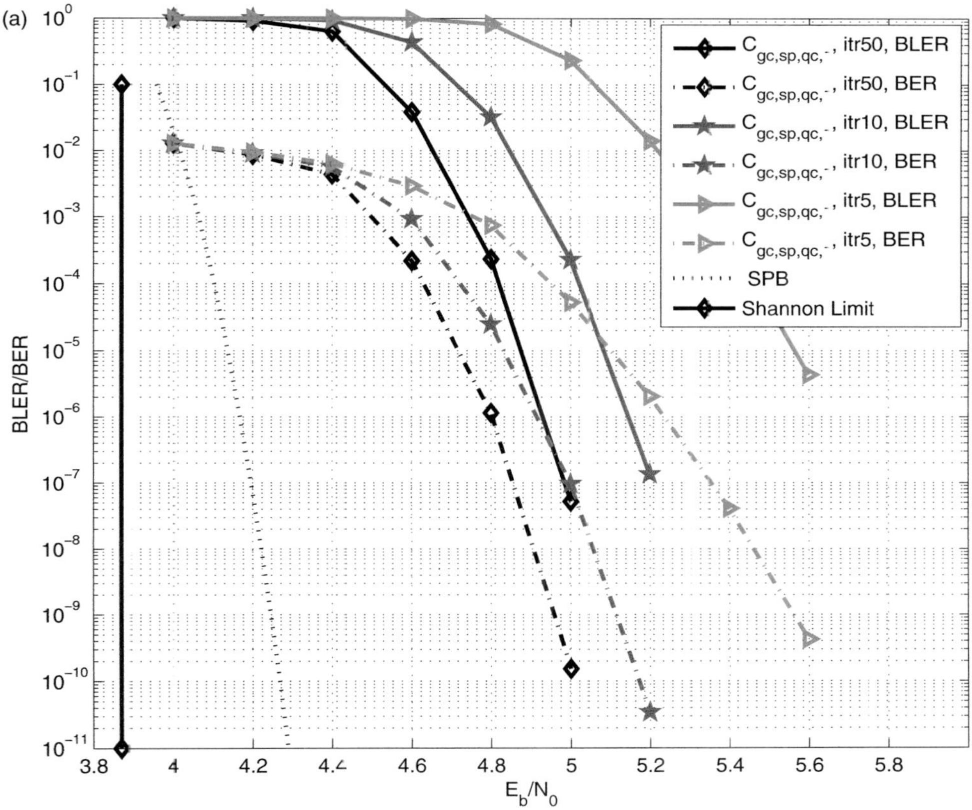

Figure 10.3. (a) The performances of the (15 876, 14 871) CN-based CPM-QC-GC-LDPC code $C_{gc,cn,sp,qc,1}$ decoded with 5, 10, and 50 iterations of the MSA given in Example 10.2 over the AWGNC; and (b) the performances of the (15 876, 15 120) CN-based CPM-QC-GC-LDPC code $C_{gc,cn,sp,qc,2}$ decoded with 5, 10, and 50 iterations of the MSA given in Example 10.2 over the AWGNC.

Code $C_{gc,cn,sp,qc,2}$ performs within 1.2 dB of the Shannon limit at a BER of 10^{-10} and 0.8 dB from the SPB at a BLER of 10^{-7}, as shown in Fig. 10.3(b). The Tanner graph of this code also contains a large number of cycles of lengths 6 and 8. Every VN also has a large connectivity, and on average, each VN is connected to 328 other VNs with paths of length 2. This large VN connectivity again results in a fast rate of decoding convergence as shown in Fig. 10.3(b). This code has almost the same rate as the (16 560, 15 737) doubly CPM-QC-SP-LDPC code given in Example 8.2. From Fig. 8.2(a) and Fig. 10.3(b), we see that the two codes perform almost the same.

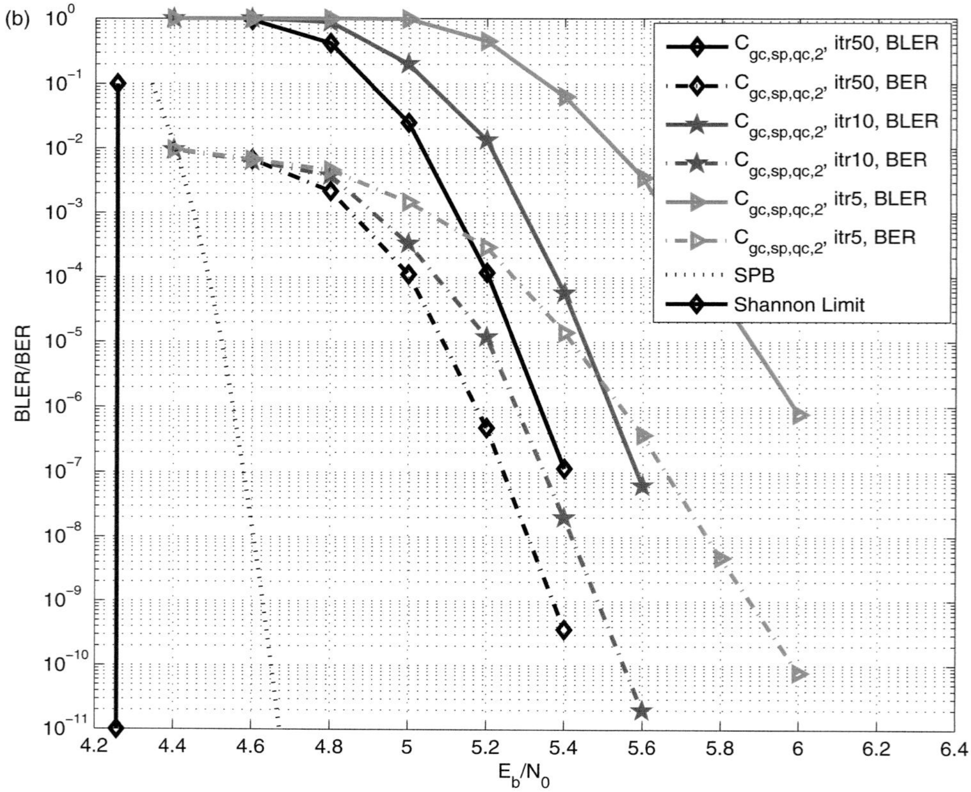

Figure 10.3. Continued

The unresolved erasure probability performances of the above two codes $C_{gc,cn,sp,qc,1}$ and $C_{gc,cn,sp,qc,2}$ over the BEC are shown in Fig. 10.4. We see that they both perform well.

The rates of the above two codes meet the high-rate requirement for high-speed optical communications and high-density flash memory.

10.2 A Local/Global Two-Phase Decoding of CN-Based CPM-QC-GC-LDPC Codes

Note that the CPM-dispersion CPM($\mathbf{R}_{0,0}$) of the $m \times n$ $\mathbf{R}_{0,0}$ is an RC-constrained $m \times n$ array of CPMs and/or ZMs of size $(q-1) \times (q-1)$. Its null space gives a QC-LDPC code $C_{0,0}$ of length $n(q-1)$. From the structure of the parity-check matrix array $\mathbf{H}_{gc,cn,sp,qc}(q-1, q-1)$ of the CN-based CPM-QC-GC-LDPC code $C_{gc,cn,sp,qc}$ displayed in (10.2), we readily see that each codeword

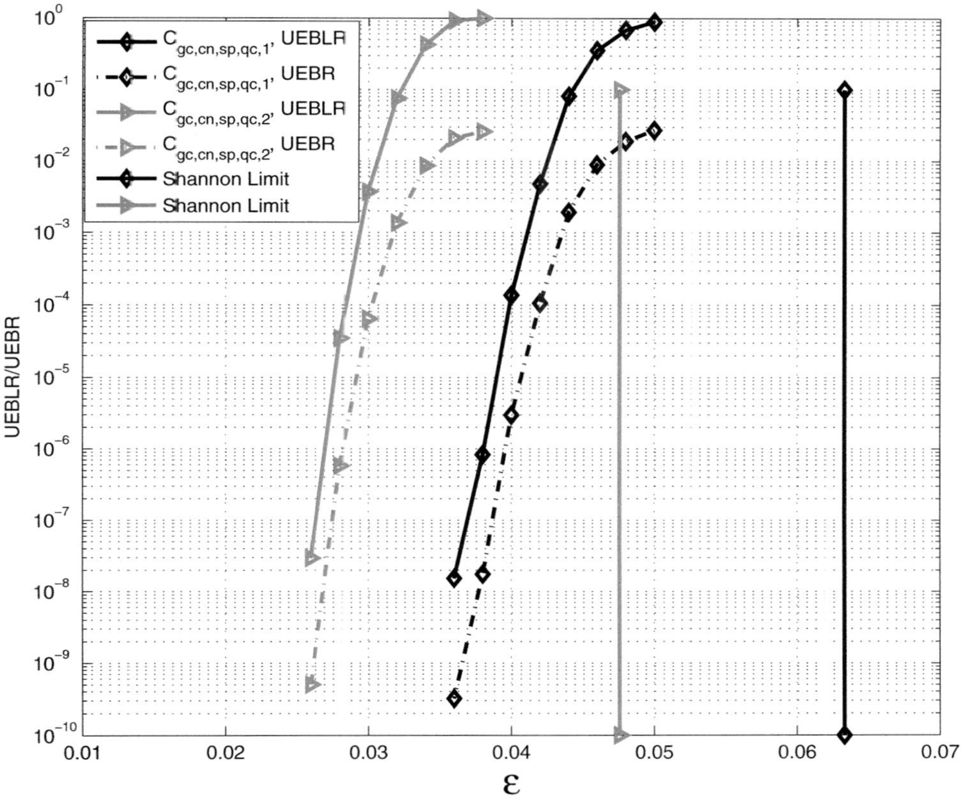

Figure 10.4. The performances of the (15 876, 14 871) and (15 876, 15 120) CN-based CPM-QC-GC-LDPC codes $C_{gc,cn,sp,qc,1}$ and $C_{gc,cn,sp,qc,2}$ given in Example 10.2 over the BEC.

$\mathbf{v} = (\mathbf{v}_0, \mathbf{v}_1, \ldots, \mathbf{v}_{t-1})$ of $C_{gc,cn,sp,qc}$ consists of t sections, each containing $n(q-1)$ code symbols. Each section $\mathbf{v}_i$ of $\mathbf{v}$ is a codeword of the code $C_{0,0}$ and the cascaded t sections of $\mathbf{v}$ in the order $\mathbf{v}_0, \mathbf{v}_1, \ldots, \mathbf{v}_{t-1}$ must satisfy the $s(q-1)$ parity-check constraints specified by the rows of the lower subarray CPM($\mathbf{X}_{gc,cn}(s,t)$) of $\mathbf{H}_{gc,cn,sp,qc}(q-1, q-1)$. Hence, we may regard the CN-based CPM-QC-GC-LDPC code $C_{gc,cn,sp,qc}$ as a *global code* which is composed of t local codes connected by $s(q-1)$ global parity-check constraints.

From the local/global point of view, a *local/global two-phase iterative decoding scheme* can be devised to correct local and global random errors. Suppose a codeword $\mathbf{v} = (\mathbf{v}_0, \mathbf{v}_1, \ldots, \mathbf{v}_{t-1})$ is transmitted. Let $\mathbf{r} = (\mathbf{r}_0, \mathbf{r}_1, \ldots, \mathbf{r}_{t-1})$ be the received word. The decoding starts with a local phase of decoding, where each received section $\mathbf{r}_i$ is decoded using a local iterative decoder based on the local code $C_{0,0}$ with a fixed number of iterations. If all the sections are successfully decoded, the t decoded sections form a locally decoded codeword $\mathbf{v}^\star = (\mathbf{v}_0^\star, \mathbf{v}_1^\star, \ldots, \mathbf{v}_{t-1}^\star)$. Then, we check whether this locally decoded codeword satisfies the $s(q-1)$ parity-check constraints imposed by the $s(q-1)$ global CNs. If it does, the locally decoded codeword is a codeword in the CN-based CPM-QC-GC-LDPC code $C_{gc,cn,sp,qc}$ and it is delivered to the user. If the locally decoded codeword $\mathbf{v}^\star$ does not satisfy the $s(q-1)$ global CN-constraints, a global decoder based on the entire code $C_{gc,cn,sp,qc}$ is activated to decode the received vector $\mathbf{r}$, with the symbol reliability measures provided by the local decoders and the channel information as inputs.

The local phase of decoding is a local error correction process that allows correction of random errors scattered among t sections of a received word, each containing a small number of random errors (or erasures). If the local decoders fail to decode the received sequence $\mathbf{r}$ into a codeword in $C_{gc,cn,sp,qc}$, the error pattern contained in $\mathbf{r}$ is regarded as a *global error pattern* which the global decoder attempts to correct.

There are two possible approaches to switch the decoding from the local phase to the global phase. The first approach is to complete the local decoding of all the received sections, and then pass the decoded information (LLRs) of all the decoded sections (successfully and unsuccessfully decoded) and the channel information to the global decoder to process the received word $\mathbf{r}$. The other approach is to switch to the global phase decoding as soon as a local decoder fails to decode a received section. In this case, the global decoder processes the received vector $\mathbf{r}$ either with only the channel information as inputs or with the combined channel information and the reliability information of successfully decoded sections as inputs (the reliability information of unsuccessfully decoded sections may also be included).

Basically, the global phase of decoding is needed only when the distribution of errors does not allow successful error correction of all the sections of the received sequence $\mathbf{r}$. Intuitively, the average number of computations should be lower than when using only a global decoder. In the decoder implementation, one could use a single local decoder based on the local code $C_{0,0}$ to process all the received sections sequentially one at a time, or use a group of (t or less) identical local decoders to process a group of (t or less) consecutive received sections in parallel. Pipeline processing is possible for high-speed decoding. Clearly, using only one local decoder results in the lowest decoder complexity, but requires the longest decoding time. Fully parallel local phase decoding using t local decoders results in the shortest decoding time, but requires the highest decoder complexity. Partially parallel local phase decoding using a group of local decoders (less than t)

allows a flexible trade-off between decoding complexity and decoding time. The design of such a local/global decoder would be an interesting problem for further investigation.

10.3 Construction of CN-Based GC-LDPC Codes: Method-2

Let $\mathbf{H}_0$ and $\mathbf{H}_1$ be $m_0 \times n_0$ and $m_1 \times n_1$ RC-constrained sparse matrices over GF(2), respectively. The null spaces of $\mathbf{H}_0$ and $\mathbf{H}_1$ give two LDPC codes, denoted by C_0 and C_1, of lengths n_0 and n_1, respectively. Let $\mathbf{h}_{1,0}, \mathbf{h}_{1,1}, \ldots, \mathbf{h}_{1,n_1-1}$ be the n_1 columns of $\mathbf{H}_1$. Then, $\mathbf{H}_1$ can be expressed in terms of its columns as follows:

$$\mathbf{H}_1 = [\mathbf{h}_{1,0}\ \mathbf{h}_{1,1}\ \ldots\ \mathbf{h}_{1,n_1-1}]. \tag{10.5}$$

For $0 \le j < n_1$, we form an $m_1 n_0 \times n_0$ matrix $\mathbf{H}_{1,j}$ based on the j-th column $\mathbf{h}_{1,j}$ of $\mathbf{H}_1$ as follows:

$$\mathbf{H}_{1,j} = \begin{bmatrix} \mathbf{h}_{1,j} & \mathbf{0} & \cdots & \mathbf{0} \\ \mathbf{0} & \mathbf{h}_{1,j} & \cdots & \mathbf{0} \\ \vdots & \vdots & \ddots & \vdots \\ \mathbf{0} & \mathbf{0} & \cdots & \mathbf{h}_{1,j} \end{bmatrix}, \tag{10.6}$$

where $\mathbf{0}$ is a column of m_1 zeros. It is clear that $\mathbf{H}_{1,j}$ satisfies the RC-constraint and the matrices $\mathbf{H}_{1,0}, \mathbf{H}_{1,1}, \ldots, \mathbf{H}_{1,n_1-1}$ formed by the n_1 columns of $\mathbf{H}_1$ satisfy the PW-RC-constraint. If we permute the rows of $\mathbf{H}_{1,j}$ using the row permutation π_{row} defined by (4.3) with m_1 replacing m, we obtain an $m_0 \times 1$ array (or a column) of *identity matrices* and ZMs of size $n_0 \times n_0$, denoted by $\mathbf{H}_{1,j,\pi}$, where the subscript "π" in $\mathbf{H}_{1,j,\pi}$ stands for "*row permutation*."

In the SP-construction of LDPC codes, we can use the matrices $\mathbf{H}_{1,0,\pi}, \mathbf{H}_{1,1,\pi}, \ldots, \mathbf{H}_{1,n_1-1,\pi}$ together with $\mathbf{H}_0$ as constituent matrices of a replacement set $\boldsymbol{R}$. In this case, the member matrices in $\boldsymbol{R}$ have two different sizes. The size of $\mathbf{H}_0$ is $m_0 \times n_0$ and the size of $\mathbf{H}_{1,j,\pi}$ for $0 \le j < n_1$ is $m_1 n_0 \times n_0$. All the matrices in $\boldsymbol{R}$ have the same number of columns, i.e., n_0, and the number of rows of $\mathbf{H}_{1,j,\pi}$ is n_0 times that of $\mathbf{H}_1$. If the base matrix $\mathbf{B}_{sp}$ for the SP-construction of an LDPC code is designed properly and the replacement of the entries by the member matrices of $\boldsymbol{R}$ is carried out properly, the constructed SP-LDPC code can perform well. Note that for $0 \le j < n_1$, $\mathbf{H}_{1,j,\pi}$ and $\mathbf{H}_0$ satisfy a *pair-wise row (PW-R) constraint*, i.e., no two rows, one in $\mathbf{H}_{1,j,\pi}$ and the other in $\mathbf{H}_0$ have more than one position where they both have nonzero components. This simply follows from the structure of $\mathbf{H}_{1,j,\pi}$ (or $\mathbf{H}_{1,j}$) in which each row contains at most one 1-entry.

In the following, we use $\mathbf{H}_0$ and $\mathbf{H}_{1,0,\pi}, \mathbf{H}_{1,1,\pi}, \ldots, \mathbf{H}_{1,n_1-1,\pi}$ to construct a CN-based GC-LDPC code. First, we construct an $(n_1+1) \times n_1$ binary matrix $\mathbf{B}_{sp}$

in the form of (10.3). Next, we replace the n_1 1-entries on the main diagonal of the upper $n_1 \times n_1$ submatrix of $\mathbf{B}_{sp}$ by the parity-check matrix $\mathbf{H}_0$ of the LDPC code C_0 and the 0-entries by ZMs of size $m_0 \times n_0$. Then, we replace the n_1 1-entries of the bottom row of $\mathbf{B}_{sp}$ by $\mathbf{H}_{1,0,\pi}, \mathbf{H}_{1,1,\pi}, \ldots, \mathbf{H}_{1,n_1-1,\pi}$. The above replacements result in the following $(n_1+1) \times n_1$ array of matrices over GF(2) of two different sizes, $m_0 \times n_0$ and $m_1 n_0 \times n_0$:

$$\mathbf{H}_{gc,cn,sp}(C_0, C_1) = \left[\begin{array}{cccc} \mathbf{H}_0 & & & \\ & \mathbf{H}_0 & & \\ & & \ddots & \\ & & & \mathbf{H}_0 \\ \hdashline \mathbf{H}_{1,0,\pi} & \mathbf{H}_{1,1,\pi} & \cdots & \mathbf{H}_{1,n_1-1,\pi} \end{array}\right]. \tag{10.7}$$

The array $\mathbf{H}_{gc,cn,sp}(C_0, C_1)$ is an $(m_0 n_1 + m_1 n_0) \times n_0 n_1$ matrix over GF(2). Since $\mathbf{H}_0$ and $\mathbf{H}_1$ satisfy the RC-constraint and $\mathbf{H}_0$ and $\mathbf{H}_{1,j,\pi}$ satisfy the PW-R-constraint, it follows from the structure of $\mathbf{H}_{gc,cn,sp}(C_0, C_1)$ displayed in (10.7) that $\mathbf{H}_{gc,cn,sp}(C_0, C_1)$ satisfies the RC-constraint. Hence, the null space of $\mathbf{H}_{gc,cn,sp}(C_0, C_1)$ gives a CN-based GC-LDPC code of length $n_0 n_1$, denoted by $C_{gc,cn,sp}(C_0, C_1) = C_0 \times C_1$, whose Tanner graph has girth at least 6.

Note that the lower submatrix $[\mathbf{H}_{1,0,\pi} \;\; \mathbf{H}_{1,1,\pi} \;\; \ldots \;\; \mathbf{H}_{1,n_1-1,\pi}]$ consists of n_0 copies of the parity-check matrix $\mathbf{H}_1$ of C_1. Therefore, the parity-check matrix $\mathbf{H}_{gc,cn,sp}(C_0, C_1)$ of $C_{gc,cn,sp}(C_0, C_1)$ consists of n_1 copies of $\mathbf{H}_0$ and n_0 copies of $\mathbf{H}_1$ (interleaved). Let $\mathbf{v} = (v_0, v_1, \ldots, v_{n_0 n_1 - 1})$ be a codeword in $C_{gc,cn,sp}(C_0, C_1)$. Divide this codeword into n_1 sections, each consisting of n_0 consecutive components of $\mathbf{v}$. For $0 \le i < n_1$, let $\mathbf{v}_i$ be the i-th section of $\mathbf{v}$. Then, $\mathbf{v} = (\mathbf{v}_0, \mathbf{v}_1, \ldots, \mathbf{v}_{n_1-1})$. From the structure of $\mathbf{H}_{gc,cn,sp}(C_0, C_1)$ displayed by (10.7), we see that each section $\mathbf{v}_i$ of $\mathbf{v}$ is a codeword in C_0. We can also divide the codeword into n_0 sections, denoted by $\mathbf{v}_0^\star, \mathbf{v}_1^\star, \ldots, \mathbf{v}_{n_0-1}^\star$, each consisting of n_1 components of $\mathbf{v}$, where for $0 \le j < n_0$, $\mathbf{v}_j^\star = (v_j, v_{n_0+j}, v_{2n_0+j}, \ldots, v_{(n_1-1)n_0+j})$. It follows from the structure of $\mathbf{H}_{1,0}, \mathbf{H}_{1,1}, \ldots, \mathbf{H}_{1,n_1-1}$ displayed by (10.6) that each section $\mathbf{v}_j^\star$ is a codeword in C_1. Let $\mathbf{v}^\star = (\mathbf{v}_0^\star, \mathbf{v}_1^\star, \ldots, \mathbf{v}_{n_0-1}^\star)$. Then, $\mathbf{v}^\star$ is simply a permutation of $\mathbf{v}$. From the structure of the codewords described above in $C_{gc,cn,sp}(C_0, C_1)$, the code $C_{gc,cn,sp}(C_0, C_1)$ is simply the *direct product* of C_0 and C_1 [74, 110, 111] with $\mathbf{H}_{gc,cn,sp}(C_0, C_1)$ as its parity-check matrix. The codes C_0 and C_1 are referred to as the *horizontal* and *vertical component codes* of the CN-based product GC-LDPC code $C_{gc,cn,sp}(C_0, C_1)$, respectively. The two matrices $\mathbf{H}_0$ and $\mathbf{H}_1$ are called the *horizontal* and *vertical component parity-check matrices* of CN-based product GC-LDPC code $C_{gc,cn,sp}(C_0, C_1)$, respectively.

Let d_0 and d_1 be the minimum distances of C_0 and C_1, respectively. Then the minimum distance of $C_{gc,cn,sp}(C_0, C_1)$ is $d_0 d_1$, the product of d_0

and d_1. Decoding of $C_{gc,cn,sp}(C_0, C_1)$ can be performed either based on the parity-check matrix $\mathbf{H}_{gc,cn,sp}(C_0, C_1)$ given by (10.7) or based on the two-phase decoding presented in Section 10.2. The two-phase decoding can be carried out iteratively, with the local-phase decoding and global-phase decoding performed alternatively.

If $\mathbf{H}_0$ is an array of circulants (or CPMs) and/or ZMs of size $k \times k$ over GF(2), then $\mathbf{H}_{gc,cn,sp}(C_0, C_1)$ is an array of circulants and ZMs of sizes $k \times k$ and $n_0 \times n_0$. If $n_0 = \delta k$, then an $n_0 \times n_0$ identity matrix can be viewed as a $\delta \times \delta$ array of identity matrices of size $k \times k$, that consists of δ $k \times k$ identity matrices lying on its main diagonal. In this case, $\mathbf{H}_{gc,cn,sp}(C_0, C_1)$ is an array of binary circulants and ZMs of size $k \times k$ and $C_{gc,cn,sp}(C_0, C_1)$ is a CN-based QC-GC-LDPC code.

The minimum distance of a CN-based product GC-LDPC code can be relatively large even if the minimum distances of its two component codes are small. A large minimum distance may result in a very low error-floor if the Tanner graph of the code does not contain trapping sets with size smaller than its minimum distance. LDPC codes constructed based on finite geometries have relatively large minimum distances [58, 74, 97] and their Tanner graphs contain no small trapping sets with sizes smaller than their minimum distances [28, 29]. Furthermore, the parity-check matrix of a finite geometry (Euclidean or projective) LDPC code has large row-redundancy [58, 46, 28]. The above three structural properties make the error-floor of a finite geometry LDPC code very low. In the product construction of a CN-based GC-LDPC code, we can use a finite geometry LDPC code as the horizontal component code C_0 and a simple LDPC code, such as a *single parity-check code* (SPC), as the vertical component code C_1. In this case, the resultant CN-based product GC-LDPC code has a minimum distance which is twice the minimum distance of the horizontal component code C_0.

Example 10.3. Suppose we use the two-dimensional Euclidean geometry EG$(2, 2^4)$ over GF(2^4) to construct the horizontal component code C_0 of a CN-based product GC-LDPC code $C_{gc,cn,sp}(C_0, C_1)$. Using the lines not passing through the origin of EG$(2, 2^4)$, we can construct an RC-constrained binary circulant $\mathbf{H}_0$ of size 255×255 with both column and row weights 16. The null space of $\mathbf{H}_0$ gives a $(255, 175)$ cyclic LDPC code C_0 with minimum distance 17 whose Tanner graph contains no trapping sets with size smaller than 17 [28]. Next, we choose the $(16, 15)$ SPC code as the vertical component code C_1 of $C_{gc,cn,sp}(C_0, C_1)$. The SPC code C_1 has minimum distance 2. The parity-check matrix $\mathbf{H}_1$ of C_1 is a row of 16 1-entries. Based on $\mathbf{H}_1$, we construct a single row-block

$[\mathbf{H}_{1,0,\pi}\ \mathbf{H}_{1,1,\pi}\ \ldots\ \mathbf{H}_{1,15,\pi}]$ of 16 identity matrices of size 255×255, which is a 255×4080 matrix over GF(2) with column and row weights 1 and 16, respectively.

Using $\mathbf{H}_0$ and $\mathbf{H}_{1,0,\pi}, \mathbf{H}_{1,1,\pi}, \ldots, \mathbf{H}_{1,15,\pi}$, we construct the parity-check matrix $\mathbf{H}_{gc,cn,sp}(C_0, C_1)$ of the CN-based product GC-LDPC code $C_{gc,cn,sp}(C_0, C_1)$ in the form given by (10.7). Since every submatrix in $\mathbf{H}_{gc,cn,sp}(C_0, C_1)$ is a 255×255 circulant, $\mathbf{H}_{gc,cn,sp}(C_0, C_1)$ is a 17×16 array of circulants and ZMs of size 255×255. It is a 4335×4080 matrix with column and row weights 17 and 16, respectively. The null space of $\mathbf{H}_{gc,cn,sp}(C_0, C_1)$ gives a (4080, 2625) CN-based product QC-GC-LDPC code $C_{gc,cn,sp}(C_0, C_1)$ with rate 0.6434 and minimum distance at least 34, a large minimum distance for an LDPC code. Note that $\mathbf{H}_0$ has 80 redundant rows, a large row redundancy. (It is typical that an EG-LDPC code has large row redundancy [58, 46, 28].) Using the local/global two-phase decoding scheme, any 8 or fewer random errors are guaranteed to be corrected by the local decoder.

If we put $\mathbf{H}_0$ as a 17×17 array of CPMs and ZMs of size 15×15 (see Appendix A or [46]) and regard each 255×255 identity matrix $\mathbf{H}_{1,j,\pi}$ as a 17×17 diagonal array of identity matrices of size 15×15. Then, $\mathbf{H}_{gc,cn,sp}(C_0, C_1)$ is a 34×272 array of CPMs and ZMs of size 15×15.

The BER and BLER performances of the code $C_{gc,cn,sp}(C_0, C_1)$ decoded with 50 iterations of the MSA over the AWGNC are shown in Fig. 10.5(a). At a BLER of 10^{-7}, the code performs about 2.2 dB away from the SPB and it performs within 2.8 dB of the Shannon limit at a BER of 10^{-10}.

The UEBR and UEBLR performances of $C_{gc,cn,sp}(C_0, C_1)$ over the BEC are shown in Fig. 10.5(b). At a UEBR of 10^{-10}, the code performs about 0.4 away from the Shannon limit.

The CN-based product GC-LDPC code $C_{gc,cn,sp}(C_0, C_1)$ is capable of correcting any *burst of erasures* of length at least up to n_0. For such a burst of erasures, there is at least one row in the lower submatrix $[\mathbf{H}_{1,0,\pi}\ \mathbf{H}_{1,1,\pi}\ \ldots\ \mathbf{H}_{1,n_1-1,\pi}]$ of $\mathbf{H}_{gc,cn,sp}(C_0, C_1)$ which checks (or contains) *one and only one* erasure in the burst. Hence, every erased code symbol can be resolved. A small number of scattering random erasures in each block can be corrected by the horizontal component code C_0. For example, the (4080, 2625) CN-based product QC-GC-LDPC code $C_{gc,cn,sp}(C_0, C_1)$ constructed in Example 10.3 is capable of correcting any erasure-burst of length up to 255. Furthermore, any 16 or fewer random erasures confined in a block of $C_{gc,cn,sp}(C_0, C_1)$ can be corrected by C_0 [74]. This is based on the RC-constraint structure of the parity-check matrix $\mathbf{H}_0$ of C_0 – that each erasure is checked by 16 rows, and the other 15 erasures can be distributed in at most 15 of these rows. Consequently, there is at least one row among these 16 rows that checks only one erasure and hence its value can be determined.

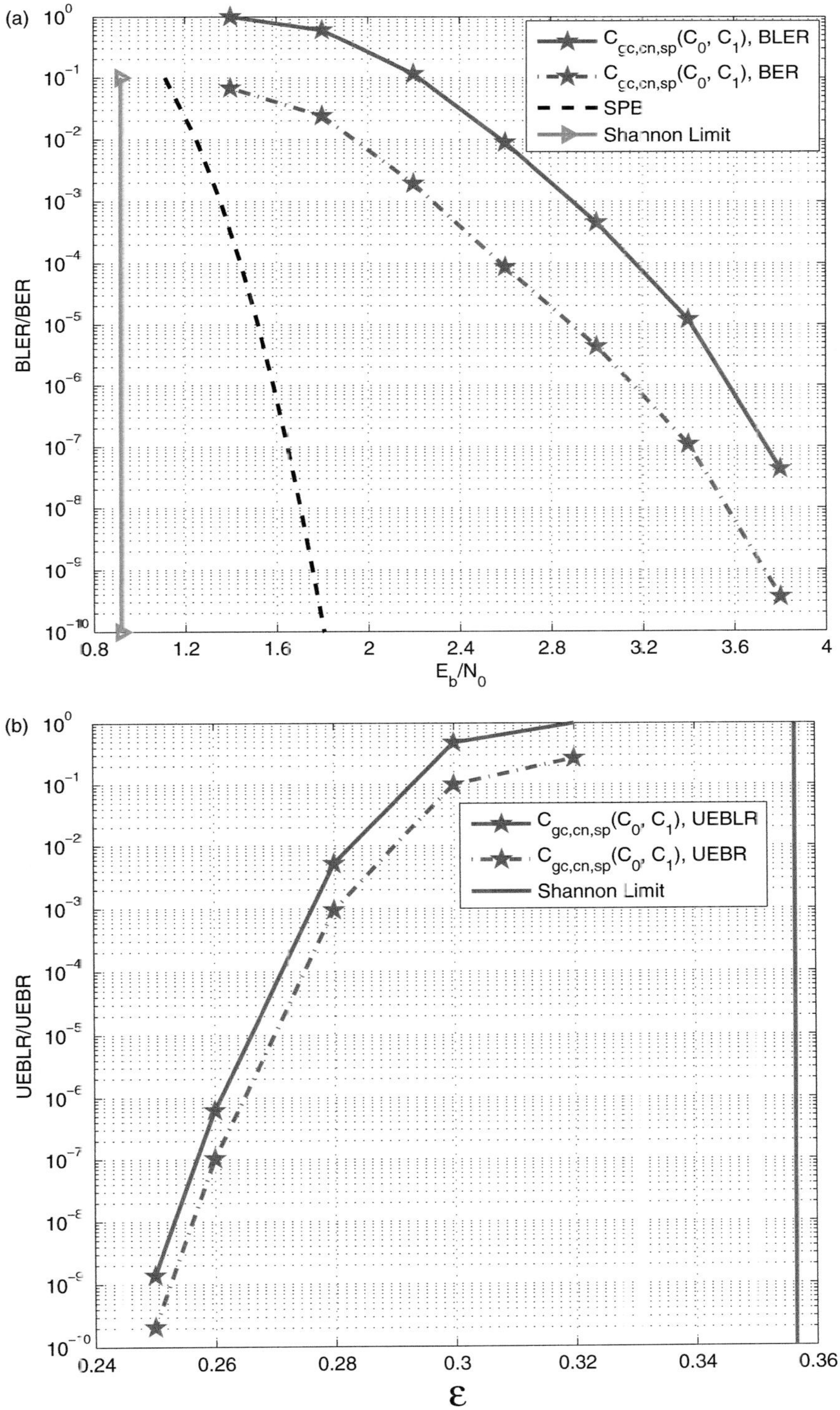

Figure 10.5. (a) The BER and BLER performances of the (4080, 2625) CN-based product GC-LDPC code $C_{gc,cn,sp}(C_0, C_1)$ given in Example 10.3 over the AWGNC; and (b) the performance of the code in (a) over the BEC.

10.4 CPM-Dispersion Construction of CN-Based Product QC-GC-LDPC Codes

A CN-based product QC-GC-LDPC code can be constructed based on 2×2 SM-constrained base matrices using the CPM-D-SP-construction. For illustration, we use the 2×2 SM-constrained base matrix $\mathbf{B}_{q,sp,s}$ over GF(q) in the form of (7.1) presented in Section 7.3 for the code construction.

Let α be a primitive element of the field GF(q) and let m_0, m_1, n_0, and n_1 be four positive integers such that $1 \leq m_0, m_1, n_0, n_1 \leq q$. Next, we form two matrices $\mathbf{B}_0$ and $\mathbf{B}_1$ with sizes $m_0 \times n_0$ and $m_1 \times n_1$, respectively, each in the form of (7.1). The matrix $\mathbf{B}_0$ is constructed based on two subsets $\boldsymbol{S}_{0,0}$ and $\boldsymbol{S}_{0,1}$ of GF(q) of sizes m_0 and n_0, respectively, and the matrix $\mathbf{B}_1$ is constructed based on two subsets $\boldsymbol{S}_{1,0}$ and $\boldsymbol{S}_{1,1}$ of GF(q) of sizes m_1 and n_1, respectively. For simplicity, we set $\eta = 1$ in the construction of $\mathbf{B}_0$ and $\mathbf{B}_1$.

The null spaces of the CPM-dispersions $\mathbf{H}_0 = \text{CPM}(\mathbf{B}_0)$ and $\mathbf{H}_1 = \text{CPM}(\mathbf{B}_1)$ of $\mathbf{B}_0$ and $\mathbf{B}_1$ give two CPM-QC-SP-LDPC codes, denoted by C_0 and C_1, of lengths $n_0(q-1)$ and $n_1(q-1)$, respectively, whose Tanner graphs have girth at least 6. Taking the direct product of C_0 and C_1 as described in Section 10.3, we obtain a CN-based product GC-LDPC code whose parity-check matrix is in the form of (10.7).

In the following, we take the CPM-dispersion approach to construct a CN-based product QC-GC-LDPC code based on the two base matrices $\mathbf{B}_0$ and $\mathbf{B}_1$. For $0 \leq j < n_1$, let $\mathbf{b}_{1,j}$ denote the j-th column of $\mathbf{B}_1$. For each column $\mathbf{b}_{1,j}$, we form an $m_1 n_0 \times n_0$ matrix over GF(q) by shifting $\mathbf{b}_{1,j}$ downward to the right $n_0 - 1$ times, each time m_1 positions, as follows:

$$\mathbf{B}_{1,j} = \begin{bmatrix} \mathbf{b}_{1,j} & & & \\ & \mathbf{b}_{1,j} & & \\ & & \ddots & \\ & & & \mathbf{b}_{1,j} \end{bmatrix}. \tag{10.8}$$

Next, we form the following $(m_0 n_1 + m_1 n_0) \times n_0 n_1$ matrix over GF(q):

$$\mathbf{B}_{gc,cn}(\mathbf{B}_0, \mathbf{B}_1) = \left[\begin{array}{cccc} \mathbf{B}_0 & & & \\ & \mathbf{B}_0 & & \\ & & \ddots & \\ & & & \mathbf{B}_0 \\ \hdashline \mathbf{B}_{1,0} & \mathbf{B}_{1,1} & \cdots & \mathbf{B}_{1,n_1-1} \end{array}\right]. \tag{10.9}$$

It follows from the structures of $\mathbf{B}_{1,j}$ and $\mathbf{B}_{gc,cn}(\mathbf{B}_0, \mathbf{B}_1)$ displayed by (10.8) and (10.9) that $\mathbf{B}_{gc,cn}(\mathbf{B}_0, \mathbf{B}_1)$ satisfies the 2×2 SM-constraint. Note that the lower submatrix $[\mathbf{B}_{1,0}\ \mathbf{B}_{1,1}\ \ldots\ \mathbf{B}_{1,n_1-1}]$ of $\mathbf{B}_{gc,cn}(\mathbf{B}_0, \mathbf{B}_1)$ is formed by taking n_0 copies of $\mathbf{B}_1$ and interleaving the copies.

Dispersing each nonzero entry of $\mathbf{B}_{gc,cn}(\mathbf{B}_0, \mathbf{B}_1)$ into a binary $(q-1) \times (q-1)$ CPM and each 0-entry into a $(q-1) \times (q-1)$ ZM, we obtain the following $(m_0 n_1 + m_1 n_0) \times n_0 n_1$ array of binary CPMs and ZMs of size $(q-1) \times (q-1)$:

$$\mathbf{H}_{b,gc,cn,sp}(C_0, C_1) = \left[\begin{array}{cccc} \mathbf{H}_0 & & & \\ & \mathbf{H}_0 & & \\ & & \ddots & \\ & & & \mathbf{H}_0 \\ \hdashline \mathbf{H}_{1,0} & \mathbf{H}_{1,1} & \cdots & \mathbf{H}_{1,n_1-1} \end{array}\right]. \qquad (10.10)$$

Note that the lower subarray $[\mathbf{H}_{1,0}\ \mathbf{H}_{1,1}\ \ldots\ \mathbf{H}_{1,n_1-1}]$ of $\mathbf{H}_{b,gc,cn,sp}(C_0, C_1)$ consists of n_0 copies of $\mathbf{H}_1$. For $0 \leq j < n_1$, each column-block of $\mathbf{H}_{1,j}$ is a column of CPMs and ZMs of size $(q-1) \times (q-1)$.

The null space of $\mathbf{H}_{b,gc,cn,sp}(C_0, C_1)$ gives a CN-based CPM-QC-GC-LDPC code $C_{b,gc,cn,sp}(C_0, C_1)$ which is the direct product of two CPM-QC-SP-LDPC codes C_0 and C_1 given by the null spaces of $\mathbf{H}_0 = \text{CPM}(\mathbf{B}_0)$ and $\mathbf{H}_1 = \text{CPM}(\mathbf{B}_1)$, respectively. The base matrix $\mathbf{B}_{gc,cn}(\mathbf{B}_0, \mathbf{B}_1)$ is referred to as the product of $\mathbf{B}_0$ and $\mathbf{B}_1$ and called the *product base matrix*. The matrices $\mathbf{B}_0$ and $\mathbf{B}_1$ are called the *horizontal* and *vertical component base matrices*, respectively.

The CN-based product CPM-QC-GC-LDPC code $C_{b,gc,cn,sp}(C_0, C_1)$ has a very interesting graphical structure. Let $\mathcal{G}_{ptg,0}$ and $\mathcal{G}_{ptg,1}$ be the Tanner graphs associated with $\mathbf{B}_0$ and $\mathbf{B}_1$, respectively. Let $\mathcal{G}_{b,gc,cn,sp}(q-1, q-1)$ be the Tanner graph of $C_{b,gc,cn,sp}(C_0, C_1)$. Then, from the structure of $\mathbf{H}_{b,gc,cn,sp}(C_0, C_1)$ and $[\mathbf{H}_{1,0}\ \mathbf{H}_{1,1}\ \ldots\ \mathbf{H}_{1,n_1-1}]$, we see that $\mathcal{G}_{b,gc,cn,sp}(q-1, q-1)$ is formed from $\mathcal{G}_{ptg,0}$ and $\mathcal{G}_{ptg,1}$ by taking the following steps: (1) Create n_1 copies of $\mathcal{G}_{ptg,0}$ and n_0 copies of $\mathcal{G}_{ptg,1}$; (2) Expand each copy of $\mathcal{G}_{ptg,0}$ and each copy of $\mathcal{G}_{ptg,1}$ by a factor of $q-1$ to form two connected Tanner graphs $\mathcal{G}_{ptg,0}(q-1, q-1)$ and $\mathcal{G}_{ptg,1}(q-1, q-1)$ through edge permutations; (3) Connect the n_1 copies of $\mathcal{G}_{ptg,0}(q-1, q-1)$ by the $m_1 n_0 (q-1)$ CNs of the n_0 copies of $\mathcal{G}_{ptg,1}(q-1, q-1)$; and (4) Connect the n_0 copies of $\mathcal{G}_{ptg,1}(q-1, q-1)$ by the $m_0 n_1 (q-1)$ CNs of the n_1 copies of $\mathcal{G}_{ptg,0}(q-1, q-1)$. Therefore, we may view the code $C_{b,gc,cn,sp}(C_0, C_1)$ as a *doubly* CN-based QC-GC-LDPC code whose Tanner graph is constructed from the copies of two protographs connected by two sets of global CNs.

We may view the Tanner graph $\mathcal{G}_{gc,cn}(\mathbf{B}_0, \mathbf{B}_1)$ associated with the base matrix $\mathbf{B}_{gc,cn}(\mathbf{B}_0, \mathbf{B}_1)$ given by (10.9) as the protograph for constructing a CN-based CPM-QC-GC-LDPC code $C_{b,gc,cn,sp}(C_0, C_1)$. In this case, we simply take $q-1$ copies of $\mathcal{G}_{gc,cn}(\mathbf{B}_0, \mathbf{B}_1)$ and connect them by permuting their edges. This copy-and-permute operation on $\mathcal{G}_{gc,cn}(\mathbf{B}_0, \mathbf{B}_1)$ gives a Tanner graph whose adjacency matrix specifies a QC-PTG-LDPC code. Therefore, a CN-based QC-GC-LDPC code can be viewed either as an SP-LDPC code or as a PTG-LDPC code.

In forming the product base matrices $\mathbf{B}_{gc,cn}(\mathbf{B}_0, \mathbf{B}_1)$, masking can be performed on either component base matrix $\mathbf{B}_0$, or $\mathbf{B}_1$, or both, to reduce the number of short cycles in the Tanner graph of the resultant masked product CN-based CPM-QC-GC-LDPC code.

Example 10.4. Let GF(331) be the field for the code construction and let α be a primitive element of GF(331). Suppose we construct a base matrix $\mathbf{B}_0$ in the form of (7.1) using the following two subsets of GF(331): $S_{0,0} = \{\alpha^0, \alpha^1, \alpha^2, \alpha^3\}$ and $S_{0,1} = \{\alpha^{130}, \alpha^{131}, \alpha^{132}, \alpha^{133}, \alpha^{134}, \alpha^{135}, \alpha^{136}, \alpha^{137}\}$. Set $\eta = -1$. The construction results in the following 4×8 matrix $\mathbf{B}_0$ over GF(331):

$$\mathbf{B}_0 = \begin{bmatrix} \alpha^{72} & \alpha^{147} & \alpha^{139} & \alpha^{218} & \alpha^{285} & \alpha^{69} & \alpha^{173} & \alpha^{189} \\ \alpha^{62} & \alpha^{73} & \alpha^{148} & \alpha^{140} & \alpha^{219} & \alpha^{286} & \alpha^{70} & \alpha^{174} \\ \alpha^{159} & \alpha^{63} & \alpha^{74} & \alpha^{149} & \alpha^{141} & \alpha^{220} & \alpha^{287} & \alpha^{71} \\ \alpha^{103} & \alpha^{160} & \alpha^{64} & \alpha^{75} & \alpha^{150} & \alpha^{142} & \alpha^{221} & \alpha^{288} \end{bmatrix}. \tag{10.11}$$

Next, we construct another base matrix $\mathbf{B}_1$ using the following two subsets of GF(331): $S_{1,0} = \{\alpha^4\}$ and $S_{1,1} = \{\alpha^{130}, \alpha^{131}, \alpha^{132}, \alpha^{133}, \alpha^{134}, \alpha^{135}\}$. Again, set $\eta = -1$. The construction results in the following 1×6 matrix $\mathbf{B}_1$ over GF(331):

$$\mathbf{B}_1 = \begin{bmatrix} \alpha^{67} & \alpha^{104} & \alpha^{161} & \alpha^{65} & \alpha^{76} & \alpha^{151} \end{bmatrix}. \tag{10.12}$$

Masking $\mathbf{B}_0$ with the masking matrix $\mathbf{Z}$ given by (7.5), we obtain the following masked base matrix:

$$\mathbf{B}_{0,mask} = \begin{bmatrix} \alpha^{72} & 0 & \alpha^{139} & 0 & \alpha^{285} & \alpha^{69} & \alpha^{173} & \alpha^{189} \\ 0 & \alpha^{73} & 0 & \alpha^{140} & \alpha^{219} & \alpha^{286} & \alpha^{70} & \alpha^{174} \\ \alpha^{159} & \alpha^{63} & \alpha^{74} & \alpha^{149} & \alpha^{141} & 0 & \alpha^{287} & 0 \\ \alpha^{103} & \alpha^{160} & \alpha^{64} & \alpha^{75} & 0 & \alpha^{142} & 0 & \alpha^{288} \end{bmatrix}. \tag{10.13}$$

Using $\mathbf{B}_{0,mask}$ and $\mathbf{B}_1$ as the horizontal and vertical component base matrices, respectively, we construct a 32×48 matrix $\mathbf{B}_{gc,cn}(\mathbf{B}_{0,mask}, \mathbf{B}_1)$ over GF(331) in the form of (10.9). The matrix $\mathbf{B}_{gc,cn}(\mathbf{B}_{0,mask}, \mathbf{B}_1)$ has column weight 4 and row weight 6. By checking, we find that $\mathbf{B}_{gc,cn}(\mathbf{B}_{0,mask}, \mathbf{B}_1)$ satisfies the 3×3 SM-constraint. Using $\mathbf{B}_{gc,cn}(\mathbf{B}_{0,mask}, \mathbf{B}_1)$ as a product base matrix, we can construct a CN-based product CPM-QC-GC-LDPC code whose Tanner graph has girth at least 8.

Dispersing each nonzero entry in $\mathbf{B}_{gc,cn}(\mathbf{B}_{0,mask}, \mathbf{B}_1)$ into a 330×330 binary CPM and each 0-entry into a 330×330 ZM, we obtain a 32×48 binary array $\mathbf{H}_{b,gc,cn,sp}(C_{0,mask}, C_1)$ of CPMs and ZMs of size 330×330. The matrix $\mathbf{H}_{b,gc,cn,sp}(C_{0,mask}, C_1)$ is a $10\,560 \times 15\,840$ matrix over GF(2) with column weight 4 and row weight 6. The null space of $\mathbf{H}_{b,gc,cn,sp}(C_{0,mask}, C_1)$ gives a (4, 6)-regular (15 840, 6600) CN-based product CPM-QC-GC-LDPC code $C_{b,gc,cn,sp}(C_{0,mask}, C_1)$, whose Tanner graph $C_{b,gc,cn,sp}(C_{0,mask}, C_1)$ has girth 8 and contains 297 000 cycles of length 8 and 79 200 cycles of length 10.

The BER and BLER performances of the code $C_{b,gc,cn,sp}(C_{0,mask}, C_1)$ over the AWGNC decoded with 50 iterations of the MSA are shown in Fig. 10.6(a). We see that the code can achieve a BER of 10^{-7} without a visible error-floor and it performs about 1.4 dB away from the SPB at a BLER of 10^{-5}.

For comparison, the BER and BLER performances of a (15 840, 6601) LDPC code C_{peg} constructed using the PEG algorithm are also included in Fig. 10.6(a). The PEG code C_{peg} has a rate close to that of the CN-based product CPM-QC-GC-LDPC code $C_{b,gc,cn,sp}(C_{0,mask}, C_1)$. The parity-check matrix of the PEG code C_{peg} is a $9240 \times 15\,840$ binary matrix and has constant column weight 4. (Note that the size of the parity-check matrix of the PEG code C_{peg} is different from that of the parity-check matrix of the code $C_{b,gc,cn,sp}(C_{0,mask}, C_1)$.) The Tanner graph of C_{peg} has girth 10 and contains 128 482 cycles of length 10 and 2 545 908 cycles of length 12. From Fig. 10.6(a), we see that the CN-based product CPM-QC-GC-LDPC code $C_{b,gc,cn,sp}(C_{0,mask}, C_1)$ outperforms the PEG code C_{peg} by about 0.6 dB.

The UEBR and UEBLR performances of the code $C_{b,gc,cn,sp}(C_{0,mask}, C_1)$ are shown in Fig. 10.6(b). We see that the code $C_{b,gc,cn,sp}(C_{0,mask}, C_1)$ also performs well over the BEC.

In forming the product base matrix $\mathbf{B}_{gc,cn}(\mathbf{B}_0, \mathbf{B}_1)$ in the form of (10.9), we can also use n_1 different $m_0 \times n_0$ matrices $\mathbf{B}_{0,0}, \mathbf{B}_{0,1}, \ldots, \mathbf{B}_{0,n_1-1}$ as the horizontal component base matrices on the main diagonal of the upper $n_1 \times n_1$ subarray of the product base matrix. This results in the following product base matrix:

$$\mathbf{B}'_{gc,cn} = \left[\begin{array}{cccc} \mathbf{B}_{0,0} & & & \\ & \mathbf{B}_{0,1} & & \\ & & \ddots & \\ & & & \mathbf{B}_{0,n_1-1} \\ \hdashline \mathbf{B}_{1,0} & \mathbf{B}_{1,1} & \cdots & \mathbf{B}_{1,n_1-1} \end{array}\right]. \tag{10.14}$$

For $0 \le i < n_1$, the null space of the CPM-dispersion of $\mathbf{B}_{0,i}$ gives a CPM-QC-SP-LDPC code $C_{0,i}$. The CN-based product CPM-QC-GC-LDPC code $C'_{b,gc,cn,sp}$ given by the null space of the binary CPM-dispersion $\mathbf{H}'_{b,gc,cn,sp}$ of the product base matrix $\mathbf{B}'_{gc,cn}$ is a product of n_1 horizontal component codes $C_{0,0}$, $C_{0,1}, \ldots, C_{0,n_1-1}$ and one vertical component code C_1.

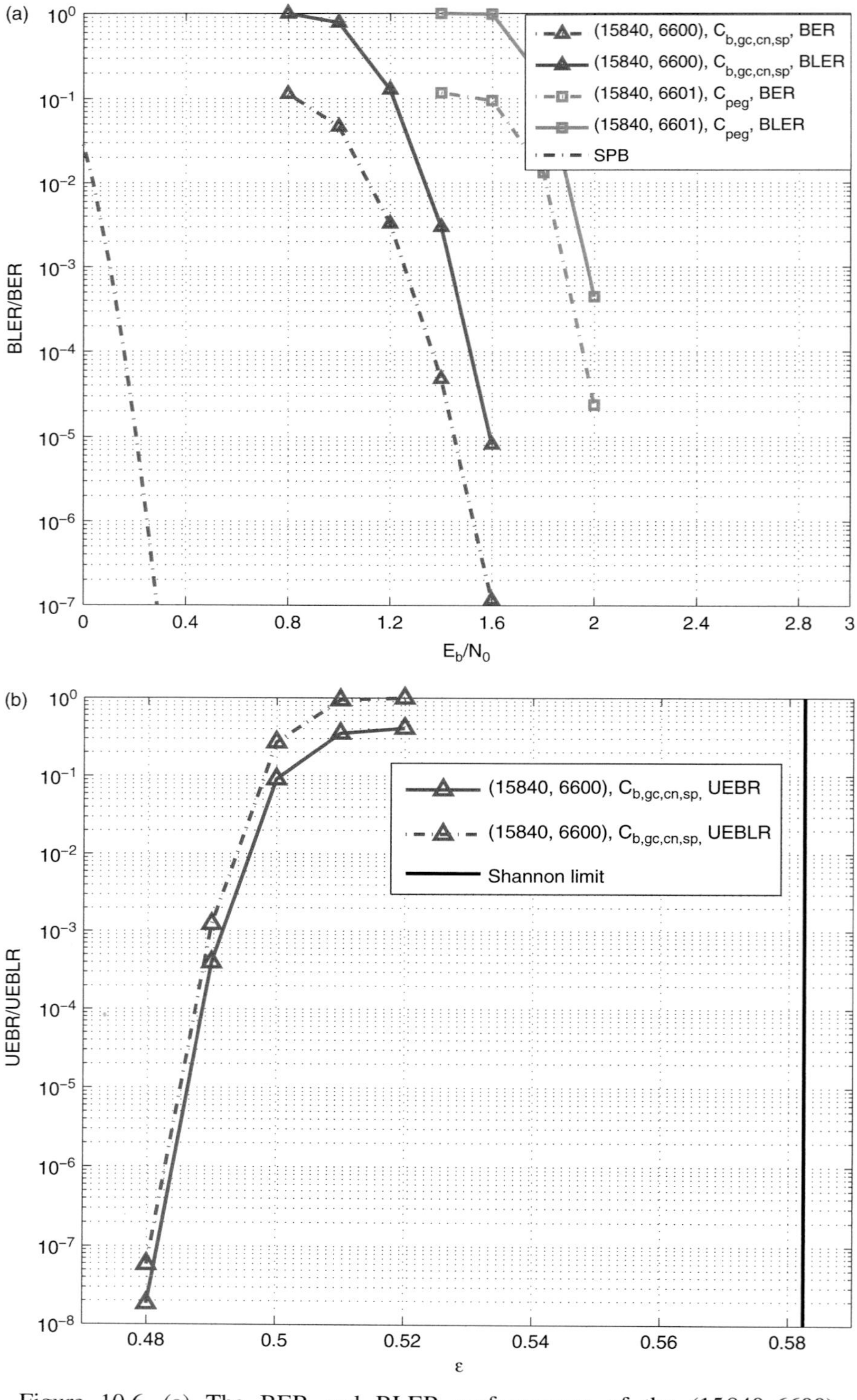

Figure 10.6. (a) The BER and BLER performances of the (15 840, 6600) CN-based product CPM-QC-GC-LDPC code $C_{b,gc,cn,sp}(C_{0,mask}, C_1)$ and the (15 840, 6601) PEG code C_{peg} given in Example 10.4 over the AWGNC; and (b) the performance of the code $C_{b,gc,cn,sp}(C_{0,mask}, C_1)$ in (a) over the BEC.

Similarly, we can use n_0 different $m_1 \times n_1$ vertical component base matrices to form the lower subarray of the product base matrix given in the form of (10.9). The null space of the CPM-dispersion of the resultant product base matrix gives a product of $n_C + 1$ CPM-QC-SP-LDPC codes, a single horizontal and n_0 vertical component codes. In forming a product base matrix in the form of (10.9), we can use n_1 different $m_0 \times n_0$ horizontal component base matrices to form the upper diagonal subarray of the product base matrix and n_0 different $m_1 \times n_1$ vertical component base matrices to form the lower subarray of the product base matrix. Then, the null space of the CPM-dispersion of the resultant product base matrix gives a product of n_1 horizontal and n_0 vertical component CPM-QC-SP-LDPC codes. A product code constructed by using more than two component codes is referred to as a *generalized product code*.

The above product construction using the CPM-D construction gives a large family of CN-based product CPM-QC-GC-LDPC codes. In the construction, any 2×2 SM-constrained base matrix can be used. Product codes are effective for correcting combinations of random and burst errors [74].

Example 10.5. Consider the code construction field GF(127). Let α be a primitive element of GF(127). In this example, we choose the following parameters: $m_0 = 2$, $n_0 = 6$, $m_1 = 1$, and $n_1 = 6$. The pairs of subsets of GF(127) for constructing the 6 horizontal component base matrices $\mathbf{B}_{0,0}$, $\mathbf{B}_{0,1}$, $\mathbf{B}_{0,2}$, $\mathbf{B}_{0,3}$, $\mathbf{B}_{0,4}$, and $\mathbf{B}_{0,5}$ are:

(1) $\boldsymbol{S}_{0,0} = \{\alpha^0, \alpha^1\}$ and $\boldsymbol{S}_{0,1} = \{\alpha^{11}, \alpha^{12}, \alpha^{13}, \alpha^{14}, \alpha^{15}, \alpha^{16}\}$;

(2) $\boldsymbol{S}_{1,0} = \{\alpha^0, \alpha^1\}$ and $\boldsymbol{S}_{1,1} = \{\alpha^{17}, \alpha^{18}, \alpha^{19}, \alpha^{20}, \alpha^{21}, \alpha^{22}\}$;

(3) $\boldsymbol{S}_{2,0} = \{\alpha^0, \alpha^1\}$ and $\boldsymbol{S}_{2,1} = \{\alpha^{23}, \alpha^{24}, \alpha^{25}, \alpha^{26}, \alpha^{27}, \alpha^{28}\}$;

(4) $\boldsymbol{S}_{3,0} = \{\alpha^0, \alpha^1\}$ and $\boldsymbol{S}_{3,1} = \{\alpha^{29}, \alpha^{30}, \alpha^{31}, \alpha^{32}, \alpha^{33}, \alpha^{34}\}$;

(5) $\boldsymbol{S}_{4,0} = \{\alpha^0, \alpha^1\}$ and $\boldsymbol{S}_{4,1} = \{\alpha^{35}, \alpha^{36}, \alpha^{37}, \alpha^{38}, \alpha^{39}, \alpha^{40}\}$;

(6) $\boldsymbol{S}_{5,0} = \{\alpha^0, \alpha^1\}$ and $\boldsymbol{S}_{5,1} = \{\alpha^{41}, \alpha^{42}, \alpha^{43}, \alpha^{44}, \alpha^{45}, \alpha^{46}\}$.

The two subsets of GF(127) for constructing the vertical component base matrix $\mathbf{B}_1$ are: $\boldsymbol{S}_{1,0} = \{\alpha^2\}$ and $\boldsymbol{S}_{1,1} = \{\alpha^{11}, \alpha^{12}, \alpha^{13}, \alpha^{14}, \alpha^{15}, \alpha^{16}\}$.

Set $\eta = -1$. Based on the above pairs of subsets of GF(127) and using the construction given by (7.1), we obtain the following horizontal and vertical component base matrices:

$$\mathbf{B}_{0,0} = \begin{bmatrix} \alpha^{8} & \alpha^{5} & \alpha^{43} & \alpha^{24} & \alpha^{79} & \alpha^{57} \\ \alpha^{102} & \alpha^{9} & \alpha^{6} & \alpha^{44} & \alpha^{25} & \alpha^{80} \end{bmatrix},$$

$$\mathbf{B}_{0,1} = \begin{bmatrix} \alpha^{86} & \alpha^{107} & \alpha^{100} & \alpha^{87} & \alpha^{105} & \alpha^{69} \\ \alpha^{58} & \alpha^{87} & \alpha^{108} & \alpha^{44} & \alpha^{88} & \alpha^{106} \end{bmatrix},$$

$$\mathbf{B}_{0,2} = \begin{bmatrix} \alpha^{54} & \alpha^{14} & \alpha^{98} & \alpha^{108} & \alpha^{59} & \alpha^{116} \\ \alpha^{70} & \alpha^{55} & \alpha^{15} & \alpha^{99} & \alpha^{109} & \alpha^{60} \end{bmatrix},$$

$$\mathbf{B}_{0,3} = \begin{bmatrix} \alpha^{7} & \alpha^{2} & \alpha^{67} & \alpha^{72} & \alpha^{83} & \alpha^{48} \\ \alpha^{117} & \alpha^{8} & \alpha^{3} & \alpha^{68} & \alpha^{73} & \alpha^{84} \end{bmatrix},$$

$$\mathbf{B}_{0,4} = \begin{bmatrix} \alpha^{96} & \alpha^{63} & \alpha^{42} & \alpha^{55} & \alpha^{122} & \alpha^{120} \\ \alpha^{49} & \alpha^{97} & \alpha^{64} & \alpha^{43} & \alpha^{56} & \alpha^{123} \end{bmatrix},$$

$$\mathbf{B}_{0,5} = \begin{bmatrix} \alpha^{97} & \alpha^{37} & \alpha^{13} & \alpha^{62} & \alpha^{64} & \alpha^{71} \\ \alpha^{121} & \alpha^{98} & \alpha^{38} & \alpha^{14} & \alpha^{63} & \alpha^{65} \end{bmatrix},$$

and

$$\mathbf{B}_1 = \begin{bmatrix} \alpha^{77} & \alpha^{103} & \alpha^{10} & \alpha^{7} & \alpha^{45} & \alpha^{26} \end{bmatrix}.$$

Based on $\mathbf{B}_{0,0}$, $\mathbf{B}_{0,1}$, $\mathbf{B}_{0,2}$, $\mathbf{B}_{0,3}$, $\mathbf{B}_{0,4}$, $\mathbf{B}_{0,5}$, and $\mathbf{B}_1$, we construct an 18×36 product base matrix $\mathbf{B}'_{gc,cn}$ over GF(127) in the form of (10.14). $\mathbf{B}'_{gc,cn}$ has column weight 3 and row weight 6 and it satisfies the 2×2 SM-constraint. By checking, we find that $\mathbf{B}'_{gc,cn}$ also satisfies the 3×3 SM-constraint. Therefore, the Tanner graph associated with the CPM-dispersion of $\mathbf{B}'_{gc,cn}$ has girth at least 8.

The binary CPM-dispersion of $\mathbf{B}'_{gc,cn}$ gives an 18×36 array $\mathbf{H}'_{b,gc,cn,sp}$ of CPMs and ZMs of size 126×126. $\mathbf{H}'_{b,gc,cn,sp}$ is a 2268×4536 matrix over GF(2) with column and row weights 3 and 6, respectively. The null space of $\mathbf{H}'_{b,gc,cn,sp}$ gives a (3, 6)-regular (4536, 2276) CN-based product CPM-QC-GC-LDPC code $C'_{b,gc,cn,sp}$ with rate 0.5018. The Tanner graph of the code $C'_{b,gc,cn,sp}$ has girth exactly 8 and contains 756 cycles of length eight, 5292 cycles of length 10, and 110 334 cycles of length 12. The BER and BLER performances of the code $C'_{b,gc,cn,sp}$ over the AWGNC decoded with 50 iterations of the MSA are shown in Fig. 10.7(a). We see that the code can achieve a BER of 10^{-9} without a visible error-floor and it performs about 1.1 dB away from the SPB at a BLER of 4×10^{-6}.

For comparison, the BER and BLER performances of a (4536, 2268) LDPC code C_{peg} constructed using the PEG algorithm are also included in Fig. 10.7(a). The PEG code C_{peg} has rate 0.5 which is slightly lower than that of the code $C'_{b,gc,cn,sp}$. The parity-check matrix of the PEG code C_{peg} is a 2268×4532 binary matrix and has constant column weight 3. The Tanner graph of C_{peg} has girth 10 and contains 3187 cycles of length 10 and 94 074 cycles of length 12. From Fig. 10.7(a), we see that the two codes perform almost the same over the AWGNC.

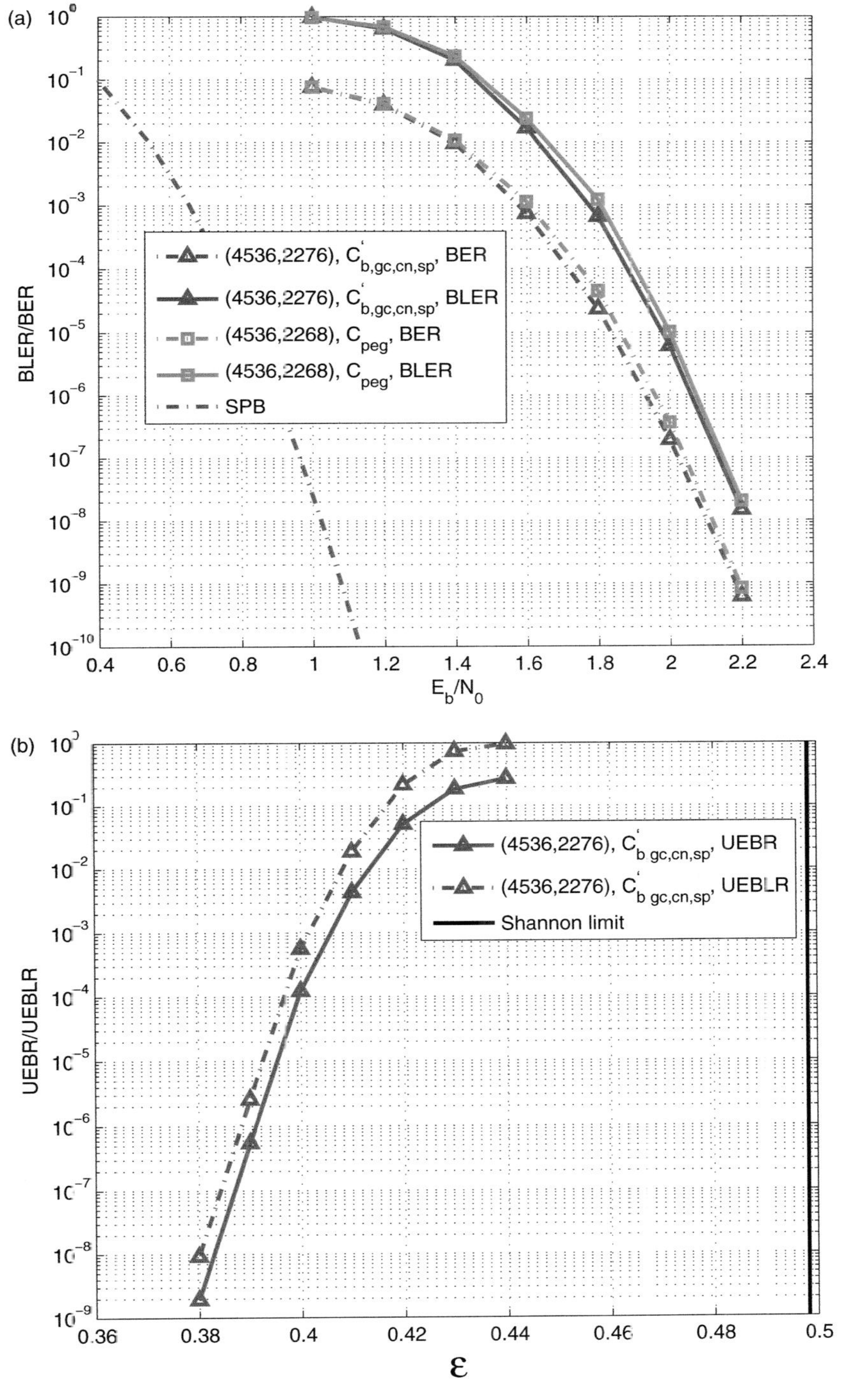

Figure 10.7. (a) The BER and BLER performances of the (4536, 2276) CN-based product CPM-QC-GC-LDPC code $C'_{b,gc,cn,sp}$ and the (4536, 2268) PEG code C_{peg} given in Example 10.5 over the AWGNC; and (b) the performance of the code $C'_{b,gc,cn,sp}$ in (a) over the BEC.

The UEBR and UEBLR performances of the code $C'_{b,gc,cn,sp}$ are shown in Fig. 10.7(b). We see that the code $C'_{b,gc,cn,sp}$ performs also well over the BEC. It can achieve a UEBR of 10^{-9} without an error-floor.

10.5 Discussion and Remarks

The counterpart of a CN-based GC-LDPC code is an LDPC code whose Tanner graph consists of a set of disjoint Tanner graphs connected by a group of *global VNs*. Such an LDPC code is called a *VN-based GC-LDPC* code. Construction of a VN-based CPM-QC-GC-LDPC code, denoted by $C_{gc,vn,qc}$, is similar to the construction of a CN-based CPM-QC-GC-LDPC code presented in Section 10.1. The base matrix $\mathbf{B}_{gc,vn}$ for constructing a VN-based CPM-QC-GC-LDPC code is of the following form:

$$\mathbf{B}_{gc,vn} = \left[\begin{array}{ccccc|c} \mathbf{R}_{0,0} & & & & & \\ & \mathbf{R}_{0,0} & & & & \\ & & \mathbf{R}_{0,0} & & & \mathbf{Y}_{gc,vn}(t,s) \\ & & & \ddots & & \\ & & & & \mathbf{R}_{0,0} & \end{array}\right], \tag{10.15}$$

where $\mathbf{Y}_{gc,vn}(t,s)$ is an $mt \times s$ matrix over GF(q) which is formed by taking and shortening s unused columns in forming the $r \times r$ cyclic array $\mathbf{B}_{q,sp,p}(m,n)$ of (8.3) from the RS-based $r \times r$ array $\mathbf{W}_{RS}$ of $l \times l$ submatrices over GF(q) given by (8.2) with $1 \leq s \leq r(l-n)$. The shortening of each unused column is done by removing the components at the locations that correspond to the rows which are not used in forming the array $\mathbf{B}_{q,sp,p}(m,n)$. Hence, each column of $\mathbf{Y}_{gc,vn}(t,s)$ consists of t sections, each with m components. $\mathbf{B}_{gc,vn}$ is a submatrix of $\mathbf{B}_{q,sp,p}(m,n)$ (after replacing some entries in $\mathbf{B}_{q,sp,p}(m,n)$ by zeros) and hence satisfies the 2×2 SM-constraint. The subscript "*vn*" in $\mathbf{B}_{gc,vn}$ and $C_{gc,vn,qc}$ stands for "*variable node*."

From the structure of $\mathbf{B}_{gc,vn}$, we see that the Tanner graph $\mathcal{G}_{gc,vn}$ associated with $\mathbf{B}_{gc,vn}$ consists of t copies of the Tanner graph associated with the $m \times n$ matrix $\mathbf{R}_{0,0}$ connected by s global VNs. The null space of the CPM-dispersion $\mathbf{H}_{gc,vn,qc}(q-1,q-1)$ of $\mathbf{B}_{gc,vn}$ gives a VN-based CPM-QC-GC-LDPC code $C_{gc,vn,qc}$ of length $(nt+s)(q-1)$ and rate at least $((n-m)t+s)/(nt+s)$. The Tanner graph $\mathcal{G}_{gc,vn}$ may be viewed as the protograph for constructing $C_{gc,vn,qc}$. In this case, $C_{gc,vn,qc}$ is QC-PTG-LDPC code.

Divide the connection matrix $\mathbf{Y}_{gc,vn}(t,s)$ into t submatrices of size $m \times s$, $\mathbf{Y}_{gc,vn,0}, \mathbf{Y}_{gc,vn,1}, \ldots, \mathbf{Y}_{gc,vn,t-1}$, each consisting of m consecutive rows of $\mathbf{Y}_{gc,vn}(t,s)$. For $0 \leq i < t$, the matrix $[\mathbf{R}_{0,0}\ \mathbf{Y}_{gc,vn,i}]$ is a 2×2 SM-constrained

$m \times (n + s)$ matrix over GF(q). The null space of the CPM-dispersion CPM($[\mathbf{R}_{0,0}\ \mathbf{Y}_{gc,vn,i}]$) gives a CPM-QC-SP-LDPC code $C_{0,i}$ of length $(n + s)(q - 1)$. From a global point of view, the code $C_{gc,vn,qc}$ given by the null space of $\mathbf{H}_{gc,vn,qc}(q - 1, q - 1)$ is a global code which is composed of t local codes $C_{0,0}, C_{0,1}, \ldots, C_{0,t-1}$. Let $\mathbf{v}$ be a codeword in $C_{gc,vn,qc}$. Divide $\mathbf{v}$ into $t + 1$ sections, $\mathbf{v}_0, \mathbf{v}_1, \ldots, \mathbf{v}_{t-1}, \mathbf{v}_t$, each of the first t sections consisting of $n(q - 1)$ consecutive components of $\mathbf{v}$ and the last section $\mathbf{v}_t$ consisting of the last $s(q - 1)$ components of $\mathbf{v}$. For $0 \leq i < t$, the vector $(\mathbf{v}_i, \mathbf{v}_t)$ is a codeword in the local code $C_{0,i}$. Note that all the t local codes are different.

A local/global two-phase iterative decoding scheme can be devised to decode the VN-based CPM-QC-GC-LDPC code $C_{gc,vn,qc}$. We leave this to the reader.

Decoding of a CN-based product GC-LDPC code $C_{gc,cn,sp}(C_0, C_1)$ based on its parity-check matrix $\mathbf{H}_{gc,cn,sp}(C_0, C_1)$ can be a serious problem. If the lengths of its two component codes C_0 and C_1 are long (even moderately long), the length of $C_{gc,cn,sp}(C_0, C_1)$ can be very long. In this case, the size of the parity-check matrix $\mathbf{H}_{gc,cn,sp}(C_0, C_1)$ is very large and the implementation of a global decoder for $C_{gc,cn,sp}(C_0, C_1)$ based on $\mathbf{H}_{gc,cn,sp}(C_0, C_1)$ will be very complex and impractical. How to devise a reduced-complexity iterative scheme to decode a long CN-based product GC-LDPC code seems to be an interesting problem.

One possible approach to resolve this high decoder complexity problem is to decode $C_{gc,cn,sp}(C_0, C_1)$ using a *turbo BP-decoder*. A turbo BP-decoder consists of two local decoders, one for decoding the horizontal component code C_0 and the other for decoding the vertical component code C_1. Each local decoder is a BP-decoder. The two local BP-decoders exchange reliability messages (LLRs) of code symbols through two interleavers in the form of turbo decoding. The two local BP-decoders can be connected either in series or in parallel.

Suppose a codeword $\mathbf{v} = (v_0, v_1, \ldots, v_{n_0 n_1 - 1})$ in $C_{gc,cn,sp}(C_0, C_1)$ is transmitted. Let $\mathbf{r} = (r_0, r_1, \ldots, r_{n_0 n_1 - 1})$ be the corresponding received word. We sectionalize $\mathbf{r}$ in two forms, $\mathbf{r} = (\mathbf{r}_0, \mathbf{r}_1, \ldots, \mathbf{r}_{n_1 - 1})$ and $\mathbf{r}^\star = (\mathbf{r}_0^\star, \mathbf{r}_1^\star, \ldots, \mathbf{r}_{n_0 - 1}^\star)$, one with n_1 equal sections and the other n_0 sections, similar to sectionalizing a codeword in $C_{gc,cn,sp}(C_0, C_1)$ as described in Section 10.3. The vector $\mathbf{r}^\star$ is a permutation of $\mathbf{r}$. For $0 \leq i < n_1$, the i-th section $\mathbf{r}_i$ of $\mathbf{r}$ is an erroneous version of a codeword in C_0. For $0 \leq j < n_0$, the j-th section $\mathbf{r}_j^\star$ of $\mathbf{r}^\star$ is an erroneous version of a codeword in C_1.

Now, we design two local BP-decoders, BP-decoder$_0$ and BP-decoder$_1$. BP-decoder$_0$ processes the n_1 sections of $\mathbf{r}$ and BP-decoder$_1$ processes the n_0 sections of the permuted received vector $\mathbf{r}^\star$ simultaneously. At the end of each decoding iteration (or a fixed number of decoding iterations), two hard-decision vectors are produced by the two BP-decoders based on the LLRs of the decoded code symbols. Both of the two hard-decision vectors are tested. There are four possible outcomes of the test:

1. The first case is that both the hard-decision vectors give the same codeword in $C_{gc,cn,sp}(C_0, C_1)$. In this case, we stop the decoding process and deliver the decoded codeword to the user.
2. The second case is that one of the two hard-decision vectors is a codeword in $C_{gc,cn,sp}(C_0, C_1)$ and the other is not. In this case, we also stop the decoding process and deliver the codeword to the user.
3. The third case is that the hard-decision vectors are two different codewords in $C_{gc,cn,sp}(C_0, C_1)$. In this case, we can either stop the decoding process and choose one at random and deliver it to the user, or continue the decoding process.
4. The fourth case is that both the hard-decision vectors are not codewords in $C_{gc,cn,sp}(C_0, C_1)$. In this case, the decoding process continues.

When another decoding iteration is needed, the two local BP-decoders exchange reliability messages and then start another decoding iteration. This decoding process continues until a codeword is delivered to the user or the preset maximum number of decoding iterations is reached. Note that, after message exchanges, each code symbol has three reliability measures which are provided by the two BP-decoders and the channel. How to use these reliability measures to start a new decoding iteration to achieve the best performance is an issue needing to be investigated. Whether the above proposed turbo BP-decoding scheme (or its variations) is effective in the trade-off between the error performance and the decoding complexity is also worthy of further study.

As pointed out in Chapter 2, a common severe weakness of LDPC codes is the error-floor. A high error-floor may preclude LDPC codes from applications where very low error rates are required. If an LDPC code has a reasonably large minimum distance and no trapping sets with size smaller than its minimum distance, then the code in general has a relatively low error-floor, e.g., a finite geometry LDPC code [58, 74, 97]. The minimum distance of a CN-based product GC-LDPC code $C_{gc,cn,sp}(C_0, C_1)$ is the product of the minimum distances of its two component codes. Even if the minimum distances of the two component codes are relatively small, their product can be relatively large. An interesting problem is how the trapping set structure of a CN-based product GC-LDPC code related to the trapping set structures of its two component codes. Is the smallest trapping set of a CN-based product GC-LDPC code much larger than the smallest trapping set of each of its two component codes? We do not have an answer to this question at this point; however, we can at least show that there is no small local trapping set in the Tanner graph $\mathcal{G}_{gc,cn,sp}(C_0, C_1)$ of $C_{gc,cn,sp}(C_0, C_1)$.

Consider the parity-check matrix $\mathbf{H}_{gc,cn,sp}(C_0, C_1)$ of $C_{gc,cn,sp}(C_0, C_1)$ given by (10.7). It consists of n_1 column-blocks, each consisting of a copy of the parity-check matrix $\mathbf{H}_0$ and an $m_1n_0 \times n_0$ matrix $\mathbf{H}_{1,j,\pi}$ with $0 \leq j < n_1$. The matrix $\mathbf{H}_{1,j,\pi}$ is formed by the j-th column of the parity-check matrix $\mathbf{H}_1$ of the vertical component code C_1 by shifting it downward to the right $n_0 - 1$ times, each time by m_1 positions. So, each row of $\mathbf{H}_{1,j,\pi}$ is either a row of zeros or a row with a single 1-entry. Suppose $\mathbf{H}_1$ has constant column weight w_1. Then each column of $\mathbf{H}_{1,j,\pi}$ has weight w_1. Suppose $\mathbf{H}_0$ has constant column weight w_0. Then, each column of the j-th column-block of $\mathbf{H}_{gc,cn,sp}(C_0, C_1)$ has constant column weight $w_0 + w_1$.

For $1 \leq \kappa \leq n_0$ and $0 \leq \tau \leq m_0$, suppose the Tanner graph $\mathcal{G}_0$ of C_0 contains a (κ, τ) trapping set $\mathbf{T}$ whose induced subgraph $\mathcal{T}$ contains κ VNs and τ CNs of odd degrees. The κ VNs correspond to κ columns of $\mathbf{H}_0$. Consider the subgraph $\mathcal{G}_{0,j}$ of the Tanner graph $\mathcal{G}_{gc,cn,sp}(C_0, C_1)$ of $C_{gc,cn,sp}(C_0, C_1)$ induced by the n_0 VNs corresponding to the n_0 columns of the j-th column-block of $\mathbf{H}_{gc,cn,sp}(C_0, C_1)$. Then, the graph $\mathcal{T}$ is a subgraph of $\mathcal{G}_{0,j}$ (and also a subgraph of $\mathcal{G}_{gc,cn,sp}(C_0, C_1)$). From the structure of the j-th column-block of $\mathbf{H}_{gc,cn,sp}(C_0, C_1)$, we see that each VN in $\mathcal{T}$ is checked by w_1 rows in $\mathbf{H}_{1,j,\pi}$. Since the two sets of 1-entries in two different columns of $\mathbf{H}_{1,j,\pi}$ are disjoint, the κ VNs in $\mathcal{T}$ are checked by a total of κw_1 rows, each checking one and only one VN in $\mathcal{T}$. As a result, the (κ, τ) trapping set $\mathbf{T}$ in $\mathcal{G}_0$ becomes a $(\kappa, \tau + \kappa w_1)$ trapping set in $\mathcal{G}_{gc,cn,sp}(C_0, C_1)$. Such a trapping set with VNs contained in the subgraph $\mathcal{G}_{0,j}$ is called a *local trapping set* in $\mathcal{G}_{gc,cn,sp}(C_0, C_1)$. We see that the product operation increases the number of CNs of odd degrees by κw_1. If $\mathbf{T}$ is a small trapping set in $\mathcal{G}_0$, it is not a small trapping set in $\mathcal{G}_{gc,cn,sp}(C_0, C_1)$ any more if $\tau + \kappa w_1 \geq 4\kappa$ (see the definition of a small trapping set in Section 2.2).

If we switch the roles of $\mathbf{H}_0$ and $\mathbf{H}_1$ in forming the parity-check matrix of a CN-based product GC-LDPC code, we can see that a (κ, τ) trapping set in the Tanner graph $\mathcal{G}_1$ of C_1 becomes a $(\kappa, \tau + \kappa w_0)$ local trapping set in $\mathcal{G}_{gc,cn,sp}(C_0, C_1)$.

Now the hard problem is the analysis of the structure of a trapping set whose VNs are spread among the subgraph $\mathcal{G}_{0,j}$ with $0 \leq j < n_1$ of $\mathcal{G}_{gc,cn,sp}(C_0, C_1)$. It is a challenging combinatorial problem and worthy of investigation.

11

SP-Construction of Nonbinary LDPC Codes

In this chapter, we consider the SP-construction of LDPC codes over NB fields. The SP-construction of NB LDPC codes is similar to the SP-construction of their binary counterparts. Most algebraic and graphical structural properties developed for binary SP-LDPC codes in earlier chapters can be applied to NB SP-LDPC codes with some modifications. In this chapter, we first give a brief description of the general SP-construction of NB LDPC codes using binary base matrices, and then present six algebraic methods for constructing these codes using both binary and NB base matrices. Two of these six construction methods are based on the conventional parity-check matrices of Reed–Solomon (RS) codes over finite fields of characteristic 2. An iterative BP-decoding algorithm, namely the *Fast-Fourier-Transform* (FFT) *q-ary* SPA (FFT-QSPA) [82, 23, 5], is presented in Appendix C. All the constructed codes in this chapter are simulated on the binary-input AWGNC.

11.1 General SP-Construction of NB LDPC Codes Using Binary Base Matrices

Let $\mathbf{B}_{sp} = [b_{i,j}]_{0 \leq i < m, 0 \leq j < n}$ be an $m \times n$ binary matrix. We use this binary matrix as the base matrix for the SP-construction of an NB LDPC code. (NB base matrices for the SP-construction of NB LDPC codes will be considered in later sections.) Let $\boldsymbol{R} = \{\mathbf{Q}_1, \mathbf{Q}_2, \ldots, \mathbf{Q}_r\}$ be a set of r *sparse* $k \times t$ matrices over an NB field GF(q) where q is a prime or a power of a prime and $q \neq 2$. We form an $m \times n$ array $\mathbf{H}_{q,sp}(k,t)$ of $k \times t$ matrices over GF(q) by replacing each 1-entry in $\mathbf{B}_{sp}$ by a member matrix in $\boldsymbol{R}$ and each 0-entry by a $k \times t$ ZM. The array $\mathbf{H}_{q,sp}(k,t)$ is an $mk \times nt$ sparse matrix over GF(q), where the subscript "q" stands for "q-*ary*."

The null space over GF(q) of $\mathbf{H}_{q,sp}(k,t)$ gives a q-ary SP-LDPC code $C_{q,sp}$. If we impose the RC-constraint on the base matrix $\mathbf{B}_{sp}$ and the RC- and

PW-RC-constraints on the member matrices of the replacement set $\boldsymbol{R}$, then the Tanner graph $\mathcal{G}_{q,sp}(k,t)$ associated with the parity-check array $\mathbf{H}_{q,sp}(k,t)$ will have girth at least 6.

The general structural properties of the Tanner graph of an NB SP-LDPC code are the same as those for the Tanner graph of a binary SP-LDPC code, except that the edges are labeled with nonzero elements in GF(q).

For a given rate R_c, we choose the parameters m, n, k, and t such that $R_c = (nk - mt)/nk$. Suppose we choose *at random* both an $m \times n$ SP-base matrix $\mathbf{B}_{sp}$ and a replacement set $\boldsymbol{R}$ of $k \times t$ sparse constituent matrices over GF(q) (i.e., the member matrices in $\boldsymbol{R}$ are chosen at random). Random replacement of the 1-entries of the SP-base matrix $\mathbf{B}_{sp}$ by the constituent matrices in $\boldsymbol{R}$ would result in an ensemble, denoted by $\xi_{q,sp}(k,t)$, of sparse arrays. The null spaces of sparse arrays in the ensemble $\xi_{q,sp}(k,t)$ give a code ensemble, denoted by $\mathbf{E}_{q,sp}(k,t)$, of q-ary SP-LDPC codes of rate R_c. When $k = t$, the ensemble $\mathbf{E}_{q,sp}(k,t)$ of q-ary SP-LDPC codes of rate R_c defined above can be viewed as an ensemble of q-ary PTG-LDPC codes of rate R_c.

11.2 SP-Construction of NB QC-LDPC Codes

Consider the special case for which $k = t$. Suppose each member matrix in the replacement set $\boldsymbol{R}$ is a circulant over GF(q) of size $k \times k$. Then, the array $\mathbf{H}_{q,sp,qc}(k,k) = [\mathbf{Q}_{i,j}]_{0 \le i < m, 0 \le j < n}$ constructed by replacing each 1-entry in $\mathbf{B}_{sp}$ by a member circulant in $\boldsymbol{R}$ and each 0-entry by a $k \times k$ ZM is an $m \times n$ array of circulants over GF(q) and/or ZMs of size $k \times k$, and its null space gives a q-ary QC-SP-LDPC code $C_{q,sp,qc}$. From the PTG point of view, $C_{q,sp,qc}$ is equivalent to a q-ary QC-PTG-LDPC code.

If we permute the rows and columns of $\mathbf{H}_{q,sp,qc}(k,k)$ based on the inverses of the row and column permutations, π_{row} and π_{col}, defined by (4.3) and (4.5), respectively, we obtain the following $k \times k$ array $\mathbf{H}_{q,sp,cyc}(m,n)$ of matrices of size $m \times n$ over GF(q) in the form of (4.1) with a block-cyclic structure:

$$\mathbf{H}_{q,sp,cyc}(m,n) = \begin{bmatrix} \mathbf{D}_0 & \mathbf{D}_1 & \cdots & \mathbf{D}_{k-1} \\ \mathbf{D}_{k-1} & \mathbf{D}_0 & \cdots & \mathbf{D}_{k-2} \\ \vdots & \vdots & \ddots & \vdots \\ \mathbf{D}_1 & \mathbf{D}_2 & \cdots & \mathbf{D}_0 \end{bmatrix}, \tag{11.1}$$

in which each constituent matrix $\mathbf{D}_i$ is an $m \times n$ matrix over GF(q). The null space of $\mathbf{H}_{q,sp,cyc}(m,n)$ gives a q-ary QC-LDPC code $C_{q,sp,cyc}$ with a block-cyclic structure.

The top (generator) row-block of $\mathbf{H}_{q,sp,cyc}(m,n)$, $[\mathbf{D}_0, \mathbf{D}_1, \ldots, \mathbf{D}_{k-1}]$, consists of k constituent matrices of size $m \times n$ over GF(q). Suppose we combine the k matrices, $\mathbf{D}_0, \mathbf{D}_1, \ldots, \mathbf{D}_{k-1}$, into a single $m \times n$ matrix, denoted by $\mathbf{B}_{q,ptg}$, in which, for $0 \leq i < m$ and $0 \leq j < n$, the entry at location (i, j) is a k-tuple $\mathbf{g}_{i,j} = (g_{i,j,0}, g_{i,j,1}, \ldots, g_{i,j,k-1})$ over GF(q) where, for $0 \leq e < k$, $g_{i,j,e}$ is the entry at the location (i, j) of the matrix $\mathbf{D}_e$. Hence, $\mathbf{B}_{q,ptg}$ is a matrix whose entries are k-tuples over GF(q). The Tanner graph $\mathcal{G}_{q,ptg}$ associated with $\mathbf{B}_{q,ptg}$ consists of n VNs and m CNs. The edge connecting the j-th VN of $\mathcal{G}_{q,ptg}$ to the i-th CN of $\mathcal{G}_{q,ptg}$ is labeled by the k-tuple $\mathbf{g}_{i,j} = (g_{i,j,0}, g_{i,j,1}, \ldots, g_{i,j,k-1})$ over GF(q). The number of nonzero components in $\mathbf{g}_{i,j}$ is the number of parallel edges that connect the j-th VN to the i-th CN of $\mathcal{G}_{q,ptg}$. If $\mathbf{g}_{i,j}$ is an all-zero k-tuple, i.e., all the components in $\mathbf{g}_{i,j}$ are zeros, then the j-th VN and the i-th CN of $\mathcal{G}_{q,ptg}$ are not connected. Following on from the replacements of the 1-entries in $\mathbf{B}_{sp}$ by the member matrices in $\boldsymbol{R}$ and the column and row permutations performed on $\mathbf{H}_{q,sp,qc}(k,k)$, we see that the vector label $\mathbf{g}_{i,j}$ for the edge (j,i) in $\mathcal{G}_{q,ptg}$ is simply the generator of the constituent circulant $\mathbf{Q}_{i,j}$ in $\mathbf{H}_{q,sp,qc}(k,k)$.

For the PTG-based construction of a q-ary QC-PTG-LDPC code $C_{q,ptg,qc}$, the graph $\mathcal{G}_{q,ptg}$ can be used as the protograph. Suppose we expand $\mathcal{G}_{q,ptg}$ by a factor of k. In the construction, we take k copies of $\mathcal{G}_{q,ptg}$ and connect the copies of $\mathcal{G}_{q,ptg}$ by permuting the edges of individual copies of $\mathcal{G}_{q,ptg}$ among the k copies of $\mathcal{G}_{q,ptg}$. The k VNs of type-j are connected to the k CNs of type-i based on the vector label $\mathbf{g}_{i,j} = (g_{i,j,0}, g_{i,j,1}, \ldots, g_{i,j,k-1})$ and their cyclic shifts for the edge (j,i) in $\mathcal{G}_{q,ptg}$. The number of nonzero components in $\mathbf{g}_{i,j}$ is equal to the number of parallel edges connecting the j-th VN and i-th CN of the protograph $\mathcal{G}_{q,ptg}$. The expansion of $\mathcal{G}_{q,ptg}$ gives a bipartite graph $\mathcal{G}_{q,ptg}(k,k)$ with nk VNs and mk CNs in which each edge is labeled with a nonzero element in GF(q). The null space of the adjacency matrix $\mathbf{H}_{q,ptg,qc}(k,k)$ of $\mathcal{G}_{q,ptg}(k,k)$ gives a q-ary QC-PTG-LDPC code $C_{q,ptg,qc}$ which is the same as the q-ary QC-SP-LDPC code $C_{q,sp,qc}$ specified by the q-ary array $\mathbf{H}_{q,sp,qc}(k,k)$ given above. This shows that for the special case $k = t$, a q-ary QC-SP-LDPC code $C_{q,sp,qc}$ constructed by the SP-construction is a q-ary QC-PTG-LDPC code $C_{q,ptg,qc}$ constructed by the PTG-based construction.

In the following, we use a simple example to illustrate the SP-construction of an NB QC-SP-LDPC code.

Example 11.1. Suppose we choose the following binary matrix as the base matrix $\mathbf{B}_{sp}$ for the SP-construction of an NB QC-SP-LDPC code:

$$\mathbf{B}_{sp} = \begin{bmatrix} 1 & 1 & 0 \\ 1 & 0 & 1 \end{bmatrix}.$$

Let α be a primitive element of GF(2^2) constructed based on the primitive polynomial $1+X+X^2$ over GF(2). The field GF(2^2) consists of four elements, 0, 1, α, and α^2. Let $\boldsymbol{R}$ be the replacement set for the SP-construction that consists of the following four circulants of size 3×3 over GF(2^2):

$$\mathbf{Q}_1 = \begin{bmatrix} 1 & \alpha & 0 \\ 0 & 1 & \alpha \\ \alpha & 0 & 1 \end{bmatrix}, \qquad \mathbf{Q}_2 = \begin{bmatrix} 0 & \alpha & 0 \\ 0 & 0 & \alpha \\ \alpha & 0 & 0 \end{bmatrix},$$

$$\mathbf{Q}_3 = \begin{bmatrix} 0 & 0 & \alpha^2 \\ \alpha^2 & 0 & 0 \\ 0 & \alpha^2 & 0 \end{bmatrix}, \qquad \mathbf{Q}_4 = \begin{bmatrix} \alpha & 0 & 0 \\ 0 & \alpha & 0 \\ 0 & 0 & \alpha \end{bmatrix}.$$

By checking, we find that the SP-base matrix $\mathbf{B}_{sp}$ and the member matrices in the replacement set $\boldsymbol{R}$ satisfy the RC- and the PW-RC-constraints, respectively.

Construct a 2×3 array $\mathbf{H}_{q,sp,qc}(3,3)$ by replacing the 1-entries of $\mathbf{B}_{sp}$ with the member circulants in $\boldsymbol{R}$ and the 0-entries by 3×3 ZMs as follows:

$$\mathbf{H}_{q,sp,qc}(3,3) = \begin{bmatrix} \mathbf{Q}_1 & \mathbf{Q}_2 & \mathbf{0} \\ \mathbf{Q}_3 & \mathbf{0} & \mathbf{Q}_4 \end{bmatrix} = \left[\begin{array}{ccc|ccc|ccc} 1 & \alpha & 0 & 0 & \alpha & 0 & 0 & 0 & 0 \\ 0 & 1 & \alpha & 0 & 0 & \alpha & 0 & 0 & 0 \\ \alpha & 0 & 1 & \alpha & 0 & 0 & 0 & 0 & 0 \\ \hline 0 & 0 & \alpha^2 & 0 & 0 & 0 & \alpha & 0 & 0 \\ \alpha^2 & 0 & 0 & 0 & 0 & 0 & 0 & \alpha & 0 \\ 0 & \alpha^2 & 0 & 0 & 0 & 0 & 0 & 0 & \alpha \end{array}\right].$$

The array $\mathbf{H}_{q,sp,qc}(3,3)$, a 6×9 matrix over GF(2^2), satisfies the RC-constraint. The null space of $\mathbf{H}_{q,sp,qc}(3,3)$ gives a 4-ary (9, 3) QC-SP-LDPC code $C_{q,sp,qc}$ with a section-wise cyclic structure. (Note that there are several other replacements of the 1-entries in the base matrix $\mathbf{B}_{sp}$ by the member circulants of $\boldsymbol{R}$ which result in RC-constrained arrays.)

Permuting the rows and the columns of $\mathbf{H}_{q,sp,qc}(3,3)$ using the inverses of the permutations π_{row} and π_{col} defined by (4.3) and (4.5), respectively, we obtain the following 3×3 array $\mathbf{H}_{q,sp,cyc}(2,3)$ of 2×3 submatrices over GF(2^2):

$$\mathbf{H}_{q,sp,cyc}(2,3) = \left[\begin{array}{ccc|ccc|ccc} 1 & 0 & 0 & \alpha & \alpha & 0 & 0 & 0 & 0 \\ 0 & 0 & \alpha & 0 & 0 & 0 & \alpha^2 & 0 & 0 \\ \hline 0 & 0 & 0 & 1 & 0 & 0 & \alpha & \alpha & 0 \\ \alpha^2 & 0 & 0 & 0 & 0 & \alpha & 0 & 0 & 0 \\ \hline \alpha & \alpha & 0 & 0 & 0 & 0 & 1 & 0 & 0 \\ 0 & 0 & 0 & \alpha^2 & 0 & 0 & 0 & 0 & \alpha \end{array}\right].$$

The null space of the array $\mathbf{H}_{q,sp,cyc}(2,3)$ gives a 4-ary $(9,3)$ QC-SP-LDPC code $C_{q,sp,cyc}$ with a block-cyclic structure. Every codeword $\mathbf{v}$ in $C_{q,sp,cyc}$ consists of three sections, each with three code symbols over GF(2^2). If we shift the codeword $\mathbf{v}$ one section (three code symbols) to the right, we obtain another codeword in $C_{q,sp,cyc}$. The code $C_{q,sp,cyc}$ is equivalent to $C_{q,sp,qc}$ in the sense that the codewords in the former are just the permutations of the codewords in the latter. By checking, we find that $\mathbf{H}_{q,sp,cyc}(2,3)$ satisfies the RC-constraint and the Tanner graph of $C_{q,sp,cyc}$ has girth 6.

The first row-block of $\mathbf{H}_{q,sp,cyc}(2,3)$ consists of three 2×3 constituent matrices over GF(2^2) which are:

$$\mathbf{D}_0 = \begin{bmatrix} 1 & 0 & 0 \\ 0 & 0 & \alpha \end{bmatrix}, \mathbf{D}_1 = \begin{bmatrix} \alpha & \alpha & 0 \\ 0 & 0 & 0 \end{bmatrix}, \mathbf{D}_2 = \begin{bmatrix} 0 & 0 & 0 \\ \alpha^2 & 0 & 0 \end{bmatrix}.$$

Combining the above three 2×3 matrices by grouping their corresponding entries into 3-tuples over GF(2^2), we obtain the following 2×3 matrix $\mathbf{B}_{q,ptg}$ in which each entry is a 3-tuple over GF(2^2):

$$\mathbf{B}_{q,ptg} = \begin{bmatrix} (1,\alpha,0) & (0,\alpha,0) & (0,0,0) \\ (0,0,\alpha^2) & (0,0,0) & (\alpha,0,0) \end{bmatrix}.$$

It is easy to check that a vector entry of $\mathbf{B}_{q,ptg}$ is the generator of a constituent circulant in the 2×3 array $\mathbf{H}_{q,sp,qc}(3,3)$.

The Tanner graphs $\mathcal{G}_{sp}$ and $\mathcal{G}_{q,ptg}$ of $\mathbf{B}_{sp}$ and $\mathbf{B}_{q,ptg}$ are shown in Fig. 11.1(a) and Fig. 11.1(b), respectively.

If we use $\mathcal{G}_{q,ptg}$ as the protograph, take three copies, and permute the edges among the copies based on their vector labels, we obtain a connected Tanner graph $\mathcal{G}_{q,ptg}(3,3)$ with 9 VNs and 6 CNs as shown in Fig. 11.1(c). The null space of the adjacency matrix of $\mathcal{G}_{q,ptg}(3,3)$ gives a 4-ary QC-PTG-LDPC code which is exactly the 4-ary QC-SP-LDPC code $C_{q,sp,qc}$ given by the null space of $\mathbf{H}_{q,sp,qc}(3,3)$.

In the following, we construct a short NB QC-SP-LDPC code and show that it performs well.

Example 11.2. Consider the following 2×4 base matrix for the SP-construction of an NB QC-SP-LDPC code:

$$\mathbf{B}_{sp} = \begin{bmatrix} 1 & 1 & 1 & 1 \\ 1 & 1 & 1 & 1 \end{bmatrix}.$$

Design a replacement set $\boldsymbol{R}$ that consists of eight weight-1 circulants of size 8×8 over GF(2^8), denoted by $\mathbf{Q}_0, \mathbf{Q}_1, \ldots, \mathbf{Q}_7$. Let α be a primitive element of

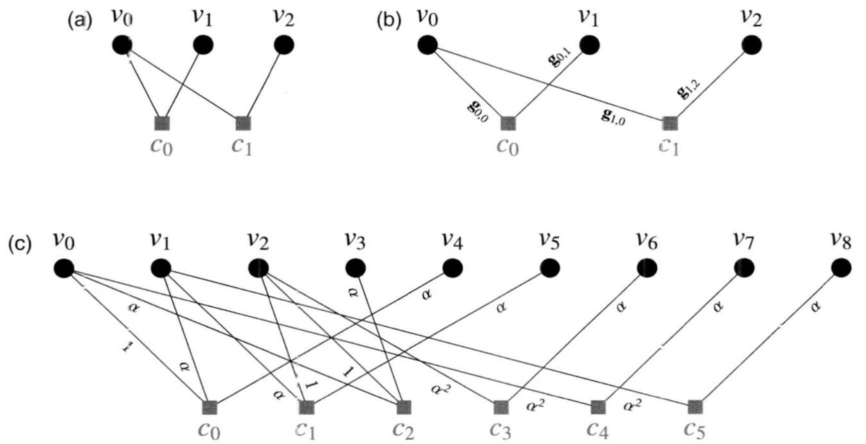

Figure 11.1. (a) The protograph $\mathcal{G}_{sp}$ associated with the base matrix $\mathbf{B}_{sp}$ given in Example 11.1; (b) the Tanner graph $\mathcal{G}_{q,ptg}$ associated with the base matrix $\mathbf{B}_{q,ptg}$ given in Example 11.1, where $\mathbf{g}_{0,0} = (1, \alpha, 0)$, $\mathbf{g}_{0,1} = (0, \alpha, 0)$, $\mathbf{g}_{1,0} = (0, 0, \alpha^2)$, and $\mathbf{g}_{1,2} = (\alpha, 0, 0)$; and (c) the Tanner graph $\mathcal{G}_{q,ptg}(3, 3)$ given in Example 11.1.

GF(2^8). The generators of these circulants over GF(2^8) in $\boldsymbol{R}$ are:

$$
\begin{aligned}
\mathbf{g}_0 &= (0,0,0,0,0,0,\alpha^{61},0), & \mathbf{g}_1 &= (0,0,0,0,0,\alpha^{106},0,0),\\
\mathbf{g}_2 &= (0,0,0,\alpha^{240},0,0,0,0), & \mathbf{g}_3 &= (0,0,0,0,0,0,0,\alpha^{125}),\\
\mathbf{g}_4 &= (0,0,0,\alpha^{229},0,0,0,0), & \mathbf{g}_5 &= (0,0,0,0,0,0,0,\alpha^{94}),\\
\mathbf{g}_6 &= (0,0,0,0,0,0,\alpha^{199},0), & \mathbf{g}_7 &= (0,0,0,0,0,\alpha^{99},0,0).
\end{aligned} \tag{11.2}
$$

These 8 member circulants satisfy both the RC- and the PW-RC-constraints. Replacing the 1-entries in $\mathbf{B}_{sp}$ by the member matrices in $\boldsymbol{R}$ according to the replacement constraint given in Section 5.1, we obtain a 2×4 array of circulants of size 8×8 over GF(2^8). There are many possible replacements that result in codes whose Tanner graphs have girth at least 6. One such replacement gives the following 2×4 array $\mathbf{H}_{q,sp,qc}(8, 8)$ of weight-1 circulants:

$$
\mathbf{H}_{q,sp,qc}(8,8) = \begin{bmatrix} \mathbf{Q}_0 & \mathbf{Q}_1 & \mathbf{Q}_2 & \mathbf{Q}_3 \\ \mathbf{Q}_4 & \mathbf{Q}_5 & \mathbf{Q}_6 & \mathbf{Q}_7 \end{bmatrix}.
$$

The array $\mathbf{H}_{q,sp,qc}(8, 8)$ is a 16×32 matrix over GF(2^8) with constant column and row weights 2 and 4, respectively, and its null space gives a $(2, 4)$-regular 256-ary (32, 16) QC-SP-LDPC code $C_{q,sp,qc}$. Using the cycle counting algorithm presented in [71], we find that the Tanner graph of the code $C_{q,sp,qc}$ has girth 8 and contains 20 cycles of length 8, no cycle of length 10, and 160 cycles of length 12.

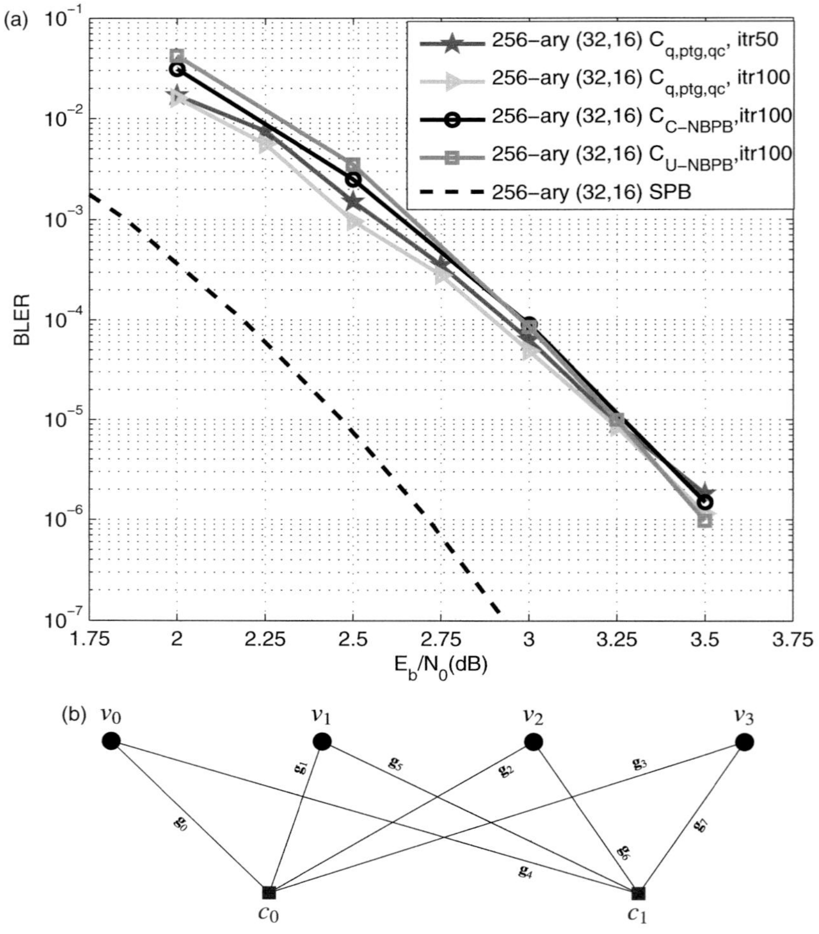

Figure 11.2. (a) The BLER performances of the 256-ary (32, 16) QC-SP-LDPC code $C_{q,sp,qc}$ given in Example 11.2 and the PTG-based codes $C_{C\text{-}NBPB}$ and $C_{U\text{-}NBPB}$ reported in [15, 36] over the binary-input AWGNC; and (b) the protograph associated with the base matrix $\mathbf{B}_{q,ptg}$ in Example 11.2.

The BLER performances of the code $C_{q,sp,qc}$ decoded with 50 and 100 iterations of the FFT-QSPA [5] (see Appendix C) are shown in Fig. 11.2(a). This code was simulated on the binary-input AWGNC. Each 256-ary code symbol is represented by a binary 8-tuple. We see that the decoding converges quickly. The gap between 50 and 100 iterations is less than 0.05 dB. At a BLER around 10^{-6}, we see that the code $C_{q,sp,qc}$ performs 0.75 dB away from the SPB.

For comparison, we also include in Fig. 11.2(b) the BLER performances of two PTG-based 256-ary LDPC codes reported in [15, 36] of the same length and rate as the code constructed above, which we denote by $C_{C\text{-}NBPB}$ and

$C_{U\text{-}NBPB}$, respectively. The subscripts "*C-NBPB*" and "*U-NBPB*" stand for "*constrained NB PTG-based*" and "*unconstrained NB PTG-based*," respectively. The code $C_{C\text{-}NBPB}$ was constructed through scale-copy-permute operations based on a scaled protograph, and the code $C_{U\text{-}NBPB}$ was constructed through copy-scale-permute operations based on an unlabeled protograph. The Tanner graphs of these two codes have the same cycle distributions and contain the same number of cycles of length 8, 10, and 12 as the code $C_{q,sp,qc}$ constructed above. The code $C_{C\text{-}NBPB}$ is a QC code, but the code $C_{U\text{-}NBPB}$ is not. Both $C_{C\text{-}NBPB}$ and $C_{U\text{-}NBPB}$ were decoded with 100 iterations of FFT-QSPA in [15, 36]. From Fig. 11.2(a), we see that, among the three 256-ary (32, 16) LDPC codes, the QC-SP-LDPC code $C_{q,sp,qc}$ slightly outperforms the codes $C_{C\text{-}NBPB}$ and $C_{U\text{-}NBPB}$.

From the PTG point of view, the 256-ary (32, 16) QC-SP-LDPC code constructed above can be viewed as a QC-PTG-LDPC code with the protograph shown in Fig. 11.2(b) whose associated base matrix is given below:

$$\mathbf{B}_{q,ptg} = \begin{bmatrix} \mathbf{g}_0 & \mathbf{g}_1 & \mathbf{g}_2 & \mathbf{g}_3 \\ \mathbf{g}_4 & \mathbf{g}_5 & \mathbf{g}_6 & \mathbf{g}_7 \end{bmatrix}$$

in which the entries are 8-tuples over GF(q) given by (11.2).

11.3 Construction of NB QC-SP-LDPC Codes Using q-ary CPM-Dispersion

A weight-1 circulant over GF(q) is called a q-ary CPM. A q-ary CPM $\mathbf{Q}$ can be obtained from a binary CPM $\mathbf{A}$ by replacing all the 1-entries in $\mathbf{A}$ by the same nonzero element chosen from GF(q). Two q-ary CPMs of the same size are mutually disjoint if the single nonzero components of their generators are located at two different positions.

In the following, we present the SP-construction of NB QC-SP-LDPC codes using the CPM-dispersion construction over an NB field. The presentation is in analogous to that of the binary case given Section 7.1.

Let α be a primitive element of the NB field GF(q). Then, the $q-1$ powers of α, namely, $\alpha^0 = 1, \alpha, \alpha^2, \ldots, \alpha^{q-2}$, give all the nonzero elements of GF(q). For $0 \leq l < q-1$, we represent the element α^l by a q-ary CPM of size $(q-1) \times (q-1)$ (with both rows and columns labeled from 0 to $q-2$) whose generator has α^l as its single nonzero component at position l. We denote this q-ary CPM by $\mathbf{Q}(\alpha^l)$. All the nonzero entries in $\mathbf{Q}(\alpha^l)$ are α^l. It is clear that the representation of the element α^l by the q-ary CPM $\mathbf{Q}(\alpha^l)$ of size $(q-1) \times (q-1)$ is unique and the mapping between α^l and $\mathbf{Q}(\alpha^l)$ is one-to-one. This matrix representation of α^l

is referred to as the *q-ary CPM-dispersion* of α^l. Note that if we replace all the nonzero entries in $\mathbf{Q}(\alpha^l)$ by the unit element "1" of GF(q), we obtain a binary CPM of size $(q-1)\times(q-1)$.

Let $\mathbf{B}_{q,sp} = [b_{i,j}]_{0\le i<m, 0\le j<n}$ be an $m\times n$ matrix over GF(q). Now, we use this NB matrix as the base matrix for the SP-construction of an NB QC-SP-LDPC code. We form an $m\times n$ array $\mathbf{H}_{q,sp,qc}(q-1,q-1)$ of q-ary CPMs and/or ZMs of size $(q-1)\times(q-1)$ from $\mathbf{B}_{q,sp}$ as follows: (1) If $b_{i,j}\neq 0$ and $b_{i,j}=\alpha^l$, we replace $b_{i,j}$ with its q-ary CPM-dispersion $\mathbf{Q}(\alpha^l)$; and (2) If $b_{i,j}=0$, we replace $b_{i,j}$ by a ZM of size $(q-1)\times(q-1)$, i.e., $\mathbf{Q}(0)$ is a $(q-1)\times(q-1)$ ZM. The array $\mathbf{H}_{q,sp,qc}(q-1,q-1)$ is called the *q-ary CPM-dispersion* of the base matrix $\mathbf{B}_{q,sp}$. The null space of $\mathbf{H}_{q,sp,qc}(q-1,q-1)$ gives a q-ary QC-SP-LDPC code $C_{q,sp,qc}$ of length $n(q-1)$, called a q-ary CPM-QC-SP-LDPC code. The above method for constructing q-ary CPM-QC-SP-LDPC code is called the *q-ary CPM-dispersion (CPM-D) construction*, which is a *deterministic* NB SP-construction.

The above CPM-D construction of a q-ary CPM-QC-SP-LDPC code reduces to the construction of its base matrix $\mathbf{B}_{q,sp}$ over GF(q). For such a q-ary CPM-QC-SP-LDPC code, the girth of its Tanner graph is related to the choice of the base matrix $\mathbf{B}_{q,sp}$. In Section 7.2, two theorems (Theorems 7.1 and 7.2) were given, which give the necessary and sufficient conditions on the base matrix for the Tanner graph of a binary CPM-QC-SP-LDPC code to have girth at least 6 and 8. These two theorems also apply to q-ary CPM-QC-SP-LDPC codes constructed using q-ary CPM-D construction.

In the CPM-D construction of a q-ary CPM-QC-SP-LDPC code, the 2×2 and the 3×3 SM-constraints given in Theorems 7.1 and 7.2 are used to construct a base matrix. All the methods presented in Chapter 7 can be applied to construct NB SP-base matrices.

Example 11.3. Consider the field GF(2^6). Let α be a primitive element of GF(2^6). Using this field, we can construct a 63×63 matrix $\mathbf{B}^*_{q,sp,p}$ over GF(2^6) in the form given by (8.1) which satisfies the 2×2 SM-constraint. Take the following 4×8 submatrix $\mathbf{B}_{q,sp}$ (avoiding the 0-entry) from $\mathbf{B}^*_{q,sp,p}$ as an SP-base matrix for the code construction:

$$\mathbf{B}_{q,sp}=\begin{bmatrix} \alpha^{35} & \alpha^{52} & \alpha^{23} & \alpha^{33} & \alpha^{47} & \alpha^{27} & \alpha^{56} & \alpha^{59} \\ \alpha^{2} & \alpha^{35} & \alpha^{52} & \alpha^{23} & \alpha^{33} & \alpha^{47} & \alpha^{27} & \alpha^{56} \\ \alpha^{25} & \alpha^{2} & \alpha^{35} & \alpha^{52} & \alpha^{23} & \alpha^{33} & \alpha^{47} & \alpha^{27} \\ \alpha^{61} & \alpha^{25} & \alpha^{2} & \alpha^{35} & \alpha^{52} & \alpha^{23} & \alpha^{33} & \alpha^{47} \end{bmatrix}. \tag{11.3}$$

Next, we mask $\mathbf{B}_{q,sp}$ with the 4×8 masking matrix $\mathbf{Z}$ given by (7.5) (or (4.7)), i.e.,

$$\mathbf{Z} = \begin{bmatrix} 1 & 0 & 1 & 0 & 1 & 1 & 1 & 1 \\ 0 & 1 & 0 & 1 & 1 & 1 & 1 & 1 \\ 1 & 1 & 1 & 1 & 1 & 0 & 1 & 0 \\ 1 & 1 & 1 & 1 & 0 & 1 & 0 & 1 \end{bmatrix}.$$

After masking, we obtain a 4×8 masked base matrix:

$$\mathbf{B}_{q,sp\ mask} = \begin{bmatrix} \alpha^{35} & 0 & \alpha^{23} & 0 & \alpha^{47} & \alpha^{27} & \alpha^{56} & \alpha^{59} \\ 0 & \alpha^{35} & 0 & \alpha^{23} & \alpha^{33} & \alpha^{47} & \alpha^{27} & \alpha^{56} \\ \alpha^{25} & \alpha^{2} & \alpha^{35} & \alpha^{52} & \alpha^{23} & 0 & \alpha^{47} & 0 \\ \alpha^{61} & \alpha^{25} & \alpha^{2} & \alpha^{35} & 0 & \alpha^{23} & 0 & \alpha^{47} \end{bmatrix}, \qquad (11.4)$$

which has row and column weights 3 and 6, respectively. The masked base matrix $\mathbf{B}_{q,sp,mask}$ satisfies the 3×3 SM-constraint. The 64-ary CPM-dispersion of $\mathbf{B}_{q,sp,mask}$ gives a 4×8 array $\mathbf{H}_{q,sp,qc,mask}(63, 63)$ of 64-ary CPMs and ZMs of size 63×63 which is a 252×504 matrix over GF(2^6) with column weight 3 and row weight 6. The null space of the array $\mathbf{H}_{q,sp,qc,mask}(63, 63)$ gives a 64-ary $(3, 6)$-regular (504, 252) CPM-QC-SP-LDPC code $C_{q,sp,qc,mask}$. The Tanner graph of $C_{q,qc,mask}$ has girth 8 and contains 819 cycles of length 8 and 12 348 cycles of length 10.

The BLER performance of the code $C_{q,sp,qc,mask}$ decoded with 50 iterations of the FFT-QSPA over the binary-input AWGNC is shown in Fig. 11.3. We see that at a BLER of 10^{-6}, $C_{q,sp,qc,mask}$ performs about 1.6 dB away from the SPB. Also included in this figure is the BLER performance of a 64-ary LDPC code $C_{q,peg}$ constructed based on the PEG algorithm, which is not QC. The code $C_{q,peg}$ is an irregular code which has column weight 3 and average row weight 6. Its Tanner graph also has girth 8 and contains 945 cycles of length 8 and 11 655 cycles of length 10. We decode this PEG code using the FFT-QSPA with 50 iterations. Fig. 11.3 shows that the code $C_{q,sp,qc,mask}$ performs just as well as the PEG code $C_{q,peg}$.

11.4 CPM-D Construction of NB CPM-QC-SP-LDPC Codes Using Binary-to-Nonbinary Replacement

Using the CPM-D construction method to construct NB CPM-QC-SP-LDPC codes directly as presented above, there is a constraint on the size of CPMs. If we desire a large CPM size, we have to use a large field. This will result in high decoding complexity. To resolve this problem, we can combine the CPM-D

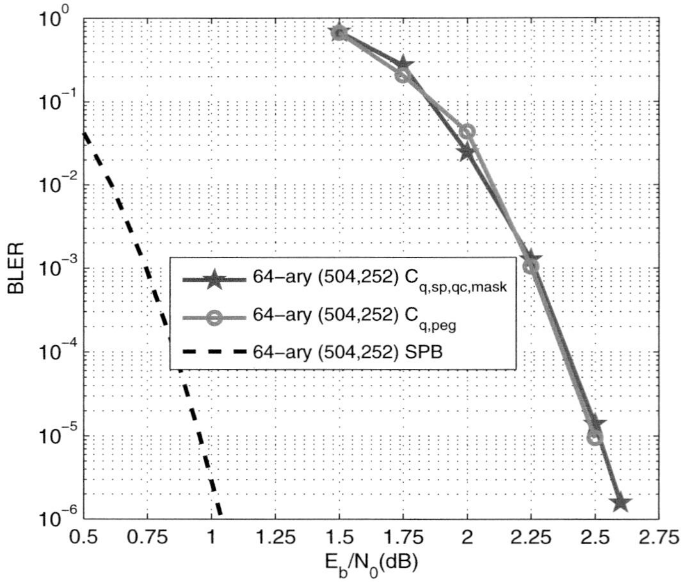

Figure 11.3. The BLER performances of the 64-ary $(504, 252)$ CPM-SP-QC-LDPC code $C_{q,sp,qc,mask}$ and the PEG code $C_{q,peg}$ given in Example 11.3 over the binary-input AWGNC.

method with a binary-to-NB (B-to-NB) replacement method to construct NB CPM-QC-SP-LDPC codes whose parity-check arrays consist of large CPMs over small fields.

First, a chosen base matrix $\mathbf{B}_{q,sp} = [b_{i,j}]_{0\leq i<m, 0\leq j<n}$ over GF(q) is dispersed into an $m \times n$ array $\mathbf{H}_{b,sp,qc}(q-1, q-1) = [\mathbf{A}_{i,j}]_{0\leq i<m, 0\leq j<n}$ of *binary* CPMs and/or ZMs of size $(q-1) \times (q-1)$. Then, for $0 \leq i < m, 0 \leq j < n$, we replace all the 1-entries in the binary CPM $\mathbf{A}_{i,j}$ by a nonzero element from a chosen *code symbol field* GF($q^\star$). The field GF($q^\star$) for code symbols can be smaller or larger than the dispersion field GF(q), or the same. This *binary to $q^\star$-ary replacement operation* transforms the binary CPM $\mathbf{A}_{i,j}$ into a $q^\star$-ary CPM $\mathbf{Q}_{i,j}$. Applying this binary to $q^\star$-ary replacement to each binary CPM in the array $\mathbf{H}_{b,sp,qc}(q-1, q-1)$, we obtain an $m \times n$ array $\mathbf{H}_{q^\star,sp,qc}(q-1, q-1) = [\mathbf{Q}_{i,j}]_{0\leq i<m, 0\leq j<n}$ of $q^\star$-ary CPMs and/or ZMs of size $(q-1) \times (q-1)$. The null space of $\mathbf{H}_{q^\star,sp,qc}(q-1, q-1)$ gives a $q^\star$-ary CPM-QC-SP-LDPC code $C_{q^\star,sp,qc}$.

Using $\mathbf{H}_{b,sp,qc}(q-1, q-1)$ as the *binary base array* and the B-to-NB replacement for the code construction, we can construct an NB CPM-QC-SP-LDPC code over any finite field, large or small. For simplicity, we call this combined construction method the *CPM-bD/B-to-NB construction*, where the symbol "bD" denotes "binary dispersion." Using the CPM-bD/B-to-NB construction method, we can construct a base array of binary CPMs of large size and then use a small code symbol field as the B-to-NB replacement field to construct a parity-check array of NB CPMs over a small field. The replacement of the 1-entries of a binary CPM by a nonzero element from another field may be considered as another superposition operation. Hence, the above construction of an NB CPM-QC-SP-LDPC code may be regarded as a *double SP-construction*. In fact, if we disperse each nonzero entry in $\mathbf{H}_{q^\star,sp,qc}(q-1,q-1)$ into either a binary or a $q^\star$-ary CPM of size $(q^\star-1)\times(q^\star-1)$, we obtain a larger array of CPMs and/or ZMs of size $(q^\star-1)\times(q^\star-1)$. The CPM-$b$D/B-to-NB construction method is very flexible for constructing both binary and NB CPM-QC-SP-LDPC codes.

Example 11.4. In this example, we use the CPM-bD/B-to-NB construction method to construct an 8-ary (2376, 1188) CPM-QC-SP-LDPC code. Based on the prime field GF(199), we can construct a 198×198 base matrix $\mathbf{B}^*_{q,sp,p}$ over GF(199) in the form given by (8.1) which satisfies the 2×2 SM-constraint. We take a 6×12 submatrix $\mathbf{B}_{q,sp}$ from $\mathbf{B}^*_{q,sp,p}$ (avoiding the 0-entries):

$$\mathbf{B}_{q,sp} =$$

$$\begin{bmatrix} \alpha^{184} & \alpha^{85} & \alpha^{71} & \alpha^{133} & \alpha^{13} & \alpha^{162} & \alpha^{140} & \alpha^{94} & \alpha^{79} & \alpha^{192} & \alpha^{42} & \alpha^{178} \\ \alpha^{98} & \alpha^{101} & \alpha^{48} & \alpha^{67} & \alpha^{130} & \alpha^{83} & \alpha^{169} & \alpha^{185} & \alpha^{34} & \alpha^{57} & \alpha^{107} & \alpha^{30} \\ \alpha^{156} & \alpha^{46} & \alpha^{26} & \alpha^{102} & \alpha^{22} & \alpha^{40} & \alpha^{64} & \alpha^{193} & \alpha^{21} & \alpha^{28} & \alpha^{137} & \alpha^{108} \\ \alpha^{43} & \alpha^{51} & \alpha^{19} & \alpha^{177} & \alpha^{194} & \alpha^{93} & \alpha^{164} & \alpha^{120} & \alpha^{4} & \alpha^{139} & \alpha^{15} & \alpha^{180} \\ \alpha^{141} & \alpha^{153} & \alpha^{150} & \alpha^{12} & \alpha^{73} & \alpha^{17} & \alpha^{110} & \alpha^{52} & \alpha^{172} & \alpha^{82} & \alpha^{69} & \alpha^{183} \\ \alpha^{131} & \alpha^{76} & \alpha^{72} & \alpha^{152} & \alpha^{181} & \alpha^{32} & \alpha^{70} & \alpha^{186} & \alpha^{77} & \alpha^{119} & \alpha^{95} & \alpha^{127} \end{bmatrix}. \tag{11.5}$$

Since $\mathbf{B}^*_{q,sp,p}$ satisfies the 2×2 SM-constraint, $\mathbf{B}_{q,sp}$, as a submatrix of $\mathbf{B}^*_{q,sp,p}$, also satisfies the 2×2 SM-constraint. Next, we design a masking matrix to mask $\mathbf{B}_{q,sp}$ to obtain a masked base matrix that additionally satisfies the 3×3 SM-constraint. To achieve this, we use the PEG algorithm given in [44] to construct a 6×12 binary matrix $\mathbf{Z}$ with average column weight 2.5 as follows:

$$\mathbf{Z} = \begin{bmatrix} 1 & 0 & 1 & 0 & 1 & 0 & 1 & 1 & 0 & 0 & 0 & 0 \\ 0 & 1 & 0 & 1 & 0 & 1 & 1 & 1 & 0 & 0 & 0 & 0 \\ 1 & 0 & 1 & 0 & 0 & 1 & 0 & 0 & 1 & 1 & 0 & 0 \\ 0 & 1 & 0 & 1 & 1 & 0 & 0 & 0 & 1 & 1 & 0 & 0 \\ 1 & 0 & 0 & 1 & 1 & 0 & 0 & 0 & 0 & 0 & 1 & 1 \\ 0 & 1 & 1 & 0 & 0 & 1 & 0 & 0 & 0 & 0 & 1 & 1 \end{bmatrix}. \tag{11.6}$$

Note that every 3×3 submatrix of $\mathbf{Z}$ contains at least one zero entry. By masking $\mathbf{B}_{q,sp}$ with the masking matrix $\mathbf{Z}$ given by (11.6), we obtain the following 6×12

masked base matrix $\mathbf{B}_{q,sp,mask}$ in which every 3×3 submatrix contains at least one zero element:

$$\mathbf{B}_{q,sp,mask} =$$

$$\begin{bmatrix} \alpha^{184} & 0 & \alpha^{71} & 0 & \alpha^{13} & 0 & \alpha^{140} & \alpha^{94} & 0 & 0 & 0 & 0 \\ 0 & \alpha^{101} & 0 & \alpha^{67} & 0 & \alpha^{83} & \alpha^{169} & \alpha^{185} & 0 & 0 & 0 & 0 \\ \alpha^{156} & 0 & \alpha^{26} & 0 & 0 & \alpha^{40} & 0 & 0 & \alpha^{21} & \alpha^{28} & 0 & 0 \\ 0 & \alpha^{51} & 0 & \alpha^{177} & \alpha^{194} & 0 & 0 & 0 & \alpha^{4} & \alpha^{139} & 0 & 0 \\ \alpha^{141} & 0 & 0 & \alpha^{12} & \alpha^{73} & 0 & 0 & 0 & 0 & 0 & \alpha^{69} & \alpha^{183} \\ 0 & \alpha^{76} & \alpha^{72} & 0 & 0 & \alpha^{32} & 0 & 0 & 0 & 0 & \alpha^{95} & \alpha^{127} \end{bmatrix}. \tag{11.7}$$

By checking, we find that $\mathbf{B}_{q,sp,mask}$ satisfies both the 2×2 and the 3×3 SM-constraints. The binary CPM-dispersion of $\mathbf{B}_{q,sp,mask}$ gives a 6×12 array $\mathbf{H}_{b,sp,qc,mask}(198, 198)$ of binary CPMs and ZMs of size 198×198. This array $\mathbf{H}_{b,sp,qc,mask}(198, 198)$ is a 1188×2376 matrix over GF(2) with average column weight 2.5. Since the base matrix $\mathbf{B}_{q,sp,mask}$ satisfies the 2×2 and the 3×3 SM-constraints, it follows from Theorem 7.2 that the Tanner graph $\mathcal{G}_{b,sp,qc,mask}(198, 198)$ associated with the binary array $\mathbf{H}_{b,sp,qc,mask}(198, 198)$ has girth at least 8. Using the cycle counting algorithm presented in [71], we find that the Tanner graph $\mathcal{G}_{b,sp,qc,mask}(198, 198)$ has girth 8 and contains 396 cycles of length eight, 594 cycles of length 10, and 3762 cycles of length 12.

Suppose we use the array $\mathbf{H}_{b,sp,qc,mask}(198, 198)$ as the binary base array and choose the field GF(2^3) as the code symbol field for constructing an 8-ary CPM-QC-SP-LDPC code with the CPM-*b*D/B-to-NB construction. Replacing all the entries in each CPM in $\mathbf{H}_{b,sp,qc,mask}(198, 198)$ with a nonzero element which is chosen at random from GF(2^3), we obtain a 6×12 array $\mathbf{H}_{q^\star,sp,qc,mask}(198, 198)$ of 8-ary CPMs and ZMs of size 198×198. Each row-block of $\mathbf{H}_{q^\star,sp,qc,mask}(198, 198)$ consists of five CPMs and seven ZMs. The generators of the 8-ary CPMs in each row-block of $\mathbf{H}_{q^\star,sp,qc,mask}(198, 198)$ are given in Table 11.1, each specified by a pair (j, β^s) with $0 \leq j < 198$ and $0 \leq s < 7$. The first component j of (j, β^s) is a number which specifies the location of the single nonzero component of the generator of a CPM and the second component β^s gives the value of this single nonzero component which is a nonzero element of GF(2^3) where β is a primitive element of GF(2^3).

The null space over GF(2^3) of the array $\mathbf{H}_{q^\star,sp,qc,mask}(198, 198)$ gives an 8-ary (2376, 1188) CPM-QC-SP-LDPC code $C_{q^\star,sp,qc,mask}$ with rate 0.5. Its Tanner graph $\mathcal{G}_{q^\star,sp,qc,mask}(198, 198)$ has the same girth and cycle distributions as the Tanner graph $\mathcal{G}_{b,sp,qc,mask}(198, 198)$ associated with the binary array $\mathbf{H}_{b,sp,qc,mask}(198, 198)$. The BLER performance of $C_{q^\star,sp,qc,mask}$ decoded with

Table 11.1. *The generators of the 8-ary CPMs of the parity-check arrays* $\mathbf{H}_{q^\star,sp,qc,mask}(198,198)$ *and* $\mathbf{H}^\star_{q^\star,sp,qc,mask}(198,198)$ *given in Examples 11.4 and 11.5, respectively*

	Row	Generators of CPMs				
$\mathbf{H}_{q^\star,sp,qc,mask}$	0	$(184,\beta^0)$	$(71,\beta^6)$	$(13,\beta^4)$	$(140,\beta^0)$	$(94,\beta^1)$
	1	$(101,\beta^3)$	$(67,\beta^6)$	$(83,\beta^6)$	$(169,\beta^1)$	$(185,\beta^6)$
	2	$(156,\beta^6)$	$(26,\beta^3)$	$(40,\beta^5)$	$(21,\beta^0)$	$(28,\beta^2)$
	3	$(51,\beta^6)$	$(177,\beta^4)$	$(194,\beta^6)$	$(4,\beta^4)$	$(139,\beta^0)$
	4	$(141,\beta^5)$	$(12,\beta^6)$	$(73,\beta^4)$	$(69,\beta^5)$	$(183,\beta^5)$
	5	$(76,\beta^2)$	$(72,\beta^4)$	$(32,\beta^1)$	$(95,\beta^4)$	$(127,\beta^0)$
$\mathbf{H}^\star_{q^\star,sp,qc,mask}$	0	$(184,\beta^5)$	$(71,\beta^1)$	$(13,\beta^3)$	$(140,\beta^4)$	$(94,\beta^6)$
	1	$(48,\beta^6)$	$(130,\beta^3)$	$(169,\beta^0)$	$(185,\beta^1)$	$(34,\beta^1)$
	2	$(26,\beta^5)$	$(22,\beta^1)$	$(193,\beta^5)$	$(137,\beta^1)$	$(108,\beta^6)$
	3	$(194,\beta^2)$	$(164,\beta^1)$	$(120,\beta^1)$	$(180,\beta^4)$	$(74,\beta^3)$
	4	$(73,\beta^2)$	$(52,\beta^5)$	$(172,\beta^4)$	$(176,\beta^3)$	$(53,\beta^6)$
	5	$(70,\beta^2)$	$(186,\beta^5)$	$(95,\beta^5)$	$(179,\beta^2)$	$(141,\beta^3)$

80 iterations of the FFT-QSPA over the binary-input AWGNC is shown in Fig. 11.4. At a BLER of 3×10^{-5}, the code $C_{q^\star,sp,qc,mask}$ performs about 0.75 dB away from the SPB.

For comparison, we construct an 8-ary $(2376, 1188)$ LDPC code $C_{q,peg}$ using the PEG algorithm and also decode it with 80 iterations of the FFT-QSPA. The parity-check matrix of $C_{q,peg}$ has the same column and row weight distributions as the CPM-QC-SP-LDPC code $C_{q^\star,sp,qc,mask}$. The Tanner graph of the code $C_{q,peg}$ has girth 12 and contains 4462 cycles of length 12 and 31 382 cycles of length 14. Its BLER performance is also included in Fig. 11.4. We see that the code $C_{q^\star,sp,qc,mask}$ constructed using the CPM-*b*D/B-to-NB method and the PEG code $C_{q,peg}$ perform almost the same, although the PEG code has no structure.

If every symbol in GF(2^3) is represented by a 3-tuple over GF(2), then the 8-ary $(2376, 1188)$ CPM-QC-SP-LDPC code $C_{q^\star,sp,qc,mask}$ is equivalent to a binary $(7128, 3564)$ LDPC code. Using the PEG algorithm, we construct a binary $(7128, 3564)$ LDPC code $C_{b,peg}$ whose parity-check matrix has the same average column weight (2.5) as the CPM-QC-SP-LDPC code $C_{q^\star,sp,qc,mask}$ and its Tanner graph has girth 14 and contains 27 144 cycles of length 14. The BLER performance of this binary PEG code decoded with 80 iterations of the SPA is also included in Fig. 11.4. We find that the 8-ary $(2376, 1188)$ CPM-QC-SP-LDPC code $C_{q^\star,sp,qc,mask}$ outperforms the binary $(7128, 3564)$ PEG code $C_{b,peg}$.

In Example 11.4, we chose a 6×12 submatrix $\mathbf{B}_{q,sp}$ of $\mathbf{B}^*_{q,sp,p}$ over GF(199) in the form given by (8.1) as the base matrix for the CPM-*b*D/B-to-NB construction

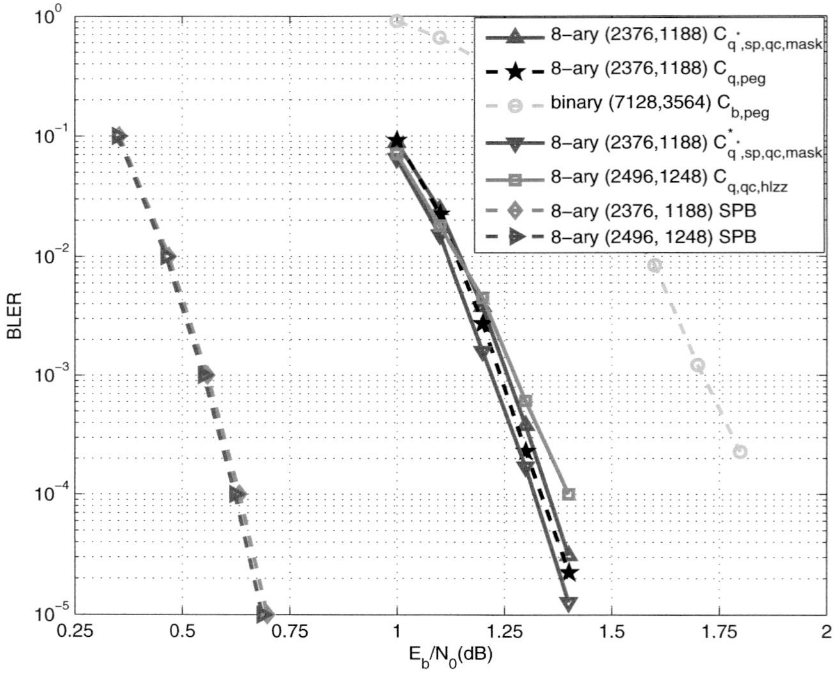

Figure 11.4. The BLER performances of the 8-ary $(2376, 1188)$ CPM-QC-SP-LDPC code $C_{q^\star,sp,qc,mask}$, the 8-ary $(2376, 1188)$ PEG code $C_{q,peg}$, the binary $(7128, 3564)$ LDPC code $C_{b,peg}$ in Example 11.4, the 8-ary $(2376, 1188)$ $C^\star_{q^\star,sp,qc,mask}$, and the 8-ary $(2496, 1248)$ code $C_{q,qc,hlzz}$ in Example 11.5 over the binary-input AWGNC.

of an NB CPM-QC-SP-LDPC code. By masking the base matrix $\mathbf{B}_{q,sp}$, we constructed an NB CPM-QC-SP-LDPC code whose Tanner graph has girth 8. There are many possible choices of 6×12 submatrices of $\mathbf{B}^*_{q,sp,p}$; a different choice with the same masking may result in a code with a larger girth or a smaller number of short cycles, or both. This will be demonstrated in the next example.

Example 11.5. Consider the 198×198 base matrix $\mathbf{B}^*_{q,sp,p}$ over GF(199) given in Example 11.4. Suppose we choose the following 6×12 submatrix of $\mathbf{B}^*_{q,sp,p}$:

$$\mathbf{B}^\star_{q,sp} =$$

$$\begin{bmatrix} \alpha^{184} & \alpha^{85} & \alpha^{71} & \alpha^{133} & \alpha^{13} & \alpha^{162} & \alpha^{140} & \alpha^{94} & \alpha^{79} & \alpha^{192} & \alpha^{42} & \alpha^{178} \\ \alpha^{101} & \alpha^{48} & \alpha^{67} & \alpha^{130} & \alpha^{83} & \alpha^{169} & \alpha^{185} & \alpha^{34} & \alpha^{57} & \alpha^{107} & \alpha^{30} & \alpha^{167} \\ \alpha^{26} & \alpha^{102} & \alpha^{22} & \alpha^{40} & \alpha^{64} & \alpha^{193} & \alpha^{21} & \alpha^{28} & \alpha^{137} & \alpha^{108} & \alpha^{5} & \alpha^{18} \\ \alpha^{177} & \alpha^{194} & \alpha^{93} & \alpha^{164} & \alpha^{120} & \alpha^{4} & \alpha^{139} & \alpha^{15} & \alpha^{180} & \alpha^{74} & \alpha^{168} & \alpha^{132} \\ \alpha^{73} & \alpha^{17} & \alpha^{110} & \alpha^{52} & \alpha^{172} & \alpha^{82} & \alpha^{69} & \alpha^{183} & \alpha^{86} & \alpha^{170} & \alpha^{176} & \alpha^{53} \\ \alpha^{32} & \alpha^{70} & \alpha^{186} & \alpha^{77} & \alpha^{119} & \alpha^{95} & \alpha^{127} & \alpha^{25} & \alpha^{158} & \alpha^{92} & \alpha^{179} & \alpha^{141} \end{bmatrix}. \tag{11.8}$$

The base matrix $\mathbf{B}^{\star}_{q,sp}$ is different from the base matrix $\mathbf{B}_{q,sp}$ chosen in Example 11.4. Masking $\mathbf{B}^{\star}_{q,sp}$ with the same 6×12 masking matrix $\mathbf{Z}$ given by (11.6), we obtain the following masked base matrix $\mathbf{B}^{\star}_{q,sp,mask}$ which also satisfies both the 2×2 and the 3×3 SM-constraints:

$$\mathbf{B}^{\star}_{q,sp,mask} =$$

$$\begin{bmatrix} \alpha^{184} & 0 & \alpha^{71} & 0 & \alpha^{13} & 0 & \alpha^{140} & \alpha^{94} & 0 & 0 & 0 & 0 \\ 0 & \alpha^{48} & 0 & \alpha^{130} & 0 & \alpha^{169} & \alpha^{185} & \alpha^{34} & 0 & 0 & 0 & 0 \\ \alpha^{26} & 0 & \alpha^{22} & 0 & 0 & \alpha^{193} & 0 & 0 & \alpha^{137} & \alpha^{108} & 0 & 0 \\ 0 & \alpha^{194} & 0 & \alpha^{164} & \alpha^{120} & 0 & 0 & 0 & \alpha^{180} & \alpha^{74} & 0 & 0 \\ \alpha^{73} & 0 & 0 & \alpha^{52} & \alpha^{172} & 0 & 0 & 0 & 0 & 0 & \alpha^{176} & \alpha^{53} \\ 0 & \alpha^{70} & \alpha^{186} & 0 & 0 & \alpha^{95} & 0 & 0 & 0 & 0 & \alpha^{179} & \alpha^{141} \end{bmatrix}. \tag{11.9}$$

The binary CPM-dispersion of $\mathbf{B}^{\star}_{q,sp,mask}$ gives a 6×12 binary array $\mathbf{H}^{\star}_{b,sp,qc,mask}(198, 198)$ of CPMs and ZMs of size 198×198. The array $\mathbf{H}^{\star}_{b,sp,qc,mask}(198, 198)$ has the same size and column and row weight distributions as the binary array $\mathbf{H}_{b,sp,qc,mask}(198, 198)$ constructed in Example 11.4, a 1188×2376 matrix over GF(2).

Using the cycle counting algorithm presented in [71], we find that the Tanner graph $\mathcal{G}^{\star}_{b,sp,qc,mask}(198, 198)$ associated with the array $\mathbf{H}^{\star}_{b,sp,qc,mask}(198, 198)$ has girth 10 and contains 396 cycles of length 10 and 4653 cycles of length 12. However, the Tanner graph $\mathcal{G}_{b,sp,qc,mask}(198, 198)$ associated with the binary array $\mathbf{H}_{b,sp,qc,mask}(198, 198)$ constructed in Example 11.4 has girth 8 and contains 396 cycles of length 8 and 594 cycles of length 10. We see that $\mathcal{G}^{\star}_{b,sp,qc,mask}(198, 198)$ has a larger girth and a smaller number of cycles of length 10 than the Tanner graph $\mathcal{G}_{b,sp,qc,mask}(198, 198)$ constructed in Example 11.4. This shows that the choice of a base matrix does affect the girth and cycle distribution of the Tanner graph of the resultant code.

Replacing the 1-entries of each CPM in $\mathbf{H}^{\star}_{b,sp,qc,mask}(198, 198)$ with nonzero elements randomly chosen from GF(2^3) (random replacement), we obtain a 6×12 array $\mathbf{H}^{\star}_{q^{\star},sp,qc,mask}(198, 198)$ of 8-ary CPM and ZMs of size 198×198. The generators of the 8-ary CPMs in each row-block of $\mathbf{H}^{\star}_{q^{\star},sp,qc,mask}(198, 198)$ are also given in Table 11.1, each specified by a pair (j, β^s) with $0 \le j < 198$ and $0 \le s < 7$ as described in Example 11.4, where β is a primitive element of GF(2^3).

The null space of $\mathbf{H}^{\star}_{q^{\star},sp,qc,mask}(198, 198)$ gives another 8-ary (2376, 1188) CPM-QC-SP-LDPC code $C^{\star}_{q^{\star},sp,qc,mask}$ with rate 0.5. The BLER performance of the code $C^{\star}_{q^{\star},sp,qc,mask}$ decoded with 80 iterations of the FFT-QSPA over the binary-input AWGNC is also included in Fig. 11.4 together with the BLER performance of the code $C_{q^\star,sp,qc,mask}$ constructed in Example 11.4. From this figure, we see that the code $C^\star_{q^\star,sp,qc,mask}$ outperforms $C_{q^\star,sp,qc,mask}$. At a BLER of 3×10^{-5}, the code $C^\star_{q^\star,sp,qc,mask}$ performs about 0.7 dB away from the SPB.

The BLER performance of the 8-ary (2496, 1248) QC-LDPC code $C_{q,qc,hlzz}$ given in [45, p.3441, Figure 2] is included in Fig. 11.4 for comparison, where the subscript "*hlzz*" stands for the initial letters of the last names of the authors. The Tanner graph of this code has girth 12 and contains 23 504 cycles of length 12. Note that the number of cycles of length 12 in the Tanner graph of $C_{q,qc,hlzz}$ is much larger than the sum of the numbers of cycles of lengths 10 and 12 in the Tanner graph of the code $C^\star_{q^\star,sp,qc,mask}$. The code $C_{q,qc,hlzz}$ is longer than the code $C^\star_{q^\star,sp,qc,mask}$ by 120 symbols. In [45], the code $C_{q,qc,hlzz}$ was decoded with 80 iterations of the FFT-QSPA. From Fig. 11.4, we see that the code $C^\star_{q^\star,sp,qc,mask}$ slightly outperforms the $C_{q,qc,hlzz}$ below a BLER of 10^{-2} even though the code $C^\star_{q^\star,sp,qc,mask}$ is shorter and the Tanner graph of the code $C_{q,qc,hlzz}$ has a larger girth. It shows that the cycle distribution plays an important role in the performance of an LDPC code.

11.5 Algebraic Construction of NB QC-PTG-LDPC Codes

Construction of NB PTG-LDPC codes can be carried out either using the conventional copy-and-permute approach presented in Section 3.1 or using the decomposition-and-replacement process presented in Section 4.1.

Let $\mathcal{G}_{q,ptg}$ be a protograph with n VNs and m CNs in which the edges are labeled with nonzero elements of an NB field GF(q). To construct a q-ary PTG-LDPC code using the conventional PTG-based construction method as presented in Section 3.1, we first design and choose an expansion factor k. Next, we create k copies of the protograph $\mathcal{G}_{q,ptg}$. Then, we permute the edges of copies and connect the copies into a connected bipartite graph $\mathcal{G}_{q,ptg}(k,k)$ under the permutation-and-connection constraint defined in Section 3.1. Since the edges of $\mathcal{G}_{q,ptg}$ are labeled with nonzero elements of GF(q), the adjacency matrix $\mathbf{H}_{q,ptg}(k,k)$ of $\mathcal{G}_{q,ptg}(k,k)$ is an $m \times n$ array of regular matrices over GF(q) of size $k \times k$. The null space of $\mathbf{H}_{q,ptg}(k,k)$ gives a q-ary PTG-LDPC code $C_{q,ptg}$ of length nk with rate at least $(n-m)/n$. If the constituent matrices in $\mathbf{H}_{q,ptg}(k,k)$ are circulants, $C_{q,ptg}$ is then a q-ary QC-PTG-LDPC code.

Study of NB PTG-LDPC codes is still in its early stage. Several specific methods for constructing NB protographs can be found in [36, 15, 16], in which some short good NB PTG-LDPC codes were given.

In the following, we give a simple example to illustrate the PTG-based construction of an NB PTG-LDPC code and show its similarity to the construction of a binary PTG-LDPC code.

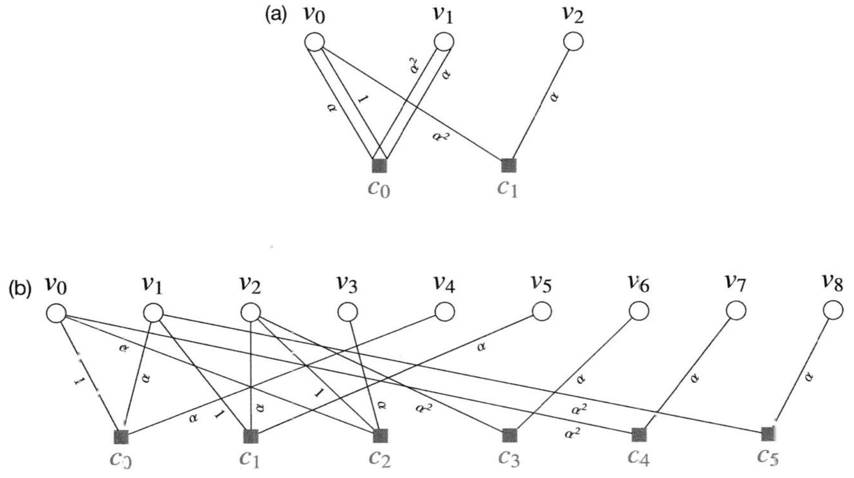

Figure 11.5. (a) The protograph $\mathcal{G}_{q,ptg}$ given in Example 11.6; and (b) the bipartite graph $\mathcal{G}_{q,ptg}(3,3)$ in Example 11.6.

Example 11.6. Consider the protograph $\mathcal{G}_{q,ptg}$ with 3 VNs and 2 CNs in which the edges are labeled with elements from GF(2^2) shown in Fig. 11.5(a), where α is a primitive element in GF(2^2). From Fig. 11.5(a), we see that VN v_0 and CN c_0 are connected by two parallel edges labeled by 1 and α, respectively. Suppose we take 3 copies of $\mathcal{G}_{q,ptg}$, permute the edges among the copies, and connect them under the permutation-and-connection constraint as presented in Section 3.1. We obtain a connected bipartite graph $\mathcal{G}_{q,ptg}(3,3)$ with 9 VNs and 6 CNs shown in Fig. 11.5(b). $\mathcal{G}_{q,ptg}(3,3)$ contains no parallel edge and has girth 6. The adjacency matrix $\mathbf{H}_{q,ptg}(3,3)$ of $\mathcal{G}_{q,ptg}(3,3)$ is a 6×9 RC-constrained matrix over GF(2^2) given below:

$$\mathbf{H}_{q,ptg}(3,3) = \begin{bmatrix} 1 & \alpha & 0 & 0 & \alpha & 0 & 0 & 0 & 0 \\ 0 & 1 & \alpha & 0 & 0 & \alpha & 0 & 0 & 0 \\ \alpha & 0 & 1 & \alpha & 0 & 0 & 0 & 0 & 0 \\ 0 & 0 & \alpha^2 & 0 & 0 & 0 & \alpha & 0 & 0 \\ \alpha^2 & 0 & 0 & 0 & 0 & 0 & 0 & \alpha & 0 \\ 0 & \alpha^2 & 0 & 0 & 0 & 0 & 0 & 0 & \alpha \end{bmatrix}.$$

Since $\mathbf{H}_{q,ptg}(3,3)$ is a 2×3 array of circulants and ZMs of size 3×3, its null space gives a 4-ary $(9,6)$ QC-PTG-LDPC code which is the same code constructed using the SP-construction in Example 11.1.

An NB QC-PTG-LDPC code can be constructed algebraically in a similar way to the algebraic construction of a binary QC-PTG-LDPC code based on the decomposition of the base matrix associated with a chosen protograph as presented in Section 4.1. The difference is that the decomposition base matrix is not a matrix over nonnegative integers, but a matrix whose entries are *sets of elements* from an NB field.

The construction starts with a protograph $\mathcal{G}_{q,ptg}$ with n VNs and m CNs in which the edges are labeled with nonzero elements from GF(q). Let $\mathbf{V} = \{v_0, v_1, \ldots, v_{n-1}\}$ and $\mathbf{C} = \{c_0, c_1, \ldots, c_{m-1}\}$ denote the sets of n VNs and m CNs, respectively. For $0 \le i < m$ and $0 \le j < n$, let $\lambda_{i,j}$ denote the number of edges connecting the CN c_i and the VN v_j. If $\lambda_{i,j} = 0$, then c_i and v_j are not connected. If $\lambda_{i,j} \ne 0$, the $\lambda_{i,j}$ parallel edges connecting c_i and v_j are labeled by a set $\mathbf{S}_{i,j}$ of $\lambda_{i,j}$ nonzero elements in GF(q). The elements in $\mathbf{S}_{i,j}$ are not necessarily different, i.e., two different parallel edges connecting c_i and v_j may be labeled by the same nonzero element in GF(q).

An adjacency matrix $\mathbf{B}_{q,ptg}$ of the protograph $\mathcal{G}_{q,ptg}$ may be constructed as follows. It is an $m \times n$ matrix in which the entry at location (i, j) is the set $\mathbf{S}_{i,j}$. Hence, $\mathbf{B}_{q,ptg}$ is an $m \times n$ matrix whose entries are sets of elements in GF(q) shown below:

$$\mathbf{B}_{q,ptg} = \begin{bmatrix} \mathbf{S}_{0,0} & \mathbf{S}_{0,1} & \cdots & \mathbf{S}_{0,n-1} \\ \mathbf{S}_{1,0} & \mathbf{S}_{1,1} & \cdots & \mathbf{S}_{1,n-1} \\ \vdots & \vdots & \ddots & \vdots \\ \mathbf{S}_{m-1,0} & \mathbf{S}_{m-1,1} & \cdots & \mathbf{S}_{m-1,n-1} \end{bmatrix}. \tag{11.10}$$

In the following, we present an algebraic construction of an NB QC-PTG-LDPC code based on the adjacency matrix $\mathbf{B}_{q,ptg}$ of the protograph $\mathcal{G}_{q,ptg}$ defined by (11.10), instead of the protograph itself, We call $\mathbf{B}_{q,ptg}$ the the decomposition base matrix.

Similar to the algebraic construction of a binary QC-PTG-LDPC code presented in Section 4.1, we first decompose $\mathbf{B}_{q,ptg}$ into k constituent matrices $\mathbf{D}_0, \mathbf{D}_1, \ldots, \mathbf{D}_{k-1}$ over GF(q) of size $m \times n$. In the decomposition of the base matrix $\mathbf{B}_{q,ptg}$, we distribute the elements of each non-empty entry-set $\mathbf{S}_{i,j}$ among the k constituent matrices $\mathbf{D}_0, \mathbf{D}_1, \ldots, \mathbf{D}_{k-1}$ in such as way that for $0 \le i < m$ and $0 \le j < n$, the $\lambda_{i,j}$ elements in the entry-set $\mathbf{S}_{i,j}$ are put in different constituent matrices at location (i, j) and the other $k - \lambda_{i,j}$ (i, j)-th locations get zeros. With this constraint on the distributions of the elements of each entry-set in $\mathbf{B}_{q,ptg}$, each constituent matrix $\mathbf{D}_e$ is an $m \times n$ matrix whose entries are elements in GF(q). The above constraint on the decomposition of the base matrix $\mathbf{B}_{q,ptg}$ is also called the *decomposition constraint* and the parameter k is called the decomposition factor. In the decomposition of $\mathbf{B}_{q,ptg}$, ZMs are allowed, i.e., there may be ZMs among the matrices $\mathbf{D}_0, \mathbf{D}_1, \ldots, \mathbf{D}_{k-1}$. Therefore, for $0 \le e < k$, each matrix $\mathbf{D}_e$ is either an $m \times n$ nonzero matrix over GF(q) or an $m \times n$ ZM. To satisfy the decomposition constraint, the decomposition factor k must be equal to or larger than the size of the *largest entry-set* $\mathbf{S}_{i,j}$ in the decomposition base matrix $\mathbf{B}_{q,ptg}$.

The base matrix $\mathbf{B}_{q,ptg}$ can be reconstructed from its constituent matrices $\mathbf{D}_0$, $\mathbf{D}_1$, ..., $\mathbf{D}_{k-1}$ by grouping the nonzero entries at location (i, j)s of these constituent matrices into the entry-set $\mathbf{S}_{i,j}$s in $\mathbf{B}_{q,ptg}$ for $0 \leq i < m$ and $0 \leq j < n$. (Note that in grouping, zero-entries in $\mathbf{D}_0$, $\mathbf{D}_1$, ..., $\mathbf{D}_{k-1}$ are not included in the reconstruction of each entry-set of $\mathbf{B}_{q,ptg}$.)

Next, we arrange the k matrices $\mathbf{D}_0$, $\mathbf{D}_1$, ..., $\mathbf{D}_{k-1}$ in a row-block $[\mathbf{D}_0\ \mathbf{D}_1\ \ldots\ \mathbf{D}_{k-1}]$, then cyclically shift this row-block $k-1$ times, each time n positions (or one constituent matrix) to the right. Placing these k row-blocks on the top of each other results in a $k \times k$ array $\mathbf{H}_{q,ptg,cyc}(m,n)$ of $m \times n$ matrices over GF(q) with a block-cyclic structure in the form given by (4.1). The row-block $[\mathbf{D}_0\ \mathbf{D}_1\ \ldots\ \mathbf{D}_{k-1}]$ is the generator row-block of the block-cyclic array $\mathbf{H}_{q,ptg,cyc}(m,n)$.

Now, we perform the row and column permutations π_{row} and π_{col} defined by (4.3) and (4.5) to convert the q-ary array $\mathbf{H}_{q,ptg,cyc}(m,n)$ into an $m \times n$ array $\mathbf{H}_{q,ptg,qc}(k,k)$ of $k \times k$ circulants over GF(q) in the form given by (4.6), duplicated as follows:

$$\mathbf{H}_{q,ptg,qc}(k,k) = \begin{bmatrix} \mathbf{Q}_{0,0} & \mathbf{Q}_{0,1} & \cdots & \mathbf{Q}_{0,n-1} \\ \mathbf{Q}_{1,0} & \mathbf{Q}_{1,1} & \cdots & \mathbf{Q}_{1,n-1} \\ \vdots & \vdots & \ddots & \vdots \\ \mathbf{Q}_{m-1,0} & \mathbf{Q}_{m-1,1} & \cdots & \mathbf{Q}_{m-1,n-1} \end{bmatrix}. \tag{11.11}$$

The generators of the constituent circulants in the q-ary array $\mathbf{H}_{q,ptg,qc}(k,k)$ can be determined directly from the constituent matrices of the generator row-block $[\mathbf{D}_0\ \mathbf{D}_1\ \ldots\ \mathbf{D}_{k-1}]$ of the q-ary block-cyclic array $\mathbf{H}_{q,ptg,cyc}(m,n)$, in the same manner as for the binary array $\mathbf{H}_{ptg,qc}(k,k)$ given by (4.6), described in Section 4.1 (or Section 11.2). Each of these generators is a k-tuple over GF(q).

Let $\mathcal{G}_{q,ptg,qc}(k,k)$ be the Tanner graph associated with $\mathbf{H}_{q,ptg,qc}(k,k)$. Then, the graph $\mathcal{G}_{q,ptg,qc}(k,k)$ is an expansion of the protograph $\mathcal{G}_{q,ptg}$ by a factor of k. In the decomposition of the base matrix $\mathbf{B}_{q,ptg}$, if we impose the RC- and the PW-RC-constraints on its constituent matrices $\mathbf{D}_0, \mathbf{D}_1, \ldots, \mathbf{D}_{k-1}$ and arrange them in a proper order (as described in Section 4.2), $\mathcal{G}_{q,ptg,qc}(k,k)$ has girth at least 6. The null space of $\mathbf{H}_{q,ptg,qc}(k,k)$ gives a q-ary QC-PTG-LDPC code, denoted by $C_{q,ptg,qc}$.

The following simple example illustrates the basic idea of the above algebraic construction method for q-ary QC-PTG-LDPC codes.

Example 11.7. Let α be a primitive element of GF(2^2). The set $\{0, 1, \alpha, \alpha^2\}$ gives all the elements of GF(2^2). Suppose we choose the following 2×4 matrix over

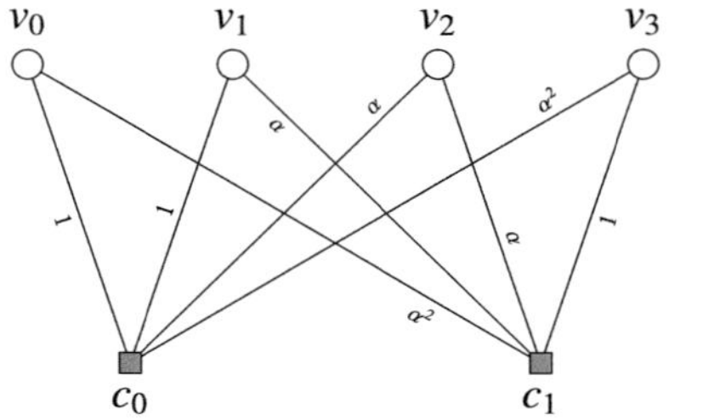

Figure 11.6. The protograph $\mathcal{G}_{q,ptg}$ of the decomposition base matrix $\mathbf{B}_{q,ptg}$ given in Example 11.7.

GF(2^2) as the decomposition base matrix:

$$\mathbf{B}_{q,ptg} = \begin{bmatrix} 1 & 1 & \alpha & \alpha^2 \\ \alpha^2 & \alpha & \alpha & 1 \end{bmatrix}.$$

The protograph $\mathcal{G}_{q,ptg}$ associated with $\mathbf{B}_{q,ptg}$ is shown in Fig. 11.6. Choose the decomposition factor $k = 3$. Suppose we decompose $\mathbf{B}_{q,ptg}$ into the following three 2×4 matrices over GF(2^2):

$$\mathbf{D}_0 = \begin{bmatrix} 1 & 0 & 0 & 0 \\ 0 & \alpha & 0 & 0 \end{bmatrix}, \mathbf{D}_1 = \begin{bmatrix} 0 & 1 & \alpha & 0 \\ \alpha^2 & 0 & 0 & 0 \end{bmatrix}, \mathbf{D}_2 = \begin{bmatrix} 0 & 0 & 0 & \alpha^2 \\ 0 & 0 & \alpha & 1 \end{bmatrix}.$$

Based on these three constituent matrices, we form the following 3×3 array $\mathbf{H}_{q,ptg,cyc}(2,4)$ of 2×4 matrices over GF(2^2) with a block-cyclic structure:

$$\mathbf{H}_{q,ptg,cyc}(2,4) = \left[\begin{array}{cccc|cccc|cccc} 1 & 0 & 0 & 0 & 0 & 1 & \alpha & 0 & 0 & 0 & 0 & \alpha^2 \\ 0 & \alpha & 0 & 0 & \alpha^2 & 0 & 0 & 0 & 0 & 0 & \alpha & 1 \\ \hline 0 & 0 & 0 & \alpha^2 & 1 & 0 & 0 & 0 & 0 & 1 & \alpha & 0 \\ 0 & 0 & \alpha & 1 & 0 & \alpha & 0 & 0 & \alpha^2 & 0 & 0 & 0 \\ \hline 0 & 1 & \alpha & 0 & 0 & 0 & 0 & \alpha^2 & 1 & 0 & 0 & 0 \\ \alpha^2 & 0 & 0 & 0 & 0 & 0 & \alpha & 1 & 0 & \alpha & 0 & 0 \end{array}\right].$$

Permuting the columns and the rows of $\mathbf{H}_{q,ptg,cyc}(2,4)$ based on the row and column permutations defined by (4.3) and (4.5), we obtain the following 2×4

array $\mathbf{H}_{q,ptg,qc}(3,3)$ of circulants of size 3×3 over GF(2^2):

$$\mathbf{H}_{q,ptg,qc}(3,3) = \begin{bmatrix} \mathbf{Q}_{0,0} & \mathbf{Q}_{0,1} & \mathbf{Q}_{0,2} & \mathbf{Q}_{0,3} \\ \mathbf{Q}_{1,0} & \mathbf{Q}_{1,1} & \mathbf{Q}_{1,2} & \mathbf{Q}_{1,3} \end{bmatrix}$$

$$= \left[\begin{array}{ccc|ccc|ccc|ccc} 1 & 0 & 0 & 0 & 1 & 0 & 0 & \alpha & 0 & 0 & 0 & \alpha^2 \\ 0 & 1 & 0 & 0 & 0 & 1 & 0 & 0 & \alpha & \alpha^2 & 0 & 0 \\ 0 & 0 & 1 & 1 & 0 & 0 & \alpha & 0 & 0 & 0 & \alpha^2 & 0 \\ \hline 0 & \alpha^2 & 0 & \alpha & 0 & 0 & 0 & 0 & \alpha & 0 & 0 & 1 \\ 0 & 0 & \alpha^2 & 0 & \alpha & 0 & \alpha & 0 & 0 & 1 & 0 & 0 \\ \alpha^2 & 0 & 0 & 0 & 0 & \alpha & 0 & \alpha & 0 & 0 & 1 & 0 \end{array}\right].$$

The null space of $\mathbf{H}_{q,ptg,qc}(3,3)$ gives a 4-ary $(12,6)$ QC-PTG-LDPC code whose Tanner graph has girth 8.

Example 11.8. Suppose we want to construct a 256-ary QC-PTG-LDPC code of symbol length 32 using the following 2×4 decomposition base matrix over GF(2^8):

$$\mathbf{B}_{q,ptg} = [b_{i,j}]_{0 \le i < 2, 0 \le j < 4} = \begin{bmatrix} \alpha^{61} & \alpha^{106} & \alpha^{240} & \alpha^{125} \\ \alpha^{229} & \alpha^{94} & \alpha^{199} & \alpha^{99} \end{bmatrix}$$

where α is a primitive element of GF(2^8). The above base matrix is chosen at random. The protograph associated with $\mathbf{B}_{q,ptg}$ is shown in Fig. 11.7(a). Choose the decomposition factor $k = 8$. Decompose $\mathbf{B}_{q,ptg}$ into the following eight 2×4 constituent matrices over GF(2^8):

$$\mathbf{D}_0 = \mathbf{D}_1 = \mathbf{D}_2 = \mathbf{D}_4 = \begin{bmatrix} 0 & 0 & 0 & 0 \\ 0 & 0 & 0 & 0 \end{bmatrix},$$

$$\mathbf{D}_3 = \begin{bmatrix} 0 & 0 & \alpha^{240} & 0 \\ 0 & 0 & \alpha^{199} & 0 \end{bmatrix}, \mathbf{D}_5 = \begin{bmatrix} 0 & \alpha^{106} & 0 & 0 \\ 0 & \alpha^{94} & 0 & 0 \end{bmatrix},$$

$$\mathbf{D}_6 = \begin{bmatrix} \alpha^{61} & 0 & 0 & 0 \\ \alpha^{229} & 0 & 0 & 0 \end{bmatrix}, \mathbf{D}_7 = \begin{bmatrix} 0 & 0 & 0 & \alpha^{125} \\ 0 & 0 & 0 & \alpha^{99} \end{bmatrix}.$$

Next, we construct a 2×4 parity-check array $\mathbf{H}_{q,ptg,qc}(8,8) = [\mathbf{Q}_{i,j}]_{0 \le i < 2, 0 \le j < 4}$ directly by finding its eight 256-ary CPMs of size 8×8. First, we determine the generators of these 256-ary CPMs from the constituent matrices, $\mathbf{D}_0$ to $\mathbf{D}_7$, which are:

$$\mathbf{g}_{0,0} = (0,0,0,0,0,0,\alpha^{61},0), \qquad \mathbf{g}_{0,1} = (0,0,0,0,0,\alpha^{106},0,0),$$

$$\mathbf{g}_{0,2} = (0,0,0,\alpha^{240},0,0,0,0), \qquad \mathbf{g}_{0,3} = (0,0,0,0,0,0,0,\alpha^{125}),$$

$$\mathbf{g}_{1,0} = (0,0,0,\alpha^{229},0,0,0,0), \qquad \mathbf{g}_{1,1} = (0,0,0,0,0,0,0,\alpha^{94}),$$

$$\mathbf{g}_{1,2} = (0,0,0,0,0,0,\alpha^{199},0), \qquad \mathbf{g}_{1,3} = (0,0,0,0,0,\alpha^{99},0,0).$$

Based on these generators, we construct eight 256-ary CPMs of size 8×8, $\mathbf{Q}_{i,j}$'s with $0 \le i < 2$ and $0 \le j < 4$. Then, we replace each entry $b_{i,j}$ in the base matrix $\mathbf{B}_{q,ptg}$ by its corresponding 256-ary CPM $\mathbf{Q}_{i,j}$. The replacement results in the following 2×4 parity-check array of 256-ary CPMs of size 8×8:

$$\mathbf{H}_{q,ptg,qc}(8,8) = \begin{bmatrix} \mathbf{Q}_{0,0} & \mathbf{Q}_{0,1} & \mathbf{Q}_{0,2} & \mathbf{Q}_{0,3} \\ \mathbf{Q}_{1,0} & \mathbf{Q}_{1,1} & \mathbf{Q}_{1,2} & \mathbf{Q}_{1,3} \end{bmatrix}.$$

The array $\mathbf{H}_{q,ptg,qc}(8,8)$ is an 18×36 matrix over GF(2^8) with column and row weights 2 and 4, respectively. The null space of $\mathbf{H}_{q,ptg,qc}(8,8)$ gives a $(2,4)$-regular (32, 16) 256-ary QC-PTG-LDPC code $C_{q,ptg,qc}$. The Tanner graph of the code $C_{q,ptg,qc}$ has girth 8 and contains 20 cycles of length 8, no cycle of length 10, and 160 cycles of length 12. The BLER performances of the code $C_{q,ptg,qc}$ over the binary-input AWGNC decoded with 50 and 100 iterations of the FFT-QSPA are shown in Fig. 11.7(b). The decoding converges quickly. The gap between 50 and 100 iterations is less than 0.05 dB. From this figure, at a BLER around 10^{-6}, we see that the code $C_{q,ptg,qc}$ performs 0.75 dB away from the SPB.

For comparison, we include two other 256-ary codes reported in [16, 36] in Fig. 11.7(b), denoted by $C_{C\text{-}NBPB}$ and $C_{U\text{-}NBPB}$, respectively. The subscripts "*C-NBPB*" and "*U-NBPB*" stand for "*constrained NB protograph-based*" and "*unconstrained NB protograph-based*," respectively. The code $C_{C\text{-}NBPB}$ is constructed through scale-copy-permute operations based on a scaled protograph [16, 36], and the code $C_{U\text{-}NBPB}$ is constructed through copy-scale-permute operations based on an unlabeled protograph [16, 36]. The Tanner graphs of these two codes have the same cycle distributions and contain the same number of cycles of lengths 8, 10, and 12 as the proposed code $C_{q,ptg,qc}$. The code $C_{C\text{-}NBPB}$ is in QC-form, but the code $C_{U\text{-}NBPB}$ is not in QC-form. The BLER performances of the two codes $C_{C\text{-}NBPB}$ and $C_{U\text{-}NBPB}$ were decoded with 100 iterations of FFT-QSPA in [16, 36]. From Fig. 11.7(b), we see that the code $C_{q,ptg,qc}$ constructed algebraically based on the base matrix $\mathbf{B}_{q,ptg}$ slightly outperforms the codes $C_{C\text{-}NBPB}$ and $C_{U\text{-}NBPB}$ in the simulation region.

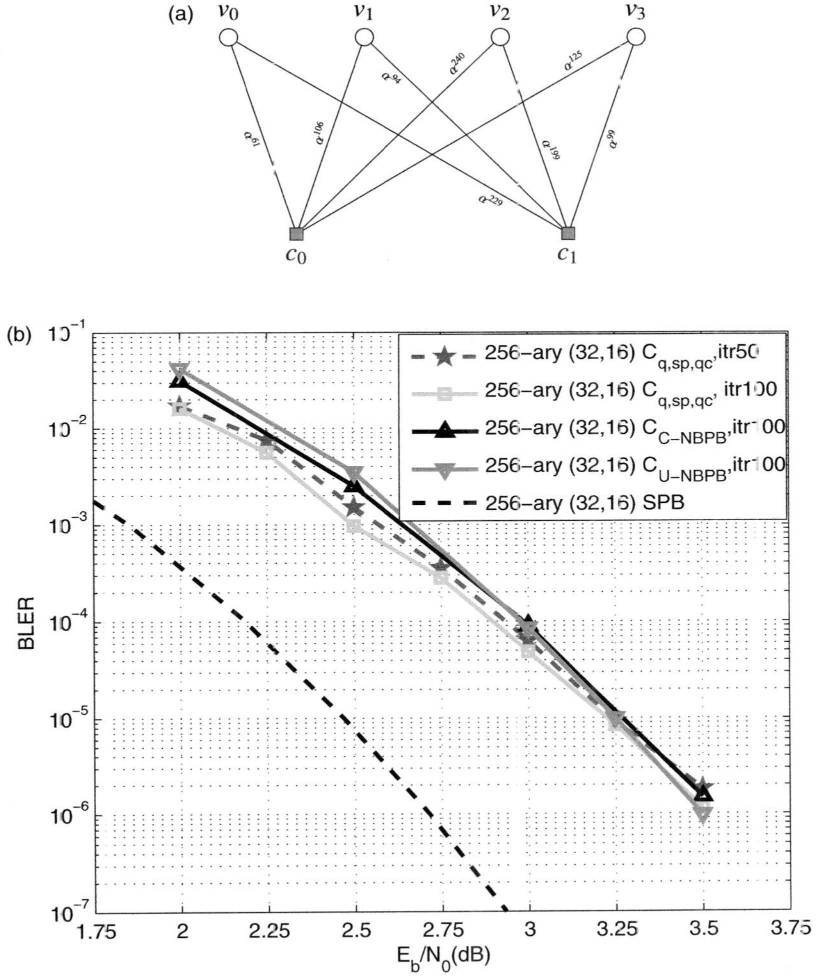

Figure 11.7. (a) The protograph $\mathcal{G}_{q,ptg}$ of the decomposition base matrix $\mathbf{B}_{q,ptg}$ given in Example 11.8; and (b) the BLER performances of the 256-ary (32, 16) QC-PTG-LDPC code $C_{q,ptg,qc}$ given in Example 11.8 and the PTG-based codes $C_{C\text{-}NBPB}$ and $C_{U\text{-}NBPB}$ reported in [16, 36] over the binary-input AWGNC.

In the above example, we consider a base matrix $\mathbf{B}_{q,ptg}$ whose entries are sets of elements of GF(q), each set consisting of only one component. In the next example, we consider a base matrix $\mathbf{B}_{q,ptg}$ consisting of entry-sets, each containing more than one element in GF(q).

Example 11.9. Consider the following 3×9 base matrix over GF(2^3) in the form of (11.10):

$$\mathbf{B}_{q,ptg} =$$

$$\begin{bmatrix} 0 & \{\alpha^5\} & \{\alpha^6\} & 0 & \{\alpha^3\} & \{\alpha^5,\alpha^6\} & \{1\} & \{\alpha^2\} & \{\alpha\} \\ \{\alpha^3\} & 0 & \{\alpha\} & \{\alpha,\alpha^3\} & 0 & \{\alpha\} & \{\alpha^5\} & \{\alpha^4\} & \{\alpha^2\} \\ \{\alpha^4\} & \{\alpha^5\} & 0 & \{\alpha^6\} & \{1,\alpha^6\} & 0 & \{\alpha\} & \{\alpha\} & \{\alpha^2\} \end{bmatrix}, \tag{11.12}$$

where α is a primitive element of GF(2^3). The protograph $\mathcal{G}_{q,ptg}$ associated with the base matrix $\mathbf{B}_{q,ptg}$ is shown in Fig. 11.8(a). There are parallel edges in $\mathcal{G}_{q,ptg}$ and each edge is labeled by a nonzero element from GF(2^3).

Choose the decomposition factor $k = 60$ and decompose the base matrix $\mathbf{B}_{q,ptg}$ into 60 RC- and PW-RC-constrained constituent matrices $\mathbf{D}_0, \mathbf{D}_1, \ldots, \mathbf{D}_{59}$, each of size 3×9. Using the row-block $[\mathbf{D}_0\ \mathbf{D}_1\ \ldots\ \mathbf{D}_{59}]$ as the generator row-block, we obtain a 60×60 block-cyclic array $\mathbf{H}_{q,ptg,cyc}(3,9)$ over GF(2^3) of submatrices of size 3×9. Perform row and column permutations π_{row} and π_{col} defined by (4.3) and (4.5) on the array $\mathbf{H}_{q,ptg,cyc}(3,9)$. Then, we obtain a 3×9 array $\mathbf{H}_{q,ptg,qc}(60,60)$ over GF(2^3) of circulants and ZMs of size 60×60 in the form of (11.11):

$$\mathbf{H}_{q,ptg,qc}(60,60) = \begin{bmatrix} \mathbf{O} & \mathbf{Q}_{0,1} & \mathbf{Q}_{0,2} & \mathbf{O} & \mathbf{Q}_{0,4} & \mathbf{Q}_{0,5} & \mathbf{Q}_{0,6} & \mathbf{Q}_{0,7} & \mathbf{Q}_{0,8} \\ \mathbf{Q}_{1,0} & \mathbf{O} & \mathbf{Q}_{1,2} & \mathbf{Q}_{1,3} & \mathbf{O} & \mathbf{Q}_{1,5} & \mathbf{Q}_{1,6} & \mathbf{Q}_{1,7} & \mathbf{Q}_{1,8} \\ \mathbf{Q}_{2,0} & \mathbf{Q}_{2,1} & \mathbf{O} & \mathbf{Q}_{2,3} & \mathbf{Q}_{2,4} & \mathbf{O} & \mathbf{Q}_{2,6} & \mathbf{Q}_{2,7} & \mathbf{Q}_{2,8} \end{bmatrix},$$

where $\mathbf{O}$ is a 60×60 ZM. The generators of circulants in $\mathbf{H}_{q,ptg,qc}(60,60)$ are given in Table 11.2, each specified by a pair (j, α^s) or two pairs, (j, α^s) and $(j', \alpha^{s'})$, with $0 \le j, j' < 60$ and $0 \le s, s' < 7$ as described in Example 11.4. The array $\mathbf{H}_{q,ptg,qc}(60,60)$ is a 180×540 matrix over GF(2^3) and it has average column weight 2.667 and constant row weight 8. Its null space gives an 8-ary (540, 360) QC-PTG-LDPC code $C_{q,ptg,qc}$ with rate 2/3. The Tanner graph of the code $C_{q,ptg,qc}$ has girth 8 and contains 3980 cycles of length 8 and 30 720 cycles of length 10. The BLER performance of the code $C_{q,ptg,qc}$ over the binary-input AWGNC decoded with 80 iterations of the FFT-QSPA is shown in Fig. 11.8(b). At a BLER of 10^{-4}, the code performs about 0.8 dB away from the SPB.

For comparison, we construct an 8-ary (540, 360) LDPC code $C_{q,peg}$ using the PEG algorithm and also decode it with 80 iterations of the FFT-QSPA. The parity-check matrix of $C_{q,peg}$ has the same average column weight as the QC-PTG-LDPC code $C_{q,ptg,qc}$. The Tanner graph of the code $C_{q,peg}$ has girth 8 and contains 3685 cycles of length 8 and 29 268 cycles of length 10. Its BLER

Table 11.2. *The generators of the circulants of the parity-check arrays* $\mathbf{H}_{q,ptg,qc}(60, 60)$ *and* $\mathbf{H}_{q,ptg,qc}(120, 120)$ *given in Example 11.9*

Generators of circulants in $\mathbf{H}_{q,ptg,qc}(60, 60)$

i	$\mathbf{Q}_{i,0}$	$\mathbf{Q}_{i,1}$	$\mathbf{Q}_{i,2}$	$\mathbf{Q}_{i,3}$	$\mathbf{Q}_{i,4}$	$\mathbf{Q}_{i,5}$	$\mathbf{Q}_{i,6}$	$\mathbf{Q}_{i,7}$	$\mathbf{Q}_{i,8}$
0	—	$(0, \alpha^5)$	$(1, \alpha^6)$	—	$(13, \alpha^3)$	$(2, \alpha^5)$ $(11, \alpha^6)$	$(15, \alpha^0)$	$(14, \alpha^2)$	$(16, \alpha)$
1	$(1, \alpha^3)$	—	$(13, \alpha)$	$(0, \alpha)$ $(2, \alpha^3)$	—	$(15, \alpha)$	$(9, \alpha^5)$	$(4, \alpha^4)$	$(7, \alpha^2)$
2	$(4, \alpha^4)$	$(14, \alpha^5)$	—	$(6, \alpha^6)$	$(1, \alpha^0)$ $(15, \alpha^6)$	—	$(16, \alpha)$	$(17, \alpha)$	$(2, \alpha^2)$

Generators of circulants in $\mathbf{H}_{q,ptg,qc}(120, 120)$

i	$\mathbf{Q}_{i,0}$	$\mathbf{Q}_{i,1}$	$\mathbf{Q}_{i,2}$	$\mathbf{Q}_{i,3}$	$\mathbf{Q}_{i,4}$	$\mathbf{Q}_{i,5}$	$\mathbf{Q}_{i,6}$	$\mathbf{Q}_{i,7}$	$\mathbf{Q}_{i,8}$
0	—	$(0, \alpha^5)$	$(1, \alpha^6)$	—	$(13, \alpha^3)$	$(2, \alpha^5)$ $(11, \alpha^6)$	$(15, \alpha^0)$	$(14, \alpha^2)$	$(16, \alpha)$
1	$(1, \alpha^3)$	—	$(13, \alpha)$	$(0, \alpha)$ $(2, \alpha^3)$	—	$(15, \alpha)$	$(9, \alpha^5)$	$(4, \alpha^4)$	$(7, \alpha^2)$
2	$(4, \alpha^4)$	$(14, \alpha^5)$	—	$(6, \alpha^6)$	$(1, \alpha^0)$ $(15, \alpha^6)$	—	$(16, \alpha)$	$(17, \alpha)$	$(2, \alpha^2)$

performance is also included in Fig. 11.8(b). We see that the QC-PTG-LDPC code $C_{q,ptg,qc}$ and the PEG code $C_{q,peg}$ perform almost the same, although the PEG code has no structure.

If we choose the decomposition factor $k = 120$ and use the same decomposition base matrix $\mathbf{B}_{q,ptg}$ given by (11.12), then we can construct an 8-ary $(1080, 720)$ QC-PTG-LDPC code $C^{\star}_{q,ptg,qc}$ with rate 2/3. The parity-check array $\mathbf{H}_{q,ptg,qc}(120, 120)$ of this code is a 3×9 array of 8-ary cirulants and ZMs of size 120×120 and has the same column and row weight distributions as the array $\mathbf{H}_{q,ptg,qc}(60, 60)$ constructed above. The Tanner graph of $C^{\star}_{q,ptg,qc}$ has girth 8 and contains 7920 cycles of length 8 and 59 400 cycles of length 10. Its BLER performance over the binary-input AWGNC decoded with 80 iterations of the FFT-QSPA is shown in Fig. 11.8(b). At a BLER of 10^{-4}, the code performs within 0.9 dB away from the SPB. An 8-ary $(1080, 720)$ LDPC code $C^{\star}_{q,peg}$ constructed using the PEG algorithm is also included in Fig. 11.8(b) for comparison. From this figure, we see that the two codes perform almost the same in the simulation region. However, the parity-check matrix of the PEG code $C^{\star}_{q,peg}$ has no QC-structure.

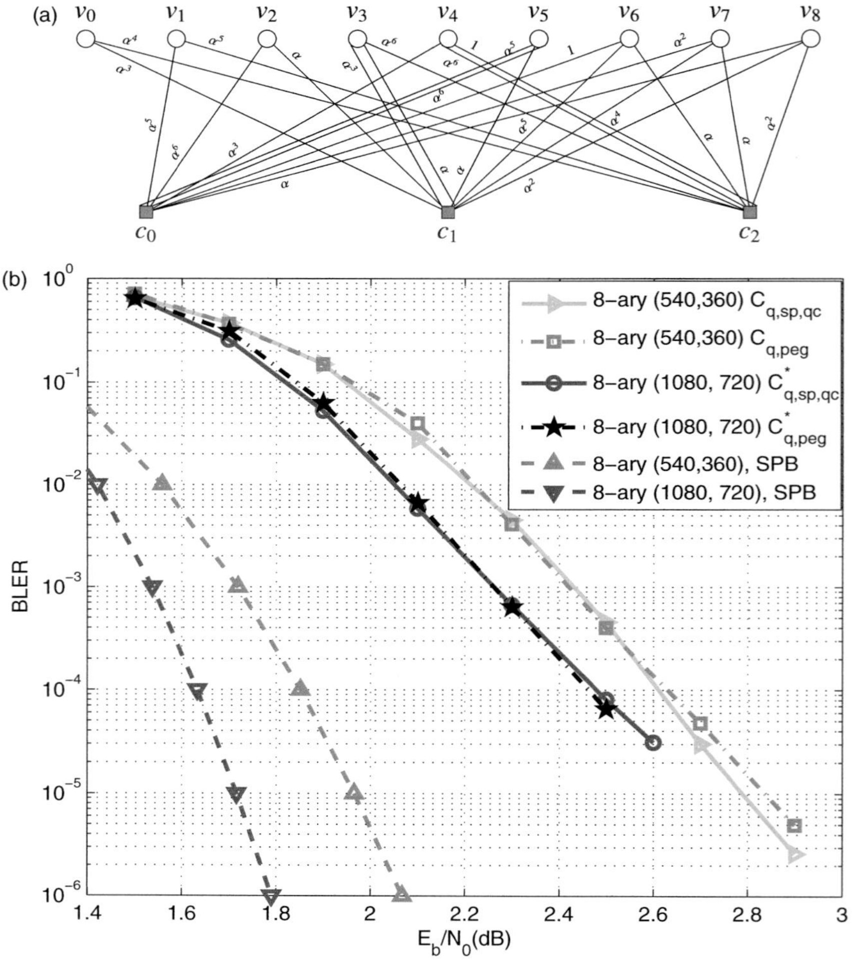

Figure 11.8. (a) The protograph $\mathcal{G}_{q,ptg}$ of the base matrix $\mathbf{B}_{q,ptg}$ given in Example 11.9; and (b) the BLER performances of the 8-ary (540, 360) QC-PTG-LDPC code $C_{q,ptg,qc}$, the 8-ary (540, 360) PEG code $C_{q,peg}$, the 8-ary (1080, 720) QC-PTG-LDPC code $C^{\star}_{q,ptg,qc}$, and the 8-ary (1080, 720) PEG code $C^{\star}_{q,peg}$ given in Example 11.9 over the binary-input AWGNC.

Construction of a base matrix $\mathbf{B}_{q,ptg}$ with set entries over an NB field GF(q) in the form of (11.10) can be achieved algebraically by using the construction method based on two arbitrary subsets $\boldsymbol{S}_0$ and $\boldsymbol{S}_1$ presented in Section 7.3. Suppose we choose $\boldsymbol{S}_0 = \boldsymbol{S}_1 = \mathrm{GF}(q)$ and a nonzero element η from GF(q). Using the sum-form given by (7.1), we obtain a $q \times q$ base matrix over GF(q), denoted by $\mathbf{B}_{q,ptg}(\eta)$, which is a Latin square of order q and satisfies the 2×2 SM-constraint. $\mathbf{B}_{q,ptg}(\eta)$ is called a Latin square associated with η.

Let α be a primitive element of GF(q). If we use each of the $q-1$ nonzero elements $1, \alpha, \alpha^2, \ldots, \alpha^{q-2}$ as ηs, we obtain $q-1$ different Latin squares, $\mathbf{B}_{q,ptg}(1)$, $\mathbf{B}_{q,ptg}(\alpha)$, $\mathbf{B}_{q,ptg}(\alpha^2)$, $\ldots$, $\mathbf{B}_{q,ptg}(\alpha^{q-2})$. Let $\mathbf{B}_{q,ptg}(\alpha^k) = [a_{i,j}]_{0\leq i,j<q}$ and $\mathbf{B}_{q,ptg}(\alpha^\ell) = [b_{i,j}]_{0\leq i,j<q}$ be two Latin squares associated with α^k and α^ℓ, respectively. Suppose we combine these two Latin squares into a $q \times q$ matrix $\mathbf{B}_{q,ptg}(\alpha^k, \alpha^\ell) = [(a_{i,j}, b_{i,j})]_{0\leq i,j<q}$ whose entries are two-tuples $(a_{i,j}, b_{i,j})$ over GF(q). All the two-tuple entries in $\mathbf{B}_{q,ptg}(\alpha^k, \alpha^\ell)$ are different. The two Latin squares $\mathbf{B}_{q,ptg}(\alpha^k)$ and $\mathbf{B}_{q,ptg}(\alpha^\ell)$ are said to be *orthogonal* [73]. The $q-1$ different Latin squares, $\mathbf{B}_{q,ptg}(1), \mathbf{B}_{q,ptg}(\alpha), \mathbf{B}_{q,ptg}(\alpha^2), \ldots, \mathbf{B}_{q,ptg}(\alpha^{q-2})$, are mutually orthogonal.

Let λ be a positive integer less than q. Suppose we take λ mutually orthogonal Latin squares of order q and form a $q \times q$ matrix by grouping their corresponding entries into sets. This results in a $q \times q$ base matrix $\mathbf{B}_{q,ptg}$ with set entries of size λ. Note that a set entry in $\mathbf{B}_{q,ptg}$ may contain a zero element. For $1 \leq m, n < q$, suppose we take an $m \times n$ submatrix $\mathbf{B}_{q,ptg}(m,n)$ from $\mathbf{B}_{q,ptg}$ and a chosen decomposition factor k. Then, based on $\mathbf{B}_{q,ptg}(m,n)$ and k, we can construct an $m \times n$ array $\mathbf{H}_{q,ptg,qc}(k,k)$ of $k \times k$ circulants in the form of (11.11) using the decomposition method presented above. The null space of $\mathbf{H}_{q,ptg,qc}(k,k)$ gives a q-ary QC-PTG-LDPC code $C_{q,ptg,qc}$. To obtain the desired column and row weight distributions of the parity-check array $\mathbf{H}_{q,ptg,qc}(k,k)$, we can set some entries in the entry-sets of the base matrix $\mathbf{B}_{q,ptg}(m,n)$ to zeros, i.e., adjust the weights of the circulants in $\mathbf{H}_{q,ptg,qc}(k,k)$.

11.6 Construction of NB LDPC Codes from Reed–Solomon Codes

Reed–Solomon (RS) codes [93, 8, 74, 97] are prevalent in many communication systems and data storage devices. They are powerful cyclic codes that can be efficiently decoded using traditional algebraic hard-decision decoding methods such as the Berlekamp–Massey algorithm (BMA) [8, 74, 97] and Euclid's algorithm [74]. Unfortunately, due to the high density of nonzero elements in their conventional parity-check matrices and the large number of short cycles, especially cycles of length 4, in their associated Tanner graphs, direct application of iterative decoding methods based on the BP is not suitable for RS codes.

In this section, we will show that powerful NB LDPC codes can be constructed from RS codes based on their conventional parity-check matrices using the masking technique presented in Section 7.4. To demonstrate this, we use two RS codes, the $(255, 239)$ and the $(255, 223)$ RS codes over GF(2^8). The first code has minimum distance 17 and can correct up to 8 random symbol errors. This code decoded with the BMA or Euclid's method has been widely used for many commercial communication systems and data storage devices over the last 40 years and is still being used. The second code has minimum distance 33 and can correct up to 16 random symbol errors. It is a CCSDS standard code for satellite and deep space communications.

Consider the $(255, 239)$ RS code over GF(2^8). It is a cyclic code whose generator polynomial has $\alpha, \alpha^2, \ldots, \alpha^{16}$ as roots where α is a primitive element of GF(2^8) [97]. Its conventional parity-check matrix $\mathbf{H}_{q,rs}$ is expressed in terms of the roots of its generator polynomial [97, pp. 124–125]:

$$\mathbf{H}_{q,rs} = \begin{bmatrix} \alpha^0 & \alpha^1 & \alpha^2 & \cdots & \alpha^{254} \\ \alpha^0 & (\alpha^2)^1 & (\alpha^2)^2 & \cdots & (\alpha^2)^{254} \\ \vdots & \vdots & \ddots & \vdots & \\ \alpha^0 & (\alpha^{16})^1 & (\alpha^{16})^2 & \cdots & (\alpha^{16})^{254} \end{bmatrix}. \tag{11.13}$$

$\mathbf{H}_{q,rs}$ is a 16×255 matrix over GF(2^8). This parity-check matrix is not suitable for iterative decoding, since its associated Tanner graph has girth 4 and contains a enormously large number of cycles of length 4. Iterative BP-decoding of the RS code based on this parity-check matrix gives a very poor error performance, much worse than the hard-decision BMA. So far, *no efficient and practically implementable iterative decoding scheme (or algorithm) has been devised for this code*. Furthermore, $\mathbf{H}_{q,rs}$ does not satisfy the 2×2 SM-constraint. Hence, the Tanner graph associated with its CPM-dispersion contains a large number of cycles of length 4 and is not suitable to construct QC-LDPC codes.

In the next example, we show that, by masking the conventional parity-check matrix $\mathbf{H}_{q,rs}$ given by (11.13) with a properly designed 16×255 masking matrix, we can construct a 256-ary $(255, 239)$ LDPC code that performs well with an iterative BP-decoding algorithm.

Example 11.10. Let $\mathbf{H}_{q,rs}$ (given in the form of (11.13)) be the conventional parity-check matrix of the $(255, 239)$ RS code C_{rs} over GF(2^8). For masking $\mathbf{H}_{q,rs}$, we need a 16×255 masking matrix. There are many ways to construct this masking matrix. Here, we use the 2-dimensional Euclidean geometry EG$(2, 5)$ over the field GF(5) for the construction. Based on the the lines in EG$(2, 5)$ not passing through the origin, we can construct a 6×6 array $\mathbf{M}$ of *binary* CPMs and ZMs of size 4×4 [46] (see Appendix A). Each row-block (or each column-block) consists of 5 CPMs and 1 ZM of size 4×4. Take 11 arbitrary 4×6 subarrays from $\mathbf{M}$, denoted by $\mathbf{M}_0, \mathbf{M}_1, \ldots, \mathbf{M}_{10}$. Each of these 4×6 subarrays of $\mathbf{M}$ consists of 4 column-blocks each having 3 CPMs and one ZM, and 2 column-blocks each having 4 CPMs. In each column-block with 4 CPMs, we choose one CPM randomly and replace it by a 4×4 ZM. By doing this, each $\mathbf{M}_j$ of these 11 subarrays becomes a 4×6 subarray $\mathbf{M}_j^\star$ in which each column-block consists of 3 CPMs and one ZM. Arranging these eleven 4×6 subarrays, $\mathbf{M}_0^\star, \mathbf{M}_1^\star, \ldots, \mathbf{M}_{10}^\star$,

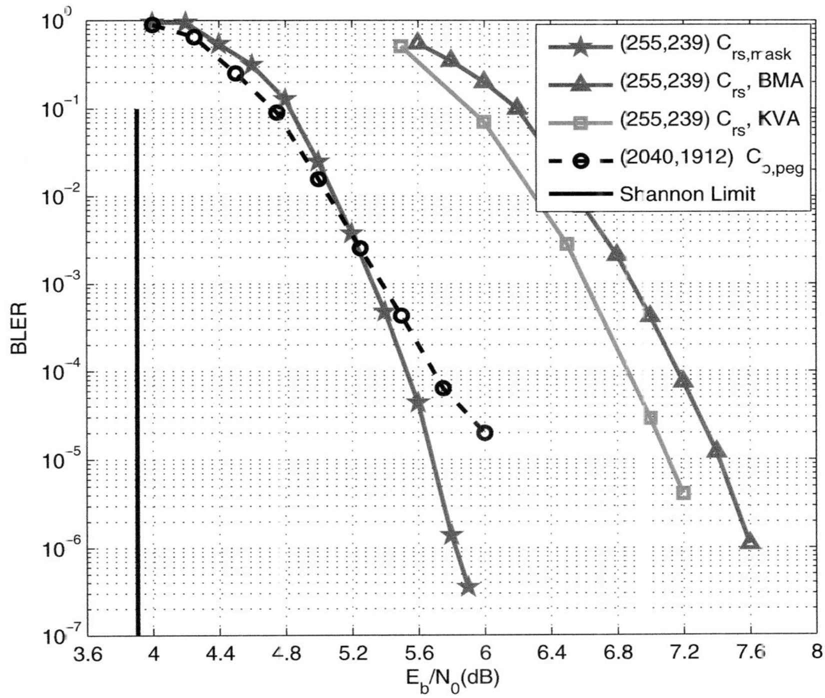

Figure 11.9. The BLER performances of the 256-ary $(255, 239)$ RS-masked LDPC code $C_{rs,mask}$, the 256-ary $(255, 239)$ RS code C_{rs} decoded with the BMA and the KVA, and the $(2040, 1912)$ binary PEG code $C_{b,peg}$ given in Example 11.10 over the binary-input AWGNC.

in a row, we obtain a 4×66 array $\mathbf{M}^\star = [\mathbf{M}_0^\star\ \mathbf{M}_1^\star\ \ldots\ \mathbf{M}_{10}^\star]$ of binary CPMs and ZMs. $\mathbf{M}^\star$ is a 16×264 matrix over GF(2). Deleting the last 9 columns from $\mathbf{M}^\star$, we obtain a 16×255 matrix $\mathbf{Z}$ with column weight 3, and 5 different row weights: 45, 46, 47, 48, and 49.

Masking $\mathbf{H}_{q,rs}$ with $\mathbf{Z}$, we obtain a masked RS matrix $\mathbf{H}_{q,rs,mask}$, whose null space gives a $(255, 239)$ LDPC code $C_{rs,mask}$ over GF(2^8), called an RS-masked LDPC code. The BLER performance of $C_{rs,mask}$ over the binary-input AWGNC is shown in Fig. 11.9. Also included in Fig. 11.9 are the BLER performances of the $(255, 239)$ RS code C_{rs} decoded with the BMA and the Kötter–Vardy algebraic soft-decision algorithm (KVA) with interpolation complexity coefficient 4.99 [54]. At a BLER of 10^{-6}, $C_{rs,mask}$ has 1.8 dB and 1.3 dB gains over the RS code C_{rs} decoded with the BMA and the KVA, respectively. The coding gains are achieved at the expense of a higher decoding complexity.

The binary image of the 256-ary $(255, 239)$ RS-masked LDPC code $C_{rs,mask}$ is a binary $(2040, 1912)$ code. For comparison, a binary $(2040, 1912)$ LDPC code

$C_{b,peg}$ is constructed using the the PEG algorithm [43, 44], which has the the same rate and same degree distributions of $C_{rs,mask}$. The PEG code has girth 6 and contains 146 250 cycles of length 6. The BLER performance of this binary LDPC code $C_{b,peg}$ over the AWGNC decoded with 50 iterations of the SPA is also included in Fig. 11.9. We see that the curves of the two codes $C_{rs,mask}$ and $C_{b,peg}$ basically overlap with each other for a BLER down to 4×10^{-4}. At a BLER lower than 4×10^{-4}, $C_{rs,mask}$ starts to outperform the binary PEG code $C_{b,peg}$ and the PEG code has an error-floor starting from a BLER of 5×10^{-6}.

Next, we consider the CCSDS standard (255, 223) cyclic RS code over GF(2^8).

Example 11.11. The (255, 223) RS code C_{rs} over GF(2^8) with minimum distance 32 has been used by international space agencies, such as US-NASA and ESA (European Space Agency), for space and satellite communications. Its conventional parity-check matrix $\mathbf{H}_{q,rs} = [(\alpha^i)^j]_{1 \leq i < \leq 32, 0 \leq j \leq 254}$ is a 32×255 matrix over GF(2^8) expressed in terms of the 32 roots of its generator polynomial, α, α^2, α^3, ..., α^{32}, where α is a primitive element of GF(2^8). To mask the parity-check matrix $\mathbf{H}_{q,rs}$ to produce a sparse matrix, we need a 32×255 masking matrix over GF(2). To do so, we construct eight 32×32 circulants $\mathbf{M}_0$, $\mathbf{M}_1$, ..., $\mathbf{M}_7$ over GF(2), each with weight 3. The generators of these eight circulants are:

$$\mathbf{g}_0 = (1,1,0,1,0,0,0,0,0,0,0,0,0,0,0,0,0,0,0,0,0,0,0,0,0,0,0,0,0,0,0,0),$$
$$\mathbf{g}_1 = (1,0,1,0,0,0,1,0,0,0,0,0,0,0,0,0,0,0,0,0,0,0,0,0,0,0,0,0,0,0,0,0),$$
$$\mathbf{g}_2 = (1,0,0,0,1,0,0,0,0,0,1,0,0,0,0,0,0,0,0,0,0,0,0,0,0,0,0,0,0,0,0,0),$$
$$\mathbf{g}_3 = (1,0,0,0,0,0,1,0,0,0,0,0,0,0,1,0,0,0,0,0,0,0,0,0,0,0,0,0,0,0,0,0),$$
$$\mathbf{g}_4 = (1,0,0,0,0,0,0,0,1,0,0,0,0,0,0,0,0,0,1,0,0,0,0,0,0,0,0,0,0,0,0,0),$$
$$\mathbf{g}_5 = (1,1,1,0,0,0,0,0,0,0,0,0,0,0,0,0,0,0,0,0,0,0,0,0,0,0,0,0,0,0,0,0),$$
$$\mathbf{g}_6 = (1,0,0,0,0,0,0,0,0,0,0,1,0,1,0,0,0,0,0,0,0,0,0,0,0,0,0,0,0,0,0,0),$$
$$\mathbf{g}_7 = (1,0,0,0,0,0,0,0,0,1,0,0,0,0,0,0,0,0,0,0,0,0,1,0,0,0,0,0,0,0,0,0).$$

Arrange these 8 circulants in a row-block to form a 32×256 matrix $\mathbf{M}^\star$ over GF(2). Deleting the last column of $\mathbf{M}^\star$, we obtain a 32×255 matrix $\mathbf{Z}$ with column weight 3 but two different row weights, 23 and 24. Masking $\mathbf{H}_{q,rs}$ with $\mathbf{Z}$, we obtain a masked 32×255 matrix $\mathbf{H}_{q,rs,mask}$. The null space of $\mathbf{H}_{q,rs,mask}$ gives a 256-ary (255, 223) RS-masked LDPC code $C_{rs,mask}$ whose BLER performance over the binary-input AWGNC decoded with 50 iterations of the FFT-QSPA is shown in Fig. 11.10. At a BLER of 10^{-5}, this code achieves almost 2 dB and 1.6 dB gains over the (255, 223) RS code C_{rs} decoded with BMA, and the KVA with interpolation complexity coefficient 4.99, respectively.

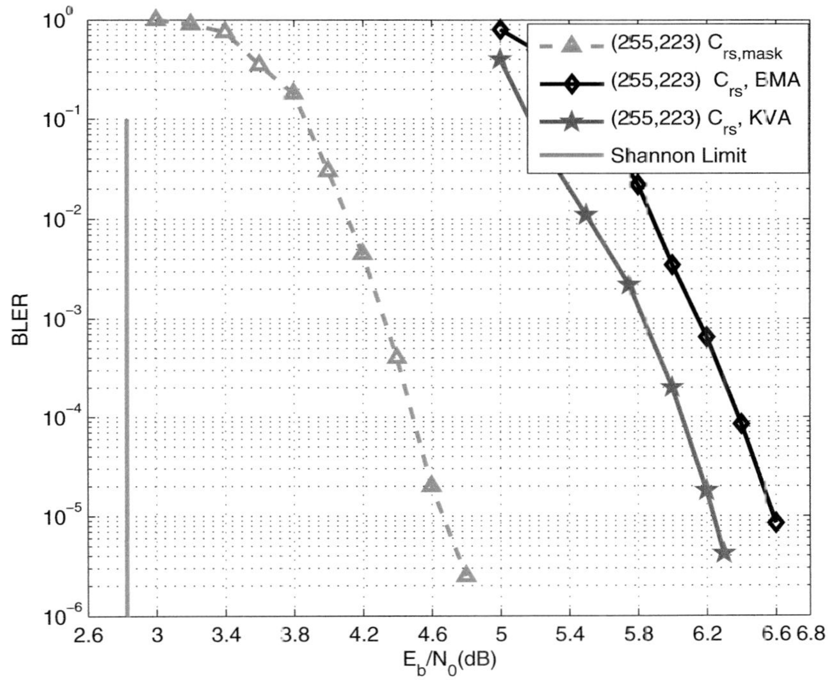

Figure 11.10. The BLER performances of the 256-ary (255, 223) RS-masked LDPC code $C_{rs,mask}$ and the 256-ary (255, 223) RS code C_{rs} decoded with BMA and the KVA given in Example 11.11 over the binary-input AWGNC.

What we showed above is that masking is a powerful technique to construct a good LDPC code from a parity-check matrix that is not suitable for iterative BP-decoding. RS-masked LDPC codes over other fields of characteristic 2 can be constructed in a similar manner.

11.7 Construction of NB QC-SP-LDPC Codes based on RS Codes

In general, the conventional parity-check matrix $\mathbf{H}_{q,rs}$ of an RS code over a field GF(2^s) of characteristic 2 does not satisfy the 2×2 SM-constraint. However, under a certain constraint, the parity-check matrix of an RS code over GF(2^s) *does* satisfy the 2×2 SM-constraint. In this case, the parity-check matrix of an RS code (or its shortened version) can be used as a base matrix to construct a 2^s-ary QC-SP-LDPC code by 2^s-ary CPM-dispersion whose Tanner graph has girth at least 6.

Let n be a factor of $2^s - 1$ such that $2^s - 1 = cn$ with $n > 1$. Let p_s be the smallest prime factor of n. Let α be a primitive element of GF(2^s) and let $\beta = \alpha^c$. The order of β is n and the set $\{1, \beta, \ldots, \beta^{n-1}\}$ forms a cyclic subgroup of GF(2^s). Let d be a positive integer less than or equal to p_s , i.e., $1 \leq d \leq p_s$. Consider the $(n, n-d)$ RS code C_{rs} over GF(2^s) of length n with minimum distance $d+1$. The generator polynomial $g(X)$ of this RS code has $\beta, \beta^2, \ldots, \beta^d$ as roots. The parity-check matrix of this RS code in terms of the roots of its generator polynomial is given below:

$$\mathbf{B}_{q,sp,rs}(d,n) = \begin{bmatrix} 1 & \beta & \beta^2 & \cdots & \beta^{n-1} \\ 1 & \beta^2 & (\beta^2)^2 & \cdots & (\beta^2)^{n-1} \\ \vdots & \vdots & \vdots & \cdots & \vdots \\ 1 & \beta^d & (\beta^d)^2 & \cdots & (\beta^d)^{n-1} \end{bmatrix}. \tag{11.14}$$

$\mathbf{B}_{q,sp,rs}(d,n)$ is a $d \times n$ matrix over GF(2^s). If $n = 2^s - 1$, then the code C_{rs} generated by $\mathbf{B}_{q,sp,rs}(d,n)$ is a primitive RS code [74, 97]. Otherwise, C_{rs} is a nonprimitive RS code.

Next, we prove that $\mathbf{B}_{q,sp,rs}(d,n)$ satisfies the 2×2 SM-constraint. Let i, j, k, and l be any four nonnegative integers such that $1 \leq i < j \leq d$ and $0 \leq k < l < n$. Consider the following 2×2 submatrix $\mathbf{B}(2,2)$ of $\mathbf{B}_{q,sp,rs}(d,n)$:

$$\mathbf{B}(2,2) = \begin{bmatrix} (\alpha^i)^k & (\alpha^i)^l \\ (\alpha^j)^k & (\alpha^j)^l \end{bmatrix}.$$

The determinant of $\mathbf{B}(2,2)$ is $\beta^{jl+ik} - \beta^{jk+il}$. Next, we show that $\beta^{jl+ik} - \beta^{jk+il} \neq 0$. Suppose $\beta^{jl+ik} - \beta^{jk+il} = 0$. Then, we must have $\beta^{(j-i)(l-k)} = 1$. Since $j - i$ is a positive integer less than $d \leq p_s$ (the smallest prime factor of n) and $l - k$ is a positive integer less than n, the product $(j-i)(l-k)$ cannot be divisible by n. Since the order of β is n and $(j-i)(l-k)$ is not divisible by n, $\beta^{(j-i)(l-k)}$ cannot be equal to 1. Hence, $\beta^{jl+ik} - \beta^{jk+il} \neq 0$ and $\mathbf{B}(2,2)$ must be a nonsingular matrix. Consequently, the matrix $\mathbf{B}_{q,sp,rs}(d,n)$ satisfies the 2×2 SM-constraint.

Since $\mathbf{B}_{q,sp,rs}(d,n)$ satisfies the 2×2 SM-constraint, it or a submatrix of it can be used as the base matrix for the SP-construction of a CPM-QC-SP-LDPC code, binary or NB, using the CPM-D method. Dispersing each entry of $\mathbf{B}_{q,sp,rs}(d,n)$ into a 2^s-ary CPM of size $(2^s - 1) \times (2^s - 1)$, we obtain a $d \times n$ array $\mathbf{H}_{q,sp,qc,rs}(2^s - 1, 2^s - 1)$ of 2^s-ary CPMs of size $(2^s - 1) \times (2^s - 1)$ which satisfies the RC-constraint. It is a $d(2^s - 1) \times n(2^s - 1)$ matrix over GF(2^s) with column and row weights d and n, respectively. The null space of $\mathbf{H}_{q,sp,qc,rs}(2^s - 1, 2^s - 1)$ gives a 2^s-ary CPM-QC-SP-LDPC code $C_{q,sp,qc,rs}$ of length $n(2^s - 1)$ whose Tanner graph has girth at least 6. We call this code an *RS-QC-SP-LDPC code* and the matrix $\mathbf{B}_{q,sp,rs}(d,n)$ an *RS-base matrix*.

Note that the entries in $\mathbf{B}_{q,sp,rs}(d,n)$ are elements of the cyclic subgroup $\{1, \beta, \beta^2, \ldots, \beta^{n-1}\}$ of GF(2^s) of order n. Dispersing each entry of $\mathbf{B}_{q,sp,rs}(d,n)$

into a 2^s-ary CPM of size $n \times n$ (as described in Section 2.1), we obtain a $d \times n$ array $\mathbf{H}_{q,sp,qc,rs}(n, n)$ of 2^s-ary CPMs of size $n \times n$ which also satisfies the RC-constraint. It is a $dn \times n^2$ matrix over GF(2^s) with column and row weights d and n, respectively. The null space of $\mathbf{H}_{q,sp,qc,rs}(n, n)$ gives a 2^s-ary (d, n)-regular RS-QC-SP-LDPC code $C_{q,sp,qc,rs}$ of length n^2 whose Tanner graph has girth at least 6.

Example 11.12. Let GF(2^6) be the field for the code construction and let α be a primitive element of GF(2^6). Then, $2^6 - 1 = 63$. Let $n = 63$. The smallest prime factor of 63 is 3. In this case, the parity-check matrix $\mathbf{B}_{q,sp,rs}(3, 63)$ of the $(63, 60)$ RS code over GF(2^6) is a 3×63 matrix and satisfies the 2×2 SM-constraint. The 64-ary CPM-dispersion of $\mathbf{B}_{q,sp,rs}(3, 63)$ gives a 3×63 array $\mathbf{H}_{q,sp,qc,rs}(63, 63)$ of 64-ary CPMs of size 63×63 which is a 189×3969 matrix over GF(2^6) with column and row weights 3 and 63, respectively. The null space of $\mathbf{H}_{q,sp,qc,rs}(63, 63)$ gives a 64-ary $(3, 63)$-regular $(3969, 3780)$ RS-QC-SP-LDPC code $C_{q,sp,qc,rs}$ with rate 0.9524 whose Tanner graph has girth at least 6. The BER, SER (symbol-error rate), and BLER performances of $C_{q,sp,qc,rs}$ over the binary-input AWGNC decoded with 50 iterations of the FFT-QSPA are shown in Fig. 11.11. At a BER of 10^{-7}, the code performs only 0.75 dB from the Shannon limit and at a BLER of 10^{-5}, it performs 0.5 dB from the SPB.

For comparison, we also include the BER, SER, and BLER performances of a 64-ary $(3969, 3780)$ LDPC code $C_{q,peg}$ constructed using the PEG algorithm in Fig. 11.11. The PEG code $C_{q,peg}$ has the same rate and dimension as $C_{q,sp,qc,rs}$. Its parity-check matrix has constant column weight 3, but not in QC-form. From Fig. 11.11, we see that the two codes perform the same.

Example 11.13. In this example, we use GF(2^8) as the field for the code construction. Factor $2^8 - 1 = 255$ as the product of 3 and 85. Set $n = 85$. The smallest prime factor of 85 is 5. Let α be a primitive element of GF(2^8) and let $\beta = \alpha^3$. The order of β is 85. Consider the nonprimitive $(85, 80)$ RS code over GF(2^8) whose generator polynomial has β, β^2, β^3, β^4, and β^5 as roots. The parity-check matrix $\mathbf{B}_{q,sp,rs}(5, 85)$ of this RS code is a 5×85 matrix over GF(2^8).

Take a 3×85 submatrix $\mathbf{B}_{q,sp,rs}(3, 85)$ from $\mathbf{B}_{q,sp,rs}(5, 85)$ and use it as the SP-base matrix for the construction of an RS-QC-SP-LDPC code. Dispersing each entry of $\mathbf{B}_{q,sp,rs}(3, 85)$ into a 256-ary CPM of size 85×85, we obtain a 3×85 array $\mathbf{H}_{q,sp,qc,rs}(85, 85)$ of 256-ary CPMs of size 85×85, which is a 255×7225 matrix over GF(2^8) with column and row weights 3 and 85, respectively. The null space of $\mathbf{H}_{q,sp,qc,rs}(85, 85)$ gives a 256-ary $(3, 85)$-regular $(7225, 6970)$ RS-QC-SP-LDPC code $C_{q,sp,qc,rs}$ with rate 0.9647. The BER, SER, and BLER performances of $C_{q,sp,qc,rs}$ over the binary-input AWGNC decoded with 50 iterations of the

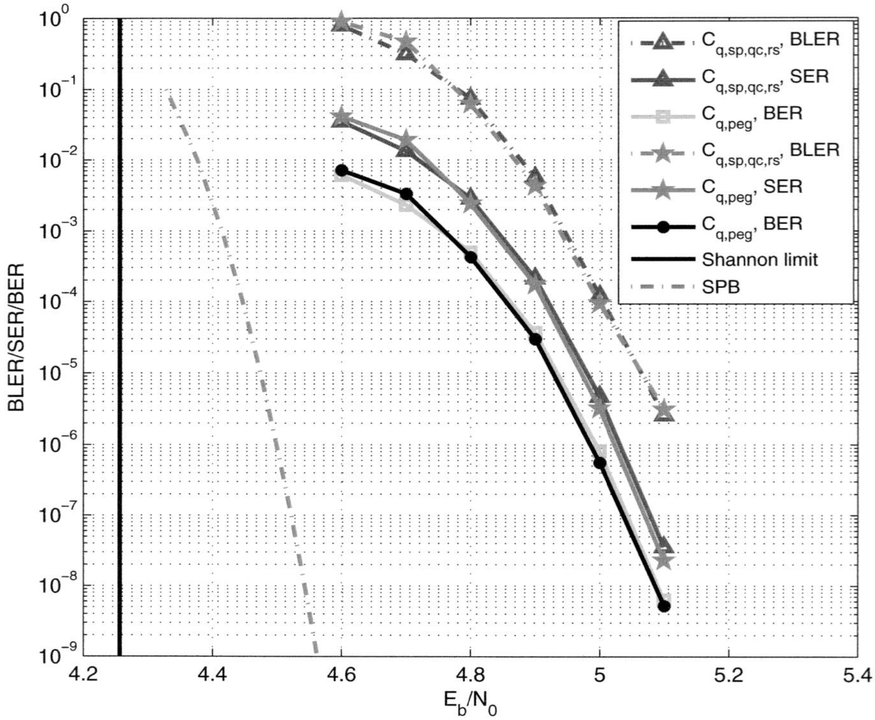

Figure 11.11. The BER, SER, and BLER performances of the 64-ary (3969, 3780) RS-QC-SP-LDPC $C_{q,sp,qc,rs}$ code and the PEG code $C_{q,peg}$ given in Example 11.12 over the binary-input AWGNC.

FFT-QSPA are shown in Fig. 11.12. At a BER of 10^{-6}, the code performs only 0.7 dB from the Shannon limit (4.64 dB) and at a BLER of 10^{-4}, it performs 0.53 dB from the SPB.

For comparison, we also include the BER, SER, and BLER performances of a 256-ary (7225, 6970) LDPC code $C_{q,peg}$ constructed using the PEG algorithm in Fig. 11.12. The PEG code $C_{q,peg}$ has the same rate and dimension as $C_{q,sp,qc,rs}$. Its parity-check matrix has constant column weight 3, but not in QC-form. From Fig. 11.12, we see that the RS-QC-SP-LDPC code $C_{q,sp,qc,rs}$ slightly outperforms the PEG code $C_{q,peg}$.

Now, we consider a special case. Let p be the largest prime factor of $2^s - 1$. Then $2^s - 1 = cp$. Let $\beta = \alpha^c$ where α is a primitive element of GF(2^s). Then, the order of β is p. Let d be a positive integer such that $d < p$. Then, the parity-check matrix $\mathbf{B}_{q,sp,rs}(d, p)$ of the $(p, p-d)$ nonprimitive RS code of length p over GF(2^s) satisfies the 2×2 SM-constraint. The generator polynomial of this RS code has

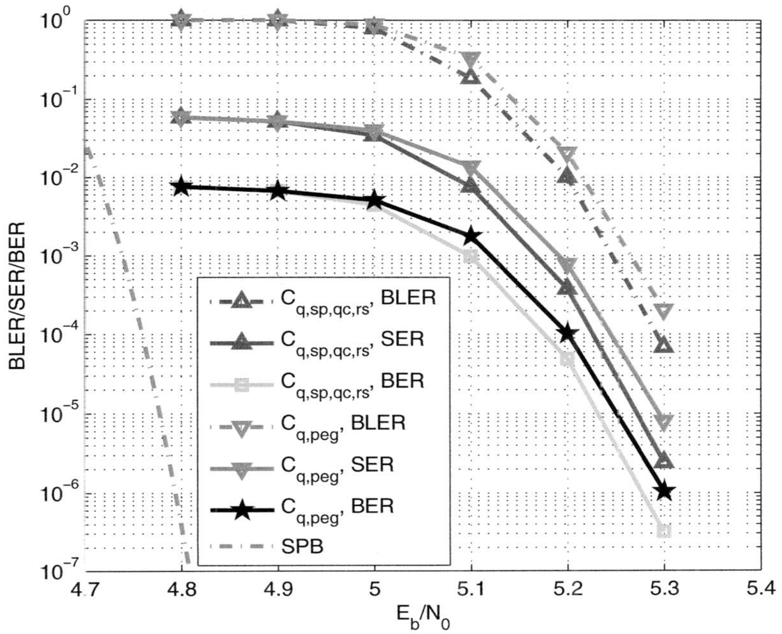

Figure 11.12. The BER, SER, and BLER performances of the 256-ary (7225, 6970) RS-QC-SP-LDPC code $C_{q,sp,qc,rs}$ and the PEG code $C_{q,peg}$ given in Example 11.13 over the binary-input AWGNC.

$\beta, \beta^2, \ldots, \beta^d$ as roots and its parity-check matrix $\mathbf{B}_{q,sp,rs}(d, p)$ is a $d \times p$ matrix over GF(2^s) specified by the d roots of the generator polynomial of the RS code.

If we disperse each entry in $\mathbf{B}_{q,sp,rs}(d, p)$ into a 2^s-ary CPM of size $(2^s - 1) \times (2^s - 1)$, we obtain a $d \times p$ array $\mathbf{H}_{q,sp,qc,rs}(2^s - 1, 2^s - 1)$ of 2^s-ary CPMs of size $(2^s - 1) \times (2^s - 1)$. The null space of $\mathbf{H}_{q,sp,qc,rs}(2^s - 1, 2^s - 1)$ gives a 2^s-ary RS-QC-SP-LDPC code $C_{q,sp,qc,rs}$ of length $p(2^s - 1)$. If we disperse each entry of $\mathbf{B}_{q,sp,rs}(d, p)$ into a 2^s-ary CPM of size $p \times p$, we then obtain a $d \times p$ array $\mathbf{H}_{q,sp,qc,rs}(p, p)$ of 2^s-ary CPMs of size $p \times p$. The null space of $\mathbf{H}_{q,sp,qc,rs}(p, p)$ gives a 2^s-ary RS-QC-SP-LDPC code $C_{q,sp,qc,rs}$ of length p^2. For various choices of $d < p$, we can construct a family of 2^s-ary RS-QC-SP-LDPC codes of various lengths and rates.

Example 11.14. Let GF(2^5) be the field for the code construction and α be a primitive element of GF(2^5). Since $2^5 - 1 = 31$ is a prime, then $p = 31$ (the largest prime factor). Set $d = 4$. The parity-check matrix of a double error-correcting RS code over GF(2^5) is a 4×31 matrix over GF(2^5) given as follows:

$$\mathbf{B}_{q,sp,rs}(4,31) = \begin{bmatrix} \alpha^0 & \alpha^1 & \alpha^2 & \cdots & \alpha^{30} \\ \alpha^0 & (\alpha^2)^1 & (\alpha^2)^2 & \cdots & (\alpha^2)^{30} \\ \alpha^0 & (\alpha^3)^1 & (\alpha^3)^2 & \cdots & (\alpha^3)^{30} \\ \alpha^0 & (\alpha^4)^1 & (\alpha^4)^2 & \cdots & (\alpha^4)^{30} \end{bmatrix}.$$

The null space of $\mathbf{B}_{q,sp,rs}(4,31)$ gives a $(31,27)$ RS code over GF(2^5).

The 32-ary CPM-dispersion of the RS-base matrix $\mathbf{B}_{q,sp,rs}(4,31)$ gives a 4×31 array $\mathbf{H}_{q,sp,qc,rs}(31,31)$ of 32-ary CPMs of size 31×31 with column and row weights 4 and 31, respectively. The null space of $\mathbf{H}_{q,sp,qc,rs}(31,31)$ gives a $(4,31)$-regular 32-ary $(961,840)$ RS-QC-SP-LDPC code of rate 0.874. Its Tanner graph has girth 6 and contains 115 320 cycles of length 6 and 8 735 490 cycles of length 8. The BER, SER, and BLER performances of the code decoded with 50 iterations of the FFT-QSPA over the binary-input AWGNC are shown in Fig. 11.13. From this figure, we see that the code can achieve a BER of 10^{-9} without a visible error-floor. At a BER of 10^{-9}, the code performs within 1.76 dB from the Shannon limit and 1.2 dB from the SPB at a BLER of 10^{-7}.

For comparison, we also include the BER, SER, and BLER performances of a 32-ary $(961,837)$ LDPC code $C_{q,peg}$ constructed using the PEG algorithm in Fig. 11.13. The rate of the PEG code $C_{q,peg}$ is slightly lower than that of the code $C_{q,sp,qc,rs}$. Its parity-check matrix has constant column weight 3, but not in QC-form. From Fig. 11.13, we see that the two codes perform the same.

Using the CPM-*b*D/B-to-NB construction presented in Section 11.4, we can construct short RS-QC-SP-LDPC codes over large symbol fields or long RS-QC-SP-LDPC codes with small symbol fields.

Example 11.15. Consider the $(7,4)$ RS code over GF(2^3) whose generator polynomial has α, α^2, and α^3 as roots where α is a primitive element of GF(2^3). Its parity-check matrix $\mathbf{B}_{q,sp,rs}(3,7)$ is a 3×7 matrix over GF(2^3) which satisfies the 2×2 SM-constraint. Suppose we take the binary CPM-dispersion of $\mathbf{B}_{q,sp,rs}(3,7)$ and use the field GF(2^8) for the B-to-NB replacement (random replacement). We obtain a 3×7 array $\mathbf{H}_{q,sp,qc,rs}(7,7)$ of CPMs of size 7×7 over GF(2^8) which is a 21×49 matrix over GF(2^8) with column and row weights 3 and 7, respectively. The null space of $\mathbf{H}_{q,sp,qc,rs}(7,7)$ gives a 256-ary $(49,28)$ RS-QC-SP-LDPC code $C_{q,sp,qc,rs,0}$, a short code over a large field. The BER, SER, and BLER performances of the code over the binary-input AWGNC decoded with 50 iterations of the FFT-QSPA are shown in Fig. 11.14. At a BLER of 10^{-6}, the code performs about 2.1 dB away from the SPB.

For comparison, we also include the BER, SER and BLER performances of a 256-ary $(49,28)$ LDPC code $C_{q,peg,0}$ constructed using the PEG algorithm in

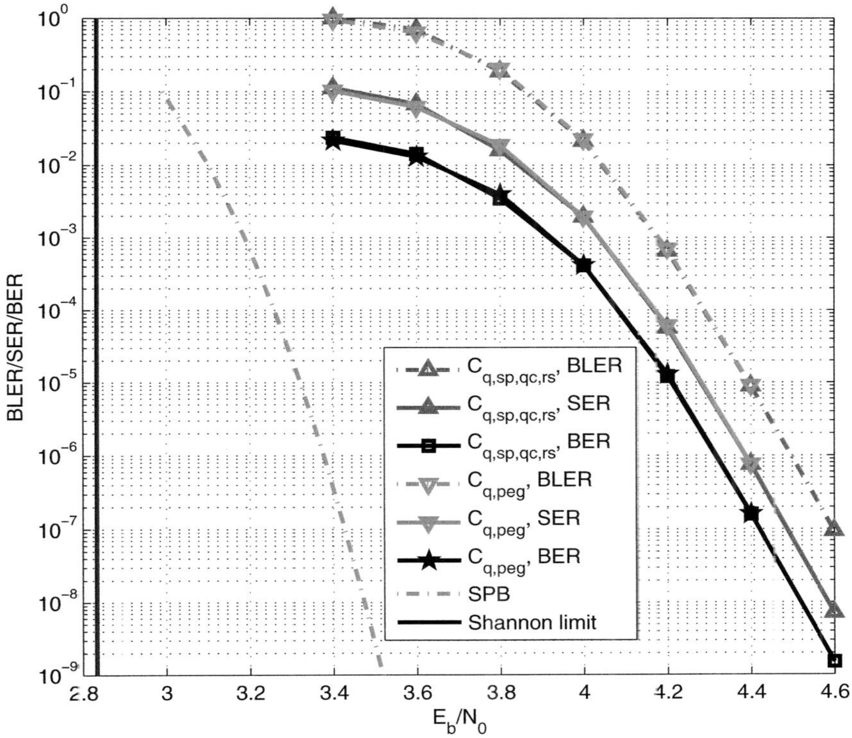

Figure 11.13. The BER, SER, and BLER performances of the 32-ary $(961, 840)$ RS-QC-SP-LDPC code $C_{q,sp,qc,rs}$ and the 32-ary $(961, 837)$ PEG code $C_{q,peg}$ given in Example 11.14 over the binary-input AWGNC.

Fig. 11.14. The rate of the PEG code $C_{q,peg,0}$ is the same as that of the code $C_{q,sp,qc,rs,0}$. Its parity-check matrix has constant column weight 3, but not in QC-form. From Fig. 11.14, we see that the two codes perform almost the same.

Suppose we disperse the matrices $\mathbf{B}_{q,sp,rs}(3, 63)$ over GF(2^6) given in Example 11.12 and $\mathbf{B}_{q,sp,rs}(3, 85)$ over GF(2^8) given in Example 11.13 into a 3×63 binary array of CPMs of size 63×63 and a 3×85 binary array of CPMs of size 85×85, respectively. Using GF(2^4) as the field for the B-to-NB replacement, we obtain a 3×63 array $\mathbf{H}_{q,sp,qc,rs}(63, 63)$ of 16-ary CPMs of size 63×63 and a 3×85 array $\mathbf{H}_{q,sp,qc,rs}(85, 85)$ of 16-ary CPMs of size 85×85. The null spaces of $\mathbf{H}_{q,sp,qc,rs}(63, 63)$ and $\mathbf{H}_{q,sp,qc,rs}(85, 85)$ give a 16-ary $(3, 63)$-regular $(3969, 3780)$ RS-QC-SP-LDPC code $C_{q,sp,qc,rs,1}$ and a 16-ary $(3, 85)$-regular $(7225, 6970)$ RS-QC-SP-LDPC code $C_{q,sp,qc,rs,2}$, respectively. The BER, SER and BLER performances of the two codes $C_{q,sp,qc,rs,1}$ and $C_{q,sp,qc,rs,2}$ over the binary-input AWGNC decoded with 50 iterations of the FFT-QSPA are shown in Fig. 11.15 and Fig. 11.16, respectively. The 16-ary $(3969, 3780)$ code $C_{q,sp,qc,rs,1}$

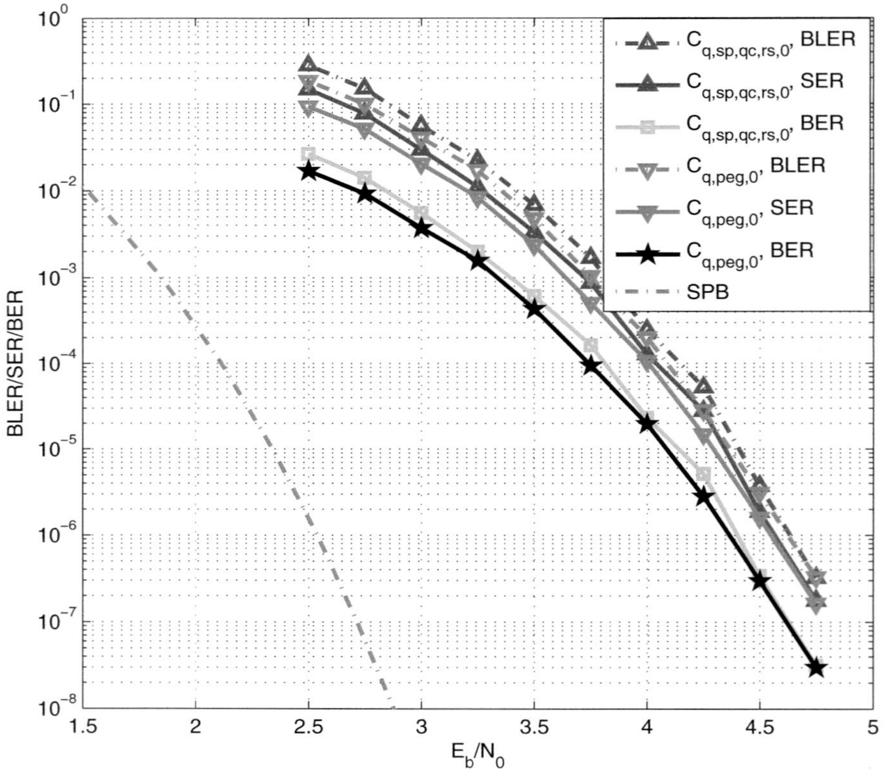

Figure 11.14. The BER, SER, and BLER performances of the 256-ary (49, 28) RS-QC-SP-LDPC code $C_{q,sp,qc,rs,0}$ and the PEG code $C_{q,peg,0}$ given in Example 11.15 over the binary-input AWGNC.

performs about 0.85 dB away from the Shannon limit at a BER of 10^{-8} and 0.55 dB away from the SPB at a BLER of 10^{-5}. The 16-ary (7225, 6970) code $C_{q,sp,qc,rs,2}$ performs about 0.67 dB away from the Shannon limit at a BER of 10^{-7} and 0.5 dB away from the SPB at a BLER of 10^{-5}.

For comparison, we also include the BER, SER, and BLER performances of a 16-ary (3969, 3780) LDPC code $C_{q,peg,1}$ constructed using the PEG algorithm in Fig. 11.15. The rate of the PEG code $C_{q,peg,1}$ is the same as that of the code $C_{q,sp,qc,rs,1}$. Its parity-check matrix has constant column weight 3, but not in QC-form. From Fig. 11.15, we see that the two codes perform almost the same.

Also included in Fig. 11.16 is the performance of a 16-ary (7225, 6970) LDPC code $C_{q,peg,2}$ whose rate and dimension are the same as those of the code $C_{q,sp,qc,rs,2}$. The PEG code $C_{q,peg,2}$ is constructed using the PEG algorithm and its parity-check matrix has constant column weight 3. From Fig. 11.16, we see that performance curves of the codes $C_{q,sp,qc,rs,2}$ and $C_{q,peg,2}$ overlap with each other.

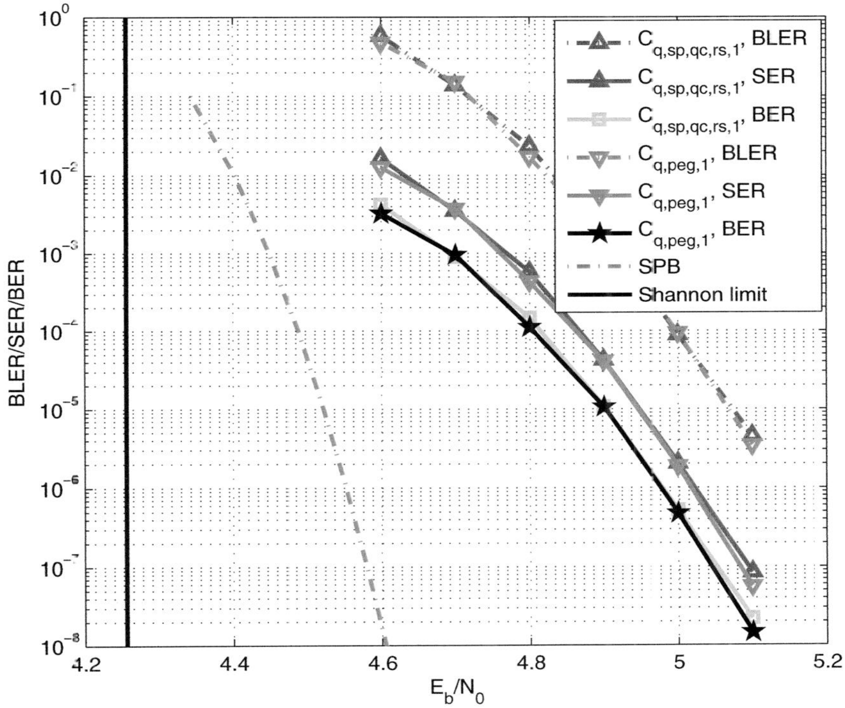

Figure 11.15. The BER, SER, and BLER performances of the 16-ary $(3969, 3780)$ RS-QC-SP-LDPC code $C_{q,sp,qc,rs,1}$ and the PEG code $C_{q,peg,1}$ given in Example 11.15 over the binary-input AWGNC.

In this and the previous sections, what we have shown is that, even though we cannot decode RS codes directly based on their conventional parity-check matrices using a iterative decoding algorithm based on BP, we can construct powerful NB LDPC codes from these conventional parity-check matrices using the masking technique, the q-ary CPM-dispersion, and the CPM-bD/B-to-NB constructions.

11.8 Discussion and Remarks

In Chapters 8 to 10, we presented various algebraic methods for constructing binary doubly QC-LDPC, QC-SC-LDPC, and QC-GC-LDPC codes. These methods of binary code construction can be generalized in a straightforward manner to construct NB doubly QC-LDPC, QC-SC-LDPC, and QC-GC-LDPC codes. This is achieved simply by dispersing a nonzero entry in a 2×2 SM-constrained base matrix into a q-ary CPM. Binary CPM-dispersion combined with B-to-NB

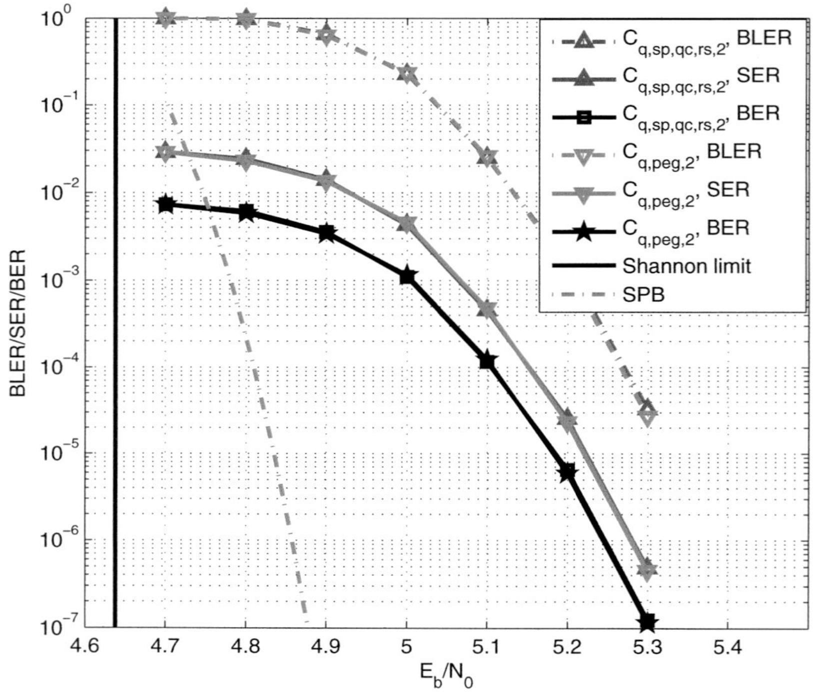

Figure 11.16. The BER, SER, and BLER performances of the 16-ary (7225, 6970) RS-QC-SP-LDPC code $C_{q,sp,qc,rs,2}$ and its comparable PEG code $C_{q,peg,2}$ given in Example 11.15 over the binary-input AWGNC.

mapping allows us to construct NB QC-LDPC codes with large size CPMs over small code symbol fields.

In general, supported by many experimental results, an NB LDPC code over a larger symbol field has a better error-floor performance than an NB LDPC codes with a smaller symbol field. However, the decoder complexity of an NB LDPC code increases greatly as the size of the symbol field increases. A reduced-complexity BP-algorithm for decoding NB LDPC codes with an efficient trade-off between error performance and decoder complexity is desired. The reduced-complexity iterative schemes proposed in [79, 69, 68, 78] for decoding binary QC-LDPC codes with a block-cyclic structure or a section-wise cyclic structure using a small submatrix of the parity-check matrix of the code may be used for decoding NB QC-LDPC codes.

Another interesting research problem is to find an effective BP-algorithm for iterative decoding of RS codes for practical application, that achieves good error performance with decoding complexity much lower than the existing algorithms presented in [54, 37, 7, 48]. Finding an efficient soft-decision decoding method (or algorithm) for RS codes has always been a challenging problem over the last 40 or more years. A breakthrough in soft-decision decoding of RS codes is needed if we want to keep these powerful error-correcting codes playing an important role in future generations of communication and data storage systems.

In the last section of this chapter, we used the conventional parity-check matrix of an RS code as the base matrix for constructing an NB CPM-QC-SP-LDPC code. Clearly, such a base matrix can be used for constructing a binary CPM-QC-SP-LDPC code if each entry is dispersed into a binary CPM as described in Chapters 7 to 10. A high-performance binary CPM-QC-SP-LDPC code was constructed using an RS parity-check matrix as the base matrix, which we would like to point out here is the $(6, 127)$-regular $(16\,129, 15\,372)$ QC-LDPC code recently reported in [29, 77]. The base matrix $\mathbf{B}_{q,sp,rs}(6, 127)$ given in the form of (11.14) for constructing this code is the conventional parity-check matrix of the triple-error-correcting $(127, 121)$ RS code of length 127 over GF(2^7) with minimum distance 7 and rate 0.9528. $\mathbf{B}_{q,sp,rs}(6, 127)$ is a 6×127 matrix over GF(2^7). The binary CPM-dispersion of $\mathbf{B}_{q,sp,rs}(6, 127)$ gives a 6×127 array $\mathbf{H}_{b,sp,qc,rs}(127, 127)$ of CPMs of size 127×127. It is a $762 \times 16\,129$ binary matrix

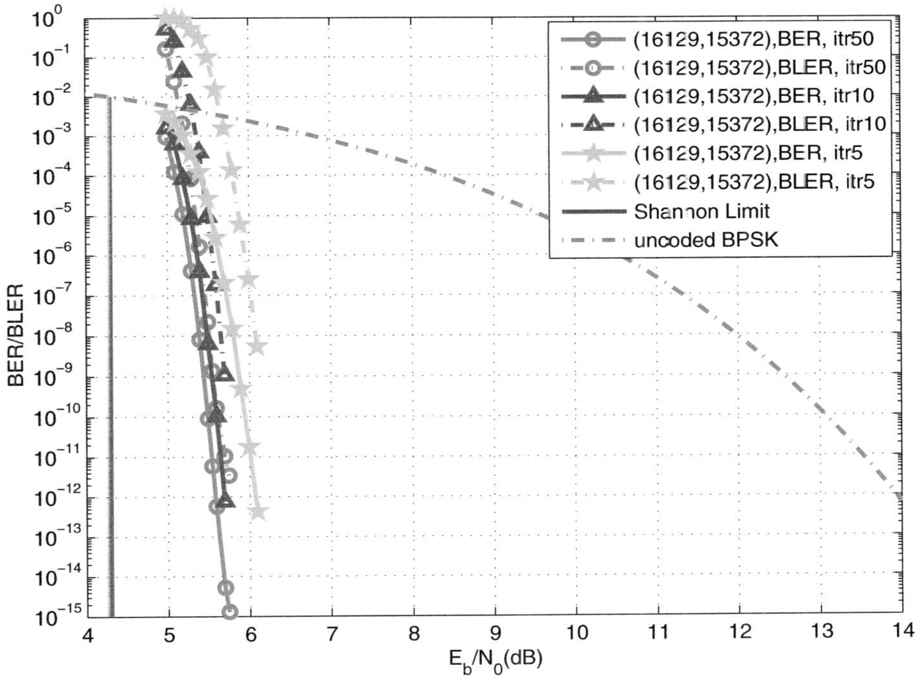

Figure 11.17. The BER and BLER performances of the binary (16 129, 15 372) QC-LDPC code reported in [29, 77] decoded with 5 , 10, and 50 iterations of the MSA over the AWGNC.

with column weight 6 and row weight 127. The null space of $\mathbf{H}_{b,sp,qc,rs}(127, 127)$ gives a (6, 127)-regular (16 129, 15 372) CPM-QC-SP-LDPC code. The BER and BLER performances of this code decoded with 5, 10, and 50 iterations of the MSA with a scaling factor 0.75 are shown in Fig. 11.17. From this figure, we see that with 50 iterations of the MSA, the code achieves a BER of 10^{-15} and a BLER of almost 10^{-12} without visible error-floors. We also see that the decoding of the code converges quickly. At a BER of 10^{-12}, the performance gap between 5 and 50 iterations of the MSA is about 0.5 dB. The performance curves of the code decoded with 10 and 50 iterations of the MSA almost overlap with each other all the way down to a BER of 10^{-12}. This code has good waterfall and error-floor performance as well as a fast rate of decoding convergence. Hence, it has an excellent overall performance as defined in Section 2.2. Using the reduced-complexity iterative decoding scheme proposed in [69, 68, 78], the code can be decoded with a matrix that consists of the 6 top rows of the 6 row-blocks of the 6×127 parity-check array $\mathbf{H}_{b,sp,qc,rs}(127, 127)$ of the code. So the decoding matrix is a 6×16129 binary matrix, a very small submatrix of $\mathbf{H}_{b,sp,qc,rs}(127, 127)$ with size 762×16129. Using this decoding matrix, the decoder complexity of this code is drastically reduced (by a factor of 127) without performance degradation.

A code with such a high rate and such a low error probability without an error-floor may find applications in high-speed optical communication and flash memory systems. No computer constructed code with the same rate and the same low error rate has been reported in published literature. This high-performance LDPC code demonstrates the art of algebraic construction of structured LDPC code.

12

Conclusion and Remarks

In this book, we unified several major types of LDPC code constructions using a single framework, namely, the SP-construction. Under this framework, the constructions of all these types of LDPC codes can be viewed as basically algebraic. In general, algebraically constructed LDPC codes have good overall performance in terms of both waterfall and error-floor performance, as well as a fast rate of decoding convergence. The unification of all these constructions may lead to better designs, and construction of high performing and more easily implementable codes for applications in next generations of communication and data storage systems.

This book consists of seven parts. In the first part (Chapter 4), we gave an algebraic interpretation of the protograph-based (PTG-based) construction of LDPC codes and presented a simple and novel algebraic method for constructing PTG-LDPC codes. The proposed algebraic method is equivalent to the graphical copy-and-permute operation and uses a simple matrix decomposition-and-replacement process to construct PTG-LDPC codes. The algebraic method is based on decomposition of a small base matrix and is very flexible in code construction. The constructed codes perform well as supported by examples and simulation results. However, how to design decomposition base matrices so that the resultant PTG-LDPC codes can perform close to their decoding thresholds is still an unresolved problem which needs further investigation.

In the second part of the book (Chapters 5 and 6), we re-interpreted the SP-construction of LDPC codes, one of the earliest methods for algebraic construction of LDPC codes, in a broader context, from both algebraic and graph-theoretic perspectives. We showed that the PTG-LDPC code construction is a special case of SP-construction. New constructions of RC-constrained base matrices and PW-RC-constrained replacement sets of matrices for the SP-construction of LDPC codes were presented.

In Chapter 5, we showed that two major ensembles of SP-LDPC codes for a given rate can be formed. One ensemble is equivalent to the ensemble of PTG-LDPC codes. In forming this ensemble, the member matrices in the replacement set $\boldsymbol{R}$ for the SP-construction of LDPC codes are regular. Since this ensemble is equivalent to the ensemble of PTG-LDPC codes, the SP-LDPC codes in this ensemble have good asymptotic performance and structural properties. For the other ensemble, the member matrices in the replacement set $\boldsymbol{R}$ for the SP-construction of LDPC codes are not regular. Whether this ensemble of SP-LDPC codes has good asymptotic properties and performance is unknown. An analysis of this ensemble is a good research problem.

The third part of the book (Chapter 7) unified most known algebraic constructions of QC-LDPC codes based on matrix dispersion under the framework of the SP-construction. This unification leads to a broader class of constructions of LDPC codes, both QC and random, allowing the possibility of finding codes with improved waterfall and error-floor performance and faster rates of decoding convergence. Necessary and sufficient conditions, namely the 2×2 and the 3×3 SM-constraints, were given for constructing base matrices whose matrix dispersions give QC-LDPC codes with girth at least 6 and 8, respectively. A powerful method was presented to construct base matrices which satisfy the 2×2 SM-constraint. This construction is based on two arbitrary subsets of a finite field. We showed that at least six algebraic methods developed over the last 11 years to construct 2×2 SM-constrained base matrices are special cases of this method. So far, there is no specific algebraic method for constructing 3×3 SM-constrained base matrices. To construct 3×3 SM-constrained base matrices, we rely on masking. A test algorithm and a cycle counting algorithm are needed to check whether a masked base matrix satisfies the 3×3 SM-constraint. Such a test algorithm was developed (see Appendix B). Therefore, to find an algebraic method to construct 3×3 SM-constrained base matrices would be an interesting problem to look into.

In Chapter 7, we also demonstrated that masking is a very effective technique to enlarge the girth and reduce the number of short cycles of the Tanner graph of an algebraically constructed LDPC code. Masking is equivalent to constructing a masking matrix with size the same as that of the base matrix to be masked. The masking operation is performed by using the Hadamard product of a masking matrix and a base matrix. Several types of small, good masking matrices were given in Chapter 7. However, the design and construction of these masking matrices is based on experiments and experience. Some criteria need to be developed for constructing good masking matrices.

In several examples, we showed that masking matrices that enlarge girth and reduce short cycles are also good base matrices for constructing QC-PTG-LDPC codes using the matrix decomposition method presented in Chapter 4. Is it generally true that a good masking matrix for the CPM-dispersion construction of a QC-SP-LDPC code is also a good base matrix for the decomposition-construction of a QC-PTG-LDPC code? If it is, then the design of a good masking matrix is much easier than the design of a protograph.

The fourth part of the book (Chapter 8) presented an algebraic method for constructing LDPC codes with a doubly QC structure which not only perform well, as shown by examples, but also have a simplified hardware decoding implementation using the reduced-complexity decoding schemes presented recently in [79] and [68]. To construct doubly QC-LDPC codes, we need 2×2 SM-constrained cyclic base matrices. Are there other methods to construct this type of base matrix, besides the 2×2 SM-constrained cyclic base matrices constructed based on RS codes with two information symbols and the two-dimensional Euclidean geometries? As shown in Chapter 9, 2×2 SM-constrained cyclic base matrices are good for constructing spatially coupled QC-LDPC codes. Finding other methods for constructing 2×2 SM-constrained cyclic base matrices seems also to be a good problem to investigate.

The fifth contribution of the book (Chapter 9) is the presentation of an algebraic method for constructing two types of time-varying spatially coupled (SC) QC-LDPC codes: semi-infinite and terminated. These codes can be regarded as either SP-LDPC codes or PTG-LDPC codes. The Tanner graphs of these codes were presented, and examples showing code performance were given. The algebraic approach may lead to better designs, and construction of high performing and more easily implementable SC-LDPC codes for practical applications. The construction methods presented in Chapter 9 are simple, systematic, and flexible, and avoid random or pseudorandom construction of long codes to achieve good error performance close to decoding thresholds.

The sixth part of the book (Chapter 10) presented algebraic methods for constructing two types of CN-based GC-LDPC codes. The first type of CN-based QC-GC-LDPC codes is constructed based on cyclic base arrays. Examples showed that codes of the first type perform very well over both the AWGNC and the BEC. The CN-based product GC-LDPC codes of the second type are direct products of two LDPC codes, and they perform well over both the AWGNC and the BEC. A method to construct CN-based product QC-GC-LDPC codes based on component base matrices and matrix dispersion was also presented. These GC-LDPC codes were shown to be special cases of SP-LDPC codes. Also presented in this part was a local/global two-phase iterative decoding scheme for CN-based GC-LDPC codes. The proposed decoding scheme allows correction of local errors in the local phase and globally distributed errors in the global phase and it seems worthy of further investigation. Another interesting problem is whether the encoding of a CN-based GC-LDPC code can be carried out in a local/global two-phase manner. The concept of global coupling seems to be new and interesting.

Decoding of a CN-based product GC-LDPC code based on its entire parity-check matrix can be a serious problem. If the lengths of its two component codes are long (even moderately long), the length of their product can be very long. In this case, the size of the parity-check matrix of the CN-based product GC-LDPC code is very large and the implementation of a global decoder for this product code based on its parity-check matrix will be very complex and impractical. How to devise a reduced-complexity iterative scheme to decode a long CN-based product GC-LDPC code seems to be an interesting problem.

One possible approach to resolve this high decoding complexity problem is to decode the product code using a turbo BP-decoder. A turbo BP-decoder consists of two local decoders which are designed for decoding the two component codes separately. Each local decoder is a BP-decoder. The two local BP-decoders exchange reliability messages (LLRs) of code symbols through two interleavers in the form of turbo decoding after each local decoding iteration (or a fixed number of local iterations).

Another interesting problem for investigation is the trapping set structure of a CN-based product GC-LDPC code. In the discussion at the end of Chapter 10, we showed that the Tanner graph of a CN-based product GC-LDPC code does not contain small local trapping sets. The question is then whether it contains small global trapping sets?

The seventh part of the book (Chapter 11) focused on the constructions of NB LDPC codes. Several effective methods for constructing NB LDPC codes were presented, including the SP, the algebraic-PTG, the NB CPM-dispersion, the B-to-NB replacement, and the masking constructions. The algebraic approach to the construction of NB PTG-LDPC codes is quite different from the construction methods presented in the current literature. To create ensembles of NB PTG-LDPC codes using this algebraic approach and to investigate their asymptotic properties are interesting research problems and worthy of further investigation. Also presented in this part of the book was the construction of NB LDPC codes based on the conventional parity-check matrices of RS codes. Two NB LDPC codes were constructed based on two well-known RS codes, the $(255, 239)$ and $(255, 223)$ RS codes over GF(2^8), which have been widely used for commercial and deep space communications and information storage systems over the last forty years. The LDPC codes constructed have the same lengths and rates as the two RS codes. They perform very well over the binary-input AWGNC in the waterfall region and have good error-floor performance. Also presented in Chapter 11 was the construction of NB QC-LDPC codes based on the parity-check matrices of RS codes with a constraint on the smallest prime factor of the code length.

One extremely interesting and challenging problem is to develop a really efficient iterative decoding scheme for RS codes, not only practically implementable but also achieving good error performance close to maximum-likelihood decoding.

Appendices

A

RC-Constrained Arrays of CPMs Constructed Based on Partial Geometries

In this appendix, we present a method for constructing RC-constrained arrays of CPMs based on *partial geometries* (PaGs) [6, 10, 12]. These PaG-based RC-constrained arrays are very powerful for creating base matrices and replacement sets of matrices for the SP-construction of QC-SP-LDPC codes.

Consider a system composed of a set **N** of n points and a set **M** of m lines, where each line is a set of points. If a line L contains a point **p**, we say that **p** is on L and that L passes through **p**. If two points are on a line, then we say that the two points are adjacent and if two lines pass through the same point, then we say that the two lines intersect; otherwise they are parallel. The system composed of the sets **N** and **M** is a *partial geometry* [6, 10, 12] if the following conditions are satisfied for some fixed integers $\gamma \geq 2$, $\rho \geq 2$, and $\delta \geq 1$:

1. Any two points are on at most one line.
2. Each point is on γ lines.
3. Each line passes through ρ points.
4. If a point **p** is not on a line L, then there are exactly δ lines, each passing through **p** and a point on L.

Such a partial geometry will be denoted by PaG(γ, ρ, δ) and γ, ρ, and δ are called the parameters of the partial geometry. The parameter δ is called the *connection number* of the geometry.

A simple counting argument [6] shows that the partial geometry PaG(γ, ρ, δ) has exactly

$$n = \rho((\rho - 1)(\gamma - 1) + \delta)/\delta \tag{A.1}$$

points and

$$m = \gamma((\rho - 1)(\gamma - 1) + \delta)/\delta \tag{A.2}$$

lines.

It follows from Properties 1 and 2 that two lines in PaG(γ, ρ, δ) can have at most one point in common, and for every point in PaG(γ, ρ, δ), there are exactly γ lines that intersect at point $\mathbf{p}$, i.e., lines that pass through $\mathbf{p}$.

Denote the n points and the m lines in PaG(γ, ρ, δ) by $\mathbf{p}_0, \mathbf{p}_1, \ldots, \mathbf{p}_{n-1}$ and $L_0, L_1, \ldots, L_{m-1}$, respectively. For $0 \leq i < m$, for every line L_i in PaG(γ, ρ, δ), we construct an n-tuple $\mathbf{v}_i = (v_{i,0}, v_{i,1}, \ldots, v_{i,n-1})$ over GF(2). For $0 \leq j < n$ and $0 \leq i < m$, $v_{i,j} = 1$ if $\mathbf{p}_j$ is on L_i and otherwise, $v_{i,j} = 0$. This n-tuple is called the *incidence vector* of the line L_i. Since two lines in PaG(γ, ρ, δ) have at most one point in common, the incidence vectors of two different lines can have at most one position where they both have 1-components. The weight of the incidence vector is ρ.

Next, we form an $m \times n$ matrix $\mathbf{H}_{\text{PaG}} = [h_{i,j}]_{0 \leq i < m,\ 0 \leq j < n}$ with the incidence vectors of the m lines in PaG(γ, ρ, δ) as its rows. Hence, the columns and the rows of $\mathbf{H}_{\text{PaG}}$ correspond to the points and the lines in PaG(γ, ρ, δ), respectively. It follows from the fundamental properties of PaG(γ, ρ, δ) that $\mathbf{H}_{\text{PaG}}$ has the following structural properties:

1. It satisfies the RC-constraint.
2. It has constant column weight γ.
3. It has constant row weight ρ.

Hence, the Tanner graph associated with $\mathbf{H}_{\text{PaG}}$ has a girth of at least 6. Any submatrix of $\mathbf{H}_{\text{PaG}}$ can be used as a parity-check matrix to generate an LDPC code. The matrix $\mathbf{H}_{\text{PaG}}$ is called the *incidence matrix* of PaG(γ, ρ, δ).

Well-known examples of partial geometries are *Euclidean* and *projective geometries* [13, 84] over finite fields. In the following, we present two types of partial geometry whose incidence matrices can be put into RC-constrained arrays of CPMs. Hence, these arrays can be used to construct QC-LDPC codes directly, or to construct base matrices and constituent matrices of replacement sets for the SP-construction of QC-LDPC codes.

A.1 RC-Constrained Arrays of CPMs Constructed Based on Two-Dimensional Euclidean Geometries over Finite Fields

A.1.1 Two-Dimensional Euclidean Geometries over Finite Fields

A two-dimensional Euclidean geometry over a finite field GF(q), denoted by EG($2, q$) [13, 84, 74, 97], is a partial geometry PaG($q+1, q, q$) that consists of q^2 points and $q^2 + q$ lines. Each line consists of q points and each point is on $q+1$ lines. If a point $\mathbf{p}$ is not on line L, then there are exactly q lines, each passing through **p** and a point on L. This implies that any two points of EG$(2,q)$ are connected by a line. Each point **p** not on a line L is on a unique line L' which is parallel to L, i.e., L and L' do not any have any point in common. The q^2+q lines in EG$(2,q)$ can be partitioned into $q+1$ classes, each consisting of q lines, such that all the lines in each class are parallel, any two lines in two different classes intersect, and each of the $n=q^2$ points is on a unique line in each class. These classes of lines are called *parallel bundles*. Each parallel bundle consists of one line passing though the origin of the geometry and $q-1$ lines not passing though the origin. For every point **p** in EG$(2,q)$, there are exactly $q+1$ lines that intersect at **p**. These lines are said to form an *intersecting bundle* at **p**.

A simple construction of EG$(2,q)$ over GF(q) is to regard the geometry as a *two-dimensional vector space* **V** of all the q^2 two-tuples (a_0,a_1) over GF(q), commonly called vectors. Each two-tuple is regarded as a point **p** in EG$(2,q)$ and the zero two-tuple $(0,0)$ is called the *origin* of the geometry. Let GF(q^2) be an extension field of GF(q). Since every element in GF(q^2) can be represented by a two-tuple over GF(q), it presents a point in EG$(2,q)$. The field GF(q^2) may be regarded as a two-dimensional vector space over GF(q). Let α be a primitive element of GF(q^2). Then, the powers of α, $\alpha^{-\infty}=0, \alpha^0=1, \alpha, \alpha^2, \ldots, \alpha^{q^2-2}$, give all the elements of GF(q^2) and they represent the q^2 points of EG$(2,q)$. The 0-element represents the origin of EG$(2,q)$.

Let α^{j_1} be a non-origin point in EG$(2,q)$. Then, the set $\{\beta\alpha^{j_1} : \beta \in \mathrm{GF}(q)\}$ of q points constitutes a line in EG$(2,q)$. For $\beta=0$, $\beta\alpha^{j_1}=0$ is the origin of EG$(2,q)$. Hence, the line $\{\beta\alpha^{j_1} : \beta \in \mathrm{GF}(q)\}$ contains (or passes through) the origin of EG$(2,q)$. This line, regarded as a set of q vectors in **V**, is simply a *one-dimensional subspace* of the two-dimensional vector space **V** over GF(q).

Let α^{j_0} and α^{j_1} be two linearly independent non-origin points in EG$(2,q)$, i.e., $\alpha^{j_1} \neq \beta\alpha^{j_0}$ (or $\alpha^{j_0} \neq \beta\alpha^{j_1}$). Then, the set $\{\alpha^{j_0}+\beta\alpha^{j_1} : \beta \in \mathrm{GF}(q)\}$ of q points forms a line passing through the point α^{j_0}. The line $\{\alpha^{j_0}+\beta\alpha^{j_1} : \beta \in \mathrm{GF}(q)\}$, regarded as a set of q vectors in **V**, is simply a *coset* of the line $\{\beta\alpha^{j_1} : \beta \in \mathrm{GF}(q)\}$ which is a one-dimensional subspace of **V**. The line $\{\alpha^{j_0}+\beta\alpha^{j_1} : \beta \in \mathrm{GF}(q)\}$ is simply obtained by taking a point α^{j_0} in EG$(2,q)$ but not in $\{\beta\alpha^{j_1} : \beta \in \mathrm{GF}(q)\}$ and adding α^{j_0} to every point in $\{\beta\alpha^{j_1} : \beta \in \mathrm{GF}(q)\}$. The two lines $\{\beta\alpha^{j_1} : \beta \in \mathrm{GF}(q)\}$ and $\{\alpha^{j_0}+\beta\alpha^{j_1} : \beta \in \mathrm{GF}(q)\}$ do not have any point in common and hence they are parallel. Since there are q cosets of $\{\beta\alpha^{j_1} : \beta \in \mathrm{GF}(q)\}$ in **V** (including $\{\beta\alpha^{j_1} : \beta \in \mathrm{GF}(q)\}$ itself), these q cosets form a bundle of q parallel lines in EG$(2,q)$, called a parallel bundle. This parallel bundle of q lines contains all the q^2 points in EG$(2,q)$ and each point appears on a unique line.

There are $q+1$ different one-dimensional subspaces in GF(q^2) (or **V**). Based on each of these one-dimensional subspaces, we can form a parallel bundle of q

lines. Hence, the lines in EG$(2,q)$ can be partitioned into $q+1$ parallel bundles, each consisting of q parallel lines. Therefore, EG$(2,q)$ consists of q^2+q lines.

Let α^{j_0}, α^{j_1}, and α^{j_2} be three points in EG$(2,q)$ which are pair-wise linearly independent. Then, the sets $\{\alpha^{j_0}+\beta\alpha^{j_1} : \beta \in \text{GF}(q)\}$ and $\{\alpha^{j_0}+\beta\alpha^{j_2} : \beta \in \text{GF}(q)\}$ form two different lines in EG$(2,q)$, both containing the point α^{j_0}. Therefore, these two lines intersect at the point α^{j_0}. Since they are not parallel, they must be in two different parallel bundles. Since the point α^{j_0} is on a unique line in each parallel bundle of EG$(2,q)$ and there are $q+1$ parallel bundles, there are exactly $q+1$ lines that intersect at the point α^{j_0}. These $q+1$ lines are said to form an intersecting bundle. For each point in EG$(2,q)$, we can form an intersecting bundle of $q+1$ lines.

What we have shown above is that the two-dimensional Euclidean geometry EG$(2,q)$ over GF(q) can be simply constructed from the two-dimensional vector space over GF(q) or the extension field GF(q^2) of the ground field GF(q). Conversely, GF(q^2) can be regarded as a two-dimensional Euclidean geometry EG$(2,q)$ over GF(q).

A.1.2 EG-Based RC-Constrained Arrays of CPMs

In the following, we show that an RC-constrained array of CPMs of size $(q-1)\times(q-1)$ can be constructed from the lines and points of EG$(2,q)$.

Factor q^2-1 as the product of $q-1$ and $q+1$, i.e., $q^2-1=(q-1)(q+1)$. Then, the order of α^{q+1} is $q-1$, i.e., $\alpha^{(q+1)(q-1)}=1$. The set $L_{-\infty}=\{\alpha^{-\infty}=0,\alpha^0,\alpha^{(q+1)},\ldots,\alpha^{(q+1)(q-2)}\}$ of q points in EG$(2,q)$ form a line which passes through the origin. Note that the elements in $L_{-\infty}$ are simply the q elements of the subfield GF(q) of GF(q^2), a one-dimensional subspace of GF(q^2). For $0\le j\le q$, the set $\alpha^j L_{-\infty}=\{0,\alpha^j,\alpha^{(q+1)+j},\ldots,\alpha^{(q+1)(q-2)+j}\}$ of q points forms a line passing through the origin of EG$(2,q)$ [74, 97]. Hence, $L_{-\infty},\alpha L_{-\infty},\ldots,\alpha^q L_{-\infty}$ give the $q+1$ lines intersecting at the origin. Note that $\alpha^{(q+1)}L_{-\infty}=L_{-\infty}$.

Let $L=\{\alpha^{j_1},\alpha^{j_2},\ldots,\alpha^{j_q}\}$ be a line in EG$(2,q)$ not passing through the origin with $\alpha^{j_1},\alpha^{j_2},\ldots,\alpha^{j_q}$ as its points where $0\le j_1,j_2,\ldots,j_q<q^2-1$. For $0\le i<q^2-1$, let $\alpha^i L=\{\alpha^{j_1+i},\alpha^{j_2+i},\ldots,\alpha^{j_q+i}\}$. Then, for $0\le i<q^2-1$, the set $\alpha^i L$ of q points also forms a line in EG$(2,q)$ not passing through the origin, and $L,\alpha L,\alpha^2 L,\ldots,\alpha^{q^2-2}L$ give all the q^2-1 lines in EG$(2,q)$ not passing through the origin [84, 97]. The structure of lines developed above is called a *cyclic structure*. This cyclic structure allows us to construct all the lines of the Euclidean geometry EG$(2,q)$ from two lines, one passing through the origin of EG$(2,q)$ and the other not passing through the origin.

Form a $q(q+1)\times q^2$ matrix $\mathbf{H}_{EG}$ over GF(2) with the incidence vectors of the $q(q+1)$ lines in EG$(2,q)$ as rows. Let $L_{-\infty}$ and L be a line passing through the origin of EG$(2,q)$ and a line not passing through the origin, respectively. We arrange the rows of $\mathbf{H}_{EG}$ in such a way that the incidence vectors of the lines $L_{-\infty}, \alpha L_{-\infty}, \alpha^2 L_{-\infty}, \ldots, \alpha^q L_{-\infty}$ are the top $q+1$ rows of $\mathbf{H}_{EG}$ and the incidence vectors of the lines $L, \alpha L, \alpha^2 L, \ldots, \alpha^{q^2-2} L$ are the bottom q^2-1 rows of $\mathbf{H}_{EG}$. We label the rows of $\mathbf{H}_{EG}$ from 0 to $q(q+1)-1$ or, equivalently, with lines $L_{-\infty}, \alpha L_{-\infty}, \alpha^2 L_{-\infty}, \ldots, \alpha^q L_{-\infty}, L, \alpha L, \alpha^2 L, \ldots, \alpha^{q^2-2} L$ of EG$(2,q)$ in this order. We label the columns of $\mathbf{H}_{EG}$ from $-\infty$, 0 to q^2-2 or, equivalently, with points $\alpha^{-\infty}, \alpha^0, \alpha, \alpha^2, \ldots, \alpha^{q^2-2}$ of EG$(2,q)$ in this order. Let $\mathbf{H}_{EG,0}$ and $\mathbf{H}_{EG,1}$ denote the submatrices of $\mathbf{H}_{EG}$ that consist of the top $q+1$ rows and the bottom q^2-1 rows of $\mathbf{H}_{EG}$, respectively. Then, $\mathbf{H}_{EG}$ has the following form:

$$\mathbf{H}_{EG} = \begin{bmatrix} \mathbf{H}_{EG,0} \\ \hdashline \mathbf{H}_{EG,1} \end{bmatrix}. \tag{A.3}$$

The top submatrix $\mathbf{H}_{EG,0}$ of $\mathbf{H}_{EG}$ is a $(q+1) \times q^2$ matrix over GF(2) with the following structural properties: (1) The $(-\infty)$-th column consists of $q+1$ one entries; (2) Each of the other q^2-1 columns contains a single one entry and q zero entries; and (3) The 0-th to the (q^2-2)-th columns form a row-block of $q-1$ identity matrices of size $(q+1) \times (q+1)$. The bottom matrix $\mathbf{H}_{EG,1}$ of $\mathbf{H}_{EG}$ has the following structural properties: (1) The $(-\infty)$-th column consists of q^2-1 zero entries; and (2) The other q^2-1 columns form a $(q^2-1) \times (q^2-1)$ circulant with both column and row weight equal to q. Hence, the column and row weights of $\mathbf{H}_{EG}$ are $q+1$ and q, respectively. $\mathbf{H}_{EG}$ is the incidence matrix of EG$(2,q)$ and satisfies the RC-constraint. The null space of $\mathbf{H}_{EG}$ gives an LDPC code, called an EG-LDPC code [58, 74, 97].

Suppose we remove the $(-\infty)$-th column from $\mathbf{H}_{EG,1}$. Then, we obtain an RC-constrained $(q^2-1) \times (q^2-1)$ *circulant*, denoted by $\mathbf{H}_{EG,cyc}$. The null space of $\mathbf{H}_{EG,cyc}$ gives a cyclic LDPC code $C_{EG,cyc}$ of length q^2-1 with minimum distance exactly $q+1$ [58, 74, 97]. It has been proved that the Tanner graph of this cyclic EG-LDPC code contains no trapping set with size smaller than its minimum distance [46, 28].

Label the rows and columns of $\mathbf{H}_{EG,cyc}$ from 0 to q^2-2. Define the following index sequences: for $0 \le i, j \le q$, $\pi_{\text{row},i} = [i, (q-1)+i, 2(q-1)+i, \ldots, (q-1)(q-1)+i]$ and $\pi_{\text{col},j} = [j, (q-1)+j, 2(q-1)+j, \ldots, (q-1)(q-1)+j]$. Let

$$\pi_{\text{row}} = [\pi_{\text{row},0}, \pi_{\text{row},1}, \ldots, \pi_{\text{row},q}], \tag{A.4}$$

$$\pi_{\text{col}} = [\pi_{\text{col},0}, \pi_{\text{col},1}, \ldots, \pi_{\text{col},q}]. \tag{A.5}$$

Then, π_{row} and π_{col} define re-orderings of the row and the column indices, respectively.

Suppose we permute the rows and the columns of $\mathbf{H}_{EG,cyc}$ based on π_{row} and π_{col}, respectively. The row and column permutations result in a $(q^2-1)\times(q^2-1)$ matrix over GF(2):

$$\mathbf{H}_{EG,qc} = [\mathbf{A}_{i,j}]_{0\le i,j\le q}, \tag{A.6}$$

which is a $(q+1)\times(q+1)$ array of matrices of size $(q-1)\times(q-1)$. It consists of $q+1$ row-blocks and $q+1$ column-blocks of matrices of size $(q-1)\times(q-1)$. Among the $q+1$ constituent matrices in each row-block (or column-block), q of them are CPMs of size $(q-1)\times(q-1)$ and one is a ZM of size $(q-1)\times(q-1)$. Hence, $\mathbf{H}_{EG,qc}$ is an RC-constrained $(q+1)\times(q+1)$ array of CPMs and ZMs of size $(q-1)\times(q-1)$ [46]. There are exactly $q+1$ ZMs in $\mathbf{H}_{EG,qc}$. The $q-1$ rows in each row-block of $\mathbf{H}_{EG,qc}$ correspond to $q-1$ parallel lines in EG$(2,q)$ not passing through the origin.

The null space of $\mathbf{H}_{EG,qc}$ gives a QC-LDPC code $C_{EG,qc}$ which is combinatorially equivalent to $C_{EG,cyc}$. The array $\mathbf{H}_{EG,qc}$ can also be used to construct RC-constrained base matrices and replacement sets for the SP-construction of QC-LDPC code as described in Sections 6.1 and 6.3.

It was shown in [46] that $\mathbf{H}_{EG,qc}$ has an interesting block-cyclic structure. Every row-block of $\mathbf{H}_{EG,qc}$ is the cyclic-shift of the row-block above one constituent matrix (or $(q-1)$ positions) to the right; but when the rightmost constituent matrix is shifted around to the left end, all its rows are cyclically shifted one position to the right within the $q-1$ positions of the constituent matrix. Note that cyclic shifting all the rows of a CPM results in a different CPM. With the above block-cyclic structure, the entire array $\mathbf{H}_{EG,qc}$ can be formed by using its top row-block.

Let β be a primitive element of GF(q), a subfield of GF(q^2). Then, $\beta^{-\infty}=0,\beta,\beta^2,\ldots,\beta^{q-2}$ give all the elements of GF(q). Since GF(q) is a subfield of GF(q^2), $\beta=\alpha^{q+1}$, where α is a primitive element of GF(q^2). For $0\le j\le q$ and $l_j\in\{-\infty,0,1,\ldots,q-2\}$, let l_j be the location of the single 1-component in the top row of the j-th constituent matrix of the top row-block of $\mathbf{H}_{EG,qc}$. If $l_j=-\infty$, the j-th constituent matrix is a ZM in the top row-block of $\mathbf{H}_{EG,qc}$; otherwise, it is a CPM whose generator has its single 1-component at location l_j. For $0\le j\le q$, if the j-th constituent matrix $\mathbf{A}_{0,j}$ of the top row-block of $\mathbf{H}_{EG,qc}$ is a CPM, then it is the CPM-dispersion of the nonzero element β^{l_j}. If $\mathbf{A}_{0,j}$ is the ZM in the top row-block of $\mathbf{H}_{EG,qc}$, then $l_j=-\infty$. Now, we form the following $(q+1)\times(q+1)$

matrix over GF(q):

$$\mathbf{B}_{EG} = \begin{bmatrix} \beta^{l_0} & \beta^{l_1} & \beta^{l_2} & \cdots & \beta^{l_{q-1}} & \beta^{l_q} \\ \beta^{l_q+1} & \beta^{l_0} & \beta^{l_1} & \cdots & \beta^{l_{q-2}} & \beta^{l_{q-1}} \\ \beta^{l_{q-1}+1} & \beta^{l_q+1} & \beta^{l_0} & \cdots & \beta^{l_{q-3}} & \beta^{l_{q-2}} \\ \vdots & \vdots & \vdots & \ddots & \vdots & \vdots \\ \beta^{l_2+1} & \beta^{l_3+1} & \beta^{l_4+1} & \cdots & \beta^{l_0} & \beta^{l_1} \\ \beta^{l_1+1} & \beta^{l_2+1} & \beta^{l_3+1} & \cdots & \beta^{l_q+1} & \beta^{l_0} \end{bmatrix}. \tag{A.7}$$

From the structure of $\mathbf{B}_{EG}$, we see that each row is a cyclic-shift of the row above it one position to the right, but when the rightmost component is shifted around to the leftmost position, it is multiplied by β. Note that the CPM obtained by cyclically shifting all the rows of the CPM-dispersion of β^{l_j} to the right by one position is the CPM-dispersion of β^{l_j+1}. Then, it follows from the cyclic structures of $\mathbf{H}_{EG,qc}$ and $\mathbf{B}_{EG}$ that $\mathbf{H}_{EG,qc}$ is the CPM-dispersion of $\mathbf{B}_{EG}$. Since $\mathbf{H}_{EG,qc}$ satisfies the RC-constraint, $\mathbf{B}_{EG}$ satisfies the 2×2 SM-constraint.

Hence, any submatrix of $\mathbf{B}_{EG}$ can be used as a base matrix for the SP-construction of CPM-QC-SP-LDPC code. For two-dimensional Euclidean geometries over various fields, we can construct base matrices of various sizes. From these base matrices, we can construct a large family of QC-SP-LDPC codes using the CPM-dispersion construction.

A.2 RC-Constrained Arrays of CPMs Based on Partial Geometries Constructed from Prime Fields

Next, we present a construction of an RC-constrained array of CPMs based on a newly discovered partial geometry constructed based on a prime field [29].

Let p be a prime and GF(p) be a prime field that consists of the following p elements: $0, 1, 2, ..., p-1$. We disperse each element i in GF(p) into a $p \times p$ CPM, denoted by $\mathbf{A}(i)$, with both columns and rows labeled from 0 to $p-1$, whose generator has its single 1-component at location i. This representation is one-to-one. For $i = 0$, $\mathbf{A}(0)$ is a $p \times p$ identity matrix.

Form the following $p \times p$ matrix $\mathbf{B}_p$ over GF(p) with both columns and rows labeled from 0 to $p-1$:

$$\mathbf{B}_p = \begin{bmatrix} 0 \cdot 0 & 0 \cdot 1 & \cdots & 0 \cdot (p-1) \\ 1 \cdot 0 & 1 \cdot 1 & \cdots & 1 \cdot (p-1) \\ 2 \cdot 0 & 2 \cdot 1 & \cdots & 2 \cdot (p-1) \\ \vdots & \vdots & \cdots & \vdots \\ (p-1) \cdot 0 & (p-1) \cdot 1 & \cdots & (p-1) \cdot (p-1) \end{bmatrix}, \tag{A.8}$$

i.e., $\mathbf{B}_p = [b_{ij}]_{0 \le i,j < p}$ with $b_{ij} = ij$, where the multiplication of two elements in GF(p) is carried out modulo p. The matrix $\mathbf{B}_p$ has the following structural properties:

1. All the entries in the 0-th row and column are zeros.
2. All the entries in any row or column, other than the 0-th row and column, are different, and they form all the elements of GF(p).
3. Two different rows or columns have the 0-element of GF(p) in common at the 0-th column or row, and differ in all the other $p-1$ positions.
4. The i-th column of $\mathbf{B}_p$ is identical to the transpose of the i-th row of $\mathbf{B}_p$ for $0 \le i < p$.

The last property implies that the transpose $\mathbf{B}_p^T$ of $\mathbf{B}_p$ is identical to $\mathbf{B}_p$, i.e., ${\mathbf{B}_p}^T = \mathbf{B}_p$.

If each entry in $\mathbf{B}_p$ is dispersed into its corresponding $p \times p$ CPM, we obtain a $p \times p$ array $\mathbf{H}_{\text{PaG},p}$ of CPMs of size $p \times p$. It is a $p^2 \times p^2$ matrix over GF(2). Each row has weight p and each column has weight p. It was proved in a recent paper [29] that $\mathbf{H}_{\text{PaG},p}$ is the incidence matrix of a partial geometry PaG($p, p, p-1$) that consists of p^2 points and p^2 lines. Each line in PaG($p, p, p-1$) consists of p points and each point is on p lines. A point $\mathbf{p}$ that is not on a line L is connected to $p-1$ points on L by lines, i.e., the connection number of a point in PaG($p, p, p-1$) is $p-1$. $\mathbf{H}_{\text{PaG},p}$ consists of p row-blocks of CPMs and p column-blocks of CPMs. The p rows of a row-block are the incidence vectors of p parallel lines in PaG($p, p, p-1$). Therefore, the q^2 lines in PaG($p, p, p-1$) can be partitioned into p parallel bundles. Since each column of $\mathbf{H}_{\text{PaG},p}$ has weight p, each point in PaG($p, p, p-1$) is on p lines. As the incidence matrix of a partial geometry, the array $\mathbf{H}_{\text{PaG},p}$ satisfies the RC-constraint. Hence, the Tanner graph associated with any subarray of $\mathbf{H}_{\text{PaG},p}$ has a girth of at least 6. The subscript "p" of $\mathbf{B}_p$ and $\mathbf{H}_{\text{PaG},p}$ stands for "*prime*."

The matrix $\mathbf{B}_p$ over GF(p) is called the base matrix for the construction of the partial geometry PaG($p, p, p-1$). The null space of any subarray of $\mathbf{H}_{\text{PaG},p}$ gives a QC-LDPC code.

The array $\mathbf{H}_{\text{PaG},p}$ can be used to construct base matrices and replacement sets for the SP-construction of QC-SP-LDPC codes as described in Sections 6.1 and 6.4.

B

An Algorithm for Searching Compatible Masking and Base Matrices for the CPM-Dispersion Construction of QC-LDPC Codes

In this appendix, we present an algorithm to search for a 2×2 SM-constrained base matrix over a given field which is a good match to a designed masking matrix. That is, when this 2×2 SM-constrained base matrix is masked with the designed masking matrix, we obtain a masked base matrix which also satisfies the 3×3 SM-constraint, and furthermore will produce a regular CPM-QC-SP-LDPC code whose Tanner graph has the largest possible girth and a small number of short cycles.

Let $\mathbf{Z}(m,n)$ be an $m \times n$ masking matrix designed for a given rate. Let w_c and w_r be the column and the row weights of $\mathbf{Z}(m,n)$, respectively. Let $\mathbf{B}_{mother}$ be an $M \times N$ matrix over a given field GF(q) with $M > m$ and $N > n$, which satisfies the 2×2 SM-constraint. Now, we want to find an $m \times n$ submatrix $\mathbf{B}(m,n)$ of $\mathbf{B}_{mother}$ such that the masked base matrix $\mathbf{B}_{mask}(m,n) = \mathbf{Z}(m,n) \odot \mathbf{B}(m,n)$ will produce a CPM-QC-SP-LDPC code whose Tanner graph has the largest possible girth and a small number of short cycles. Such a base matrix $\mathbf{B}(m,n)$ is said to be compatible with the designed masking matrix $\mathbf{Z}(m,n)$. The matrix $\mathbf{B}_{mother}$ is called the *mother base matrix*.

The search for such an $m \times n$ submatrix $\mathbf{B}(m,n)$ from the mother base matrix $\mathbf{B}_{mother}$ can be carried out systematically. To simplify the search process, we only consider those $m \times n$ submatrices which are taken from m consecutive rows and n consecutive columns in $\mathbf{B}_{mother}$ (no end around rows and columns). The search begins from the first m rows (the 0-th to the $(m-1)$-th rows) of $\mathbf{B}_{mother}$ and moves from the leftmost edge of $\mathbf{B}_{mother}$ to the right, one position (or one column) at a time. Each time, we mask an $m \times n$ submatrix $\mathbf{B}(m,n)$ with the masking matrix $\mathbf{Z}(m,n)$ and check whether the $m \times n$ masked submatrix $\mathbf{B}_{mask}(m,n)$ satisfies the 3×3 SM-constraint. After checking the last $m \times n$ submatrix of the first m rows of $\mathbf{B}_{mother}$, we start to search across the next m consecutive rows (the 1-st to the m-th rows) of $\mathbf{B}_{mother}$ from left to right. We move one position at a time until we reach and test the last $m \times n$ submatrix. Then, we move back to the leftmost edge

Table B.1. *Numbers of* 4×8 *masked base matrices over various fields* $GF(q)$ *that give rate-1/2 CPM-QC-SP-LDPC codes with girth 8 or 10 for* $\eta = 1$

q	$\#\mathbf{B}_{mask}(4,8)_8$	$\#cyc8_{smallest}$	$\#cyc6_{unmasked}$	$\#cyc8_{unmasked}$	$\#\mathbf{B}_{mask}(4,8)_{10}$	$\#cyc10_{smallest}$
53	220	806	1,612	26,286	0	0
64	542	819	1,890	25,965	0	0
89	1,433	616	2,640	26,576	0	0
128	6,872	635	3,810	28,575	0	0
181	15,799	360	4,140	27,990	680	11,880
256	45,977	255	6,375	41,820	5	9,435
257	36,927	256	6,400	46,976	1,479	11,776
331	65,373	330	7,920	34,320	8,970	11,220

of $\mathbf{B}_{mother}$ and start across the next m consecutive rows (the 2-nd to the $(m+1)$-th rows) of $\mathbf{B}_{mother}$ from left to right. This searching process continues until we finish searching the last m consecutive rows of $\mathbf{B}_{mother}$. While we are checking whether an $m \times n$ masked submatrix $\mathbf{B}_{mask}(m,n)$ satisfies the 3×3 SM-constraint, we also check the girth g and count the numbers of short cycles of lengths g, $g+2$, and $g+4$ of the Tanner graph associated with the CPM-dispersion of $\mathbf{B}_{mask}(m,n)$ using the cycle counting algorithm presented in [71]. By the end of the search, we will find a base matrix $\mathbf{B}(m,n)$ whose masked version $\mathbf{B}_{mask}(m,n)$ produces a code whose Tanner graph has the largest girth, or the smallest number of short cycles, or both. This base matrix is said to be compatible with the designed masking matrix $\mathbf{Z}(m,n)$. If we are not satisfied with the results, we could design another masking matrix and repeat the search algorithm.

The algorithm to perform the above search of a compatible base matrix for a designed masking matrix is called a *compatible base matrix search algorithm* (CBMSA). This search algorithm is a generalization of the algorithm devised in [68].

Table B.1 gives a list of numbers of 4×8 base matrices $\mathbf{B}(4,8)$s which are submatrices of the mother matrix $\mathbf{B}_{mother}$, constructed using the method presented in Section 7.3 in the form of (7.1) by choosing the two subsets $S_0 = S_1 = \text{GF}(q)$ and $\eta = 1$ for various q. This mother matrix is just a Latin square with order q. The masking matrix used for search is the 4×8 masking matrix given by (7.5). The second and the sixth columns of Table B.1 contain the numbers of $\mathbf{B}_{mask}(4,8)$s which give codes with girths 8 and 10, denoted by $\#\mathbf{B}_{mask}(4,8)_8$ and $\#\mathbf{B}_{mask}(4,8)_{10}$, respectively. The third and the seventh columns contain the smallest numbers of cycles of lengths 8 and 10 for each category of 4×8 masked base matrices, denoted by $\#cyc8_{smallest}$ and $\#cyc10_{smallest}$, respectively. The fourth and the fifth columns of the table give the smallest numbers of cycles of lengths 6 and 8, denoted by $\#cyc6_{unmasked}$ and $\#cyc8_{unmasked}$, respectively, among all unmasked 4×8 base matrices $\mathbf{B}(4, 8)$.

For example, consider the 4×8 masked matrices constructed based on the field GF(331). Using the CBMSA devised above, we find 65 373 4×8 masked base matrices which give codes with girth 8 and 8970 masked base matrices which give codes with girth 10. Among these masked matrices, the smallest numbers of cycles of lengths 8 and 10 are 330 and 11 220, respectively. Consider an unmasked 4×8 base matrix $\mathbf{B}(4, 8)$ that produces a masked base matrix $\mathbf{B}_{mask}(4, 8)$ resulting in a code with girth 8 and the smallest number of cycles of length 8. From Table B.1, we find that the numbers of cycles of lengths 6 and 8 in the Tanner graph associated with the CPM-dispersion of an unmasked base matrix $\mathbf{B}(4, 8)$ are at least 7920 and 34 320, respectively. However, the number of cycles of length 8 in the Tanner graph associated with the CPM-dispersion of the masked base matrix $\mathbf{B}_{mask}(4, 8)$ is only 330. So, masking not only increases the girth, from 6 to 8, but also reduces the number of cycles of length 8, from 34 320 to 330.

C

Iterative Decoding Algorithm for NB LDPC Codes

C.1 Introduction

This appendix will present the most popular decoding algorithm for LDPC codes over GF(q), where $q = 2^s$, $s > 1$. This algorithm, which relies on the fast Hadamard transform, was first presented in [82, 23, 5].

We will assume that each q-ary symbol is transmitted as a group of s bits over the binary-input AWGNC, although the results can be extended to NB-input channels. We will let n and m denote the number of columns and rows, respectively, of the code's (low-density) parity-check matrix over GF(q), $\mathbf{H} = [h_{i,j}]_{0 \le i < m, 0 \le j < n}$ with $h_{i,j} \in \text{GF}(q)$.

Recall that, for binary LDPC codes, the decoder iteratively updates estimates on $P_j[0]$ and $P_j[1]$, the probabilities that code bit v_j is 0 and 1, respectively, for $j = 0, 1, \ldots, n-1$. For mathematical convenience and reduced implementation complexity, the decoder estimates these probabilities indirectly by instead iteratively estimating the log-likelihood ratio (LLR) for v_j, given by

$$L_j = \log\left[\frac{P_j[0 \mid \mathbf{y}]}{P_j[1 \mid \mathbf{y}]}\right] \tag{C.1}$$

where the dependency of these probabilities on the channel output vector $\mathbf{y}$ is made explicit in the *a posteriori* probabilities $P_j[0 \mid \mathbf{y}]$ and $P_j[1 \mid \mathbf{y}]$ for code bit v_j. An additional condition is implied: namely, the code constraints as encapsulated in the code's parity-check matrix $\mathbf{H}$.

For q-ary LDPC codes, a pair of probabilities or a single LLR for each code symbol v_j is not appropriate because v_j can take on one of q values, so q probabilities must be estimated. These q probabilities form a probability mass function (pmf). Thus, the q-ary LDPC code decoder, which consists of VNs and CNs connected by edges labeled with elements from GF(q), must propagate pmfs instead of LLRs along the edges of the code's Tanner graph representation. Once the decoder discontinues its iterative computations, its decision for v_j is that value of $v \in \mathrm{GF}(q)$ which maximizes the *a posteriori* probability $P_j[v \mid \mathbf{y}]$.

C.2 Algorithm Derivation

As with the binary LDPC code decoding algorithm, the algorithm for NB LDPC codes receives information from the channel and sends that information up to the CN processors. The CN processors then send their processed outputs to the VN processors which take these inputs along with the channel information to produce information that is sent back to the CN processors. The VN/CN processing iterations continue until a codeword is found or the preset maximum number of iterations is reached.

For the AWGNC with two-sided power spectral density $N_0/2$, the appropriate bitwise information computed from the channel output y for candidate binary input $b \in \{\pm 1\}$ is

$$\Pr(b \mid y) = p(y \mid b)\Pr(b)/p(y) = \frac{1}{1+\exp(-4yb/N_0)}. \tag{C.2}$$

These bitwise probabilities are converted by a preprocessor to symbol-wise probabilities by computing appropriate products of the former. For example, for $s = 4$, suppose $\alpha^2 \in \mathrm{GF}(2^4)$ has the binary representation [0 0 1 0], where α is a primitive element of $\mathrm{GF}(q)$. Then $\Pr(\alpha^2 \mid \overline{y}) = \Pr(0 \mid y_3)\Pr(0 \mid y_2)\Pr(1 \mid y_1)\Pr(0 \mid y_0)$, where $\overline{y} = [y_3\ y_2\ y_1\ y_0]$ is the group of channel outputs corresponding to $s = 4$ consecutive binary inputs. Each VN processor receives q such symbol-wise probabilities from the decoder preprocessor; that is, each VN processor receives a (conditional) pmf on the elements of $\mathrm{GF}(q)$. The preprocessor pmf for VN j will be denoted by $\boldsymbol{P}_j$, for $j = 0, 1, \ldots, n-1$. The q elements of $\boldsymbol{P}_j$ will be denoted by $P_j(0), P_j(1), P_j(\alpha), P_j(\alpha^2), \ldots, P_j(\alpha^{q-2})$, where

$$P_j(\beta) = \Pr(v_j = \beta \mid \overline{y}) \tag{C.3}$$

for all $\beta \in \mathrm{GF}(q)$, with $P_j(\beta)$ given by a product of s probabilities $\Pr(b \mid y)$.

Given the channel outputs and computed pmfs, we now need to develop the VN and CN update equations involved in the iterative decoding. To aid the discussion, we define the *CN neighborhood* of VN j, for $0 \le j < n$, to be

$$M_j = \{i : h_{i,j} \ne 0\}. \tag{C.4}$$

Similarly, we define the *VN neighborhood* of CN i, for $0 \le i < m$, to be

$$N_i = \{j : h_{i,j} \ne 0\}. \tag{C.5}$$

C.2.1 VN Update

As is customary, we regard each VN to be a repetition code so that all of the edges leaving a VN carry the same value in GF(q). (These values are later scaled by the edge labels on their way up to the CNs.) Moreover, we assume that the messages from the CNs to the VNs are independent. Because the optimum repetition code decoder adds LLRs, or multiplies probabilities, it follows that, for each $\beta \in \mathrm{GF}(q)$, the (extrinsic) message $m_{j \to i}(\beta)$ to be sent from VN j to CN i is given by

$$m_{j \to i}(\beta) = P_j(\beta) \prod_{k \in M_j \setminus i} m_{k \to j}(\beta) \tag{C.6}$$

where $m_{k \to j}(\beta)$ is the message sent from CN k to VN j about the probability that code symbol v_j is equal to β. Note that the independence assumption on the incoming messages at VN j actually holds only for the first $g/2$ iterations, where g is the girth of the code's Tanner graph. Note also that, because of the exclusion $M_j \setminus i$ in the product, the message $m_{j \to i}(\beta)$ does not send CN i information that it already has.

We should emphasize that, while a single computation of the form (C.6) is performed in the binary case, q such computations are performed for NB LDPC codes, one for each value of $\beta \in \mathrm{GF}(q)$.

The computation of the CN-to-VN messages is much more involved, as is its development, which we will do in stages.

C.2.2 CN Update: Complex Version

The CN update equation takes more time to develop, but the concepts are well known to students of coding. First, note that a CN is considered to be an NB single-parity check (SPC) code. Thus, the i-th CN, representing the i-th row, $[h_{i,0}\ h_{i,1}\ \ldots\ h_{i,n-1}]$, of the parity-check matrix $\mathbf{H}$, corresponds to the parity-check equation

$$\sum_{j \in N_i} v_j h_{i,j} = 0. \tag{C.7}$$

In this equation, $v_j \in \mathrm{GF}(q)$ is the code symbol corresponding to VN j, $h_{i,j}$ with $j = 0, 1, \ldots, n-1$ are the nonzero elements of the i-th row of $\mathbf{H}$, and addition and multiplication are over GF(q).

Letting $v'_j = v_j h_{i,j}$, the above sum can be rewritten as

$$\sum_{j \in N_i} v'_j = 0. \tag{C.8}$$

The previous equation leads to the CN update equation given CN inputs from neighboring VNs. In particular, consider the i-th CN, its neighborhood of VNs, N_i, and their pmfs. From these pmfs (messages), we are to compute the pmf to be sent from CN i to VN j, for all js in N_i. The pmf to be sent to VN j is the pmf of

$$v'_j = \sum_{\ell \in N_i \setminus j} v'_\ell. \tag{C.9}$$

When a discrete-valued random variable, v'_j, is the sum of independent discrete-valued random variables, v'_ℓ, then the pmf of v'_j is given by the *cyclic* convolution of the pmfs of the v'_ℓs.

It is important to highlight here an important detail regarding the convolution of pmfs of elements of GF(q). Consider two random variables, X and Y, taking values in GF(q). Let their corresponding pmfs be denoted by $\mathbf{p}_X$ and $\mathbf{p}_Y$. We are interested in determining the pmf of their sum, $Z = X + Y$, noting that the sum is performed via the modulo-2 addition of the binary s-tuple representations of X and Y. Thus, it is convenient to represent the elements of GF(q) by their binary s-tuples or their decimal equivalents. We have then that

$$\mathbf{p}_Z(z) = \sum_{x,y:x+y=z} \mathbf{p}_X(x)\mathbf{p}_Y(y) = \sum_x \mathbf{p}_X(x)\mathbf{p}_Y(z-x), \tag{C.10}$$

or, using the cyclic convolution operator shorthand,

$$\mathbf{p}_Z = \mathbf{p}_X \circledast \mathbf{p}_X. \tag{C.11}$$

This generalizes to a sum of more than two random variables. To apply this result to (C.9), we let $\mathbf{m}_{i\to j} = [m_{i\to j}(0)\ m_{i\to j}(1)\ \ldots\ m_{i\to j}(\alpha^{q-2})]$ be the conditional pmf of v'_j and $\mathbf{m}_{\ell\to i} = [m_{\ell\to i}(0)\ m_{\ell\to i}(1)\ \ldots\ m_{\ell\to i}(\alpha^{q-2})]$ the conditional pmf of v'_ℓ. It follows from the development above that the conditional pmf of v'_j is

$$\mathbf{m}_{i\to j} = \circledast_{\ell \in N_i \setminus j} \mathbf{m}_{\ell\to i}. \tag{C.12}$$

C.2.3 CN Update: Fast Hadamard Transform Version

The convolution of multiple pmfs to obtain the pmf of v'_j is obviously quite complex. The computational complexity may be vastly reduced by using Fast Hadamard Transform (FHT) techniques to perform the convolutions. To introduce the FHT-based algorithm, some preliminary information on the Hadamard transform itself is necessary. We will present the fast Hadamard transform algorithm later.

The Hadamard transform of a length-q, real-valued (row) vector $\mathbf{p} = [\ p_0\ p_1 \cdots p_{q-1}\ ]$, considered to be a pmf here, is given by

$$\mathbf{P} = \mathcal{H}(\mathbf{p}) = \mathbf{p}\mathbf{H}_q \tag{C.13}$$

where $\mathbf{H}_q$ is recursively defined as

$$\mathbf{H}_q = \frac{1}{\sqrt{2}} \begin{bmatrix} \mathbf{H}_{q/2} & \mathbf{H}_{q/2} \\ \mathbf{H}_{q/2} & -\mathbf{H}_{q/2} \end{bmatrix}, \tag{C.14}$$

with the initial condition

$$\mathbf{H}_2 = \frac{1}{\sqrt{2}} \begin{bmatrix} 1 & 1 \\ 1 & -1 \end{bmatrix}. \tag{C.15}$$

Letting $\boldsymbol{\phi}_x$ represent the x-th row of $\mathbf{H}_q$, the Hadamard transform of $\mathbf{p}$ may be written as

$$\mathbf{P} = \sum_{x=0}^{q-1} p_x \boldsymbol{\phi}_x \ . \tag{C.16}$$

Because $\mathbf{H}_q$ is recursively defined, it is not difficult to prove by induction the following properties of the Hadamard transform.

1. $\mathbf{H}_q^T = \mathbf{H}_q$ (where the superscript T denotes matrix transpose).
2. $\mathbf{H}_q \mathbf{H}_q^T = \mathbf{H}_q^T \mathbf{H}_q = \mathbf{I}_q$ ($\mathbf{I}_q$ is the $q \times q$ identity matrix). In particular, the inverse Hadamard transform is performed by $\mathbf{H}_q^T = \mathbf{H}_q^{-1}$, that is, $\mathcal{H}^{-1}(\mathbf{P}) = \mathbf{P}\mathbf{H}_q^T = \mathbf{p}\mathbf{H}_q\mathbf{H}_q^T = \mathbf{p}$. A corollary to this property is that the rows of $\mathbf{H}_q$ are orthonormal $\boldsymbol{\phi}_x \boldsymbol{\phi}_y^T = \boldsymbol{\delta}_{x-y}$, where $\boldsymbol{\delta}_n$ is the Kronecker delta function.
3. The component-wise multiplication, denoted by $\odot$, of the x-th and y-th rows of $\mathbf{H}_q$ is $\boldsymbol{\phi}_x \odot \boldsymbol{\phi}_y = \frac{1}{\sqrt{2}} \boldsymbol{\phi}_{x \oplus y}$, where $x \oplus y$ is component-wise mod-2 addition of the binary representations of x and y. For example, $13 \oplus 10 = 7$.

Observe that the component-wise mod-2 addition mentioned in Property 3 is identical to addition of two elements in GF(2^s), although, following convention, we have used $\oplus$ for the former and $+$ for the latter. Property 3 is critical to the proof of the following cyclic convolution theorem for Hadamard transforms that is central to the low-complexity computation of the cyclic convolution of pmfs of random variables that take values in GF(q). It will be seen in the proof that the indices to the rows of $\mathbf{H}_q$ are put into a one-to-one correspondence with the elements (s-tuples) of the additive group in GF(q).

Theorem C.1. *Consider two independent random variables, X and Y, defined on GF(q) with pmfs $\mathbf{p}_X$ and $\mathbf{p}_Y$. Then, the pmf $\mathbf{p}_Z$ of their sum in GF(q), $Z = X + Y$, may be computed from* $\sqrt{2}\,\mathcal{H}^{-1}\left[\mathcal{H}(\mathbf{p}_X) \odot \mathcal{H}(\mathbf{p}_Y)\right]$.

Proof. For simplicity of notation, instead of $p_X(x)$ and $p_Y(y)$, $p(x)$ and $p(y)$ will be written for individual components of $\mathbf{p}_X$ and $\mathbf{p}_Y$. All summations below are from 0 to $q-1$, or their binary s-tuple equivalents.

$$\begin{aligned}
\sqrt{2}\,\mathcal{H}^{-1}\left[\mathcal{H}(\mathbf{p}_X)\odot\mathcal{H}(\mathbf{p}_Y)\right] &= \sqrt{2}\mathcal{H}^{-1}\left[\sum_x p(x)\boldsymbol{\phi}_x \odot \sum_y p(y)\boldsymbol{\phi}_y\right] \\
&= \sqrt{2}\mathcal{H}^{-1}\left[\sum_x\sum_y p(x)\boldsymbol{\phi}_x \odot p(y)\boldsymbol{\phi}_y\right] \\
&= \sqrt{2}\mathcal{H}^{-1}\left[\sum_x\sum_y p(x)p(y)\frac{1}{\sqrt{2}}\boldsymbol{\phi}_{x\oplus y}\right] \\
&= \sum_x\sum_y\sum_z p(x)p(y)\boldsymbol{\phi}_{x\oplus y}\boldsymbol{\phi}_z^T \qquad\qquad (C.17)\\
&= \sum_x\sum_y\sum_z p(x)p(y)\delta_{(x\oplus y)-z} \\
&= \sum_{x,y:x\oplus y=z} p(x)p(y) \\
&= \sum_x p(x)p(z-x) \\
&= \mathbf{p}_X \circledast \mathbf{p}_Y\,.
\end{aligned}$$

The first two lines are obvious; the third follows from Property 3; the fourth from Properties 1 and 2; the fifth from the corollary to Property 2; and the last three are obvious. □

Example C.1. Although the development so far has focused on NB LDPC codes, the results are applicable to binary LDPC codes as well, that is, the case $q=2$. Substitution of $\mathbf{p}_X = [p_X(0)\quad p_X(1)]$ and $\mathbf{p}_Y = [p_Y(0)\quad p_Y(1)]$ into $\sqrt{2}\,\mathcal{H}^{-1}\left[\mathcal{H}(\mathbf{p}_X)\odot\mathcal{H}(\mathbf{p}_Y)\right]$ yields $\mathbf{p}_X \circledast \mathbf{p}_Y = [\ p_X(0)p_Y(0) + p_X(1)p_Y(1)\quad p_X(0)p_Y(1) + p_X(1)p_Y(0)\ ]$. Clearly, the first component of this vector is equal to $\Pr[X+Y=0]$ and the second component is equal to $\Pr[X+Y=1]$. The logarithm of the ratio of the second component to the first appears in the proof of Lemma 5.2 in [97], which shows how to compute the LLR of a sum (over GF(2)) of two independent binary random variables, given their individual LLRs.

The Fast Hadamard Transform

Efficient computation of a Hadamard transform, called the fast Hadamard transform algorithm (FHT) (which is also called the *fast Fourier transform*

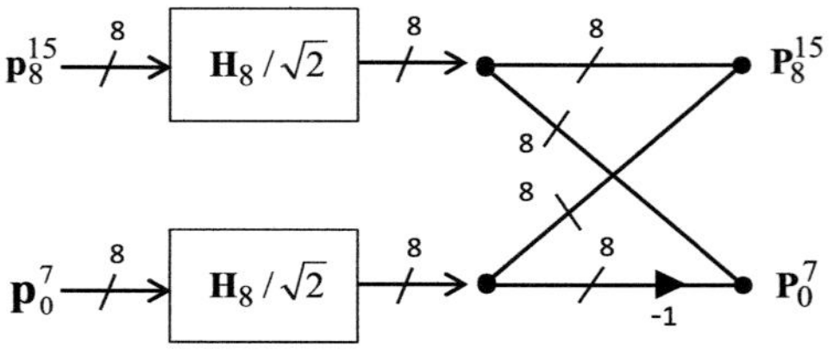

Figure C.1. Diagram of implementation of $\mathbf{P} = \mathbf{pH}_{16}$.

(FFT) in [82, 97]), is possible because of the recursive nature of Hadamard matrices. Further, from Properties 1 and 2, the Hadamard transform and the inverse Hadamard transform are identical, so we need only discuss one fast algorithm.

Presentation of the FHT is easiest by way of example. Let us consider the 16-ary case so that $q = 16$, $\mathbf{p} = [p_0\ p_1\ \dots\ p_{15}]$, and $\mathbf{P} = \mathcal{H}(\mathbf{p}) = \mathbf{pH}_{16}$. Observe that

$$\mathbf{H}_{16} = \frac{1}{\sqrt{2}}\begin{bmatrix} \mathbf{H}_8 & \mathbf{H}_8 \\ \mathbf{H}_8 & -\mathbf{H}_8 \end{bmatrix} = \frac{1}{\sqrt{2}}\begin{bmatrix} \mathbf{H}_8 & \mathbf{0} \\ \mathbf{0} & \mathbf{H}_8 \end{bmatrix}\begin{bmatrix} \mathbf{I}_8 & \mathbf{I}_8 \\ \mathbf{I}_8 & -\mathbf{I}_8 \end{bmatrix}. \tag{C.18}$$

Fig. C.1 illustrates the implementation of this equation, where $\mathbf{p}_0^7$ is the subvector $[p_0\ p_1\ \dots\ p_7]$ of $\mathbf{p}$, and similarly for the other subvectors. In the figure, quantities along edges that diverge are identical and quantities along edges that merge are added.

This implementation does not look advantageous until we observe that $\mathbf{H}_8$ decomposes as

$$\mathbf{H}_8 = \frac{1}{\sqrt{2}}\begin{bmatrix} \mathbf{H}_4 & \mathbf{0} \\ \mathbf{0} & \mathbf{H}_4 \end{bmatrix}\begin{bmatrix} \mathbf{I}_4 & \mathbf{I}_4 \\ \mathbf{I}_4 & -\mathbf{I}_4 \end{bmatrix}, \tag{C.19}$$

which is implemented as in Fig. C.2 for the input $\mathbf{p}_0^7$. The implementation for the input $\mathbf{p}_8^{15}$ is essentially identical. The setup in Fig. C.2 and the analogous one for $\mathbf{p}_8^{15}$ replace the two $\mathbf{H}_8$ blocks in Fig. C.1. Next, each of the $\mathbf{H}_4$ blocks in the implementation of $\mathbf{H}_8$ is replaced by the implementation of

$$\mathbf{H}_4 = \frac{1}{\sqrt{2}}\begin{bmatrix} \mathbf{H}_2 & \mathbf{0} \\ \mathbf{0} & \mathbf{H}_2 \end{bmatrix}\begin{bmatrix} \mathbf{I}_2 & \mathbf{I}_2 \\ \mathbf{I}_2 & -\mathbf{I}_2 \end{bmatrix}, \tag{C.20}$$

resulting in eight $\mathbf{H}_2$ blocks.

The culmination of all of these steps is Fig. C.3, which is a concatenation of $\log_2(q) = 4$ levels of "butterflies," with eight butterflies in the first level, four in the second, two in the third and one in the last. The scale factor of $1/4$ at the output accounts for the four factors of $1/\sqrt{2}$ that would occur in each of the four stages, but combined near the output.

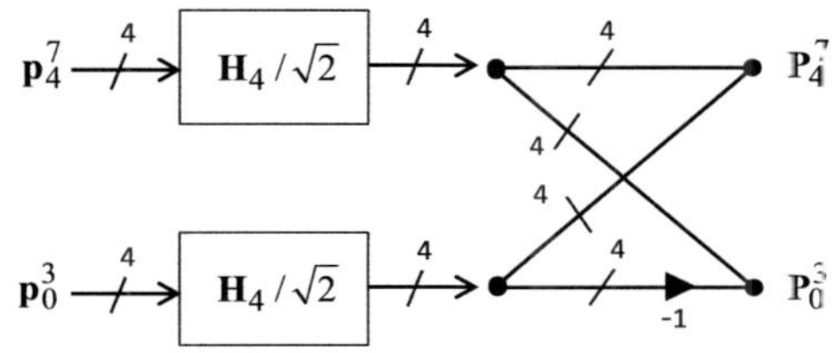

Figure C.2. Diagram of implementation of $\mathbf{P} = \mathbf{p}_0^7\mathbf{H}_8$.

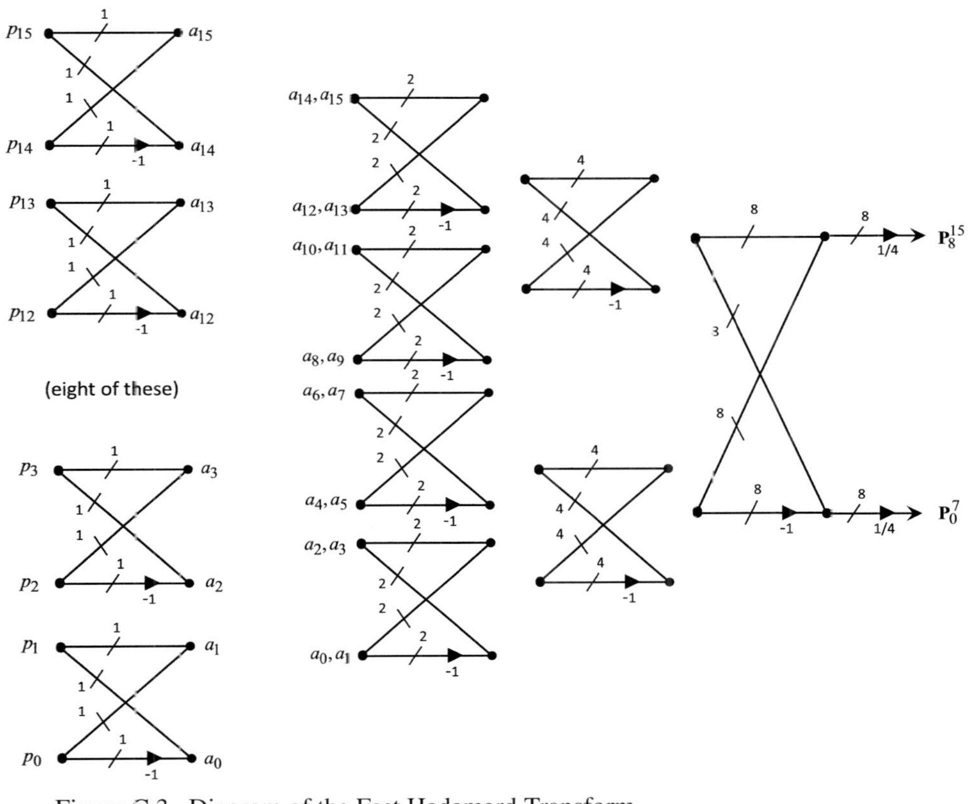

Figure C.3. Diagram of the Fast Hadamard Transform.

C.3 The NB LDPC Decoding Algorithm

Below we summarize the NB LDPC code decoding algorithm in the foregoing, called the fast Fourier transform q-ary sum-product algorithm (FFT-QSPA). The messages $m_{j\to i}(\beta)$ computed in Step 2 correspond to v_j, whereas the convolutions in (C.12) correspond to v'_j, so the translations in Steps 3 and 5 are necessary. Each of the steps is to be applied to all $\beta \in \mathrm{GF}(q)$ and all i and j for which $h_{i,j} \neq 0$, with $0 \leq i < m$ and $0 \leq j < n$.

1. Initialize the probabilities $P_j(\beta)$ according to equations (C.2) and (C.3) and the discussion in between those equations. Initialize also the CN messages $m_{i\to j}(\beta)=1$.
2. Compute the VN messages $m_{j\to i}(\beta)$ according to equation (C.6).
3. Translate the messages $m_{j\to i}(\beta)$ to the messages $m_{j\to i}(\beta')$, according to $m_{j\to i}(\beta')=m_{j\to i}(\beta)$, where $\beta'=h_{i,j}\beta$.
4. Using the Hadamard transform convolution theorem and the FHT algorithm, compute the CN messages:

$$\mathbf{m}_{i\to j}=\mathcal{H}^{-1}\left\{\prod_{\ell\in N_i\backslash j}\mathcal{H}\{\mathbf{m}_{\ell\to i}\}\right\}. \tag{C.21}$$

5. Translate the messages $m_{i\to j}(\beta')$ to the messages $m_{i\to j}(\beta)$, according to $m_{i\to j}(\beta)=m_{i\to j}(\beta')$, where $\beta'=h_{i,j}\beta$.
6. Make symbol decisions using

$$\hat{v}_j=\text{argmax}_\beta\, m_j(\beta), \tag{C.22}$$

where

$$m_j(\beta)=P_j(\beta)\prod_{k\in M_j}m_{k\to j}(\beta). \tag{C.23}$$

If a codeword found (if $\hat{\mathbf{v}}\mathbf{H}^T=\mathbf{0}$) or if the maximum number of iterations reached, then stop; otherwise, go to Step 2.

References

[1] Abu-Surra, S., Divsalar, D., and Ryan, W. E. 2010 (Jan.). On the existence of typical minimum distance for protograph-based LDPC codes. Pages 1–7 of: *Proc. IEEE Inf. Theory Applic. Workshop*. San Diego, CA, USA, January 31–February 5 2010.

[2] Abu-Surra, S., Divsalar, D., and Ryan, W. E. 2011. Enumerators for protograph-based ensembles of LDPC and generalized LDPC codes. *IEEE Trans. Inf. Theory*, **57**(2), 858–886.

[3] Ammar, B., Honary, B., Kou, Y., Xu, J., and Lin, S. 2004. Construction of low-density parity-check codes based on balanced incomplete block designs. *IEEE Trans. Inf. Theory*, **50**(6), 1257–1269.

[4] Bahl, L., Cocke, J., Jelinek, F., and Raviv, J. 1974. Optimal decoding of linear codes for minimizing symbol error rate. *IEEE Trans. Inf. Theory*, **20**(2), 284–287.

[5] Barnault, L., and Declercq, D. 2003 (Mar.). Fast decoding algorithm for LDPC over GF(2^q). Pages 70–73 of: *Proc. IEEE Inf. Theory Workshop*. La Sorbonne, Paris, France, March 31–April 4, 2003.

[6] Batten, L. M. 1997. *Combinatorics of Finite Geometries, 2nd ed.* Cambridge, UK: Cambridge University Press.

[7] Bellorado, J., and Kavcic, A. 2010. Low-complexity soft-decoding algorithms for Reed–Solomon codes part I: An algebraic soft-in hard-out Chase decoder. *IEEE Trans. Inf. Theory*, **56**(3), 945–959.

[8] Berlekamp, E. R. 1984. *Algebraic Coding Theory*. Laguna Hills, CA: Aegean Park Press.

[9] Bose, R. C. 1939. On the construction of balanced incomplete block designs. *Ann. Eugenics*, **9**(4), 353–399.

[10] Bose, R. C. 1963. Strongly regular graphs, partial geometries and partially balanced designs. *Pacific J. Math.*, **13**(2), 389–419.

[11] Butler, B. K., and Siegel, P. H. 2010 (Jun.). On distance properties of quasi-cyclic protograph-based LDPC codes. Pages 809–813 of: *Proc. IEEE Int. Symp. Inf. Theory*. Austin, TX, USA, June 13–18, 2010.

[12] Cameron, P. J., and Van Lint, J. H. 1991. *Designs, Graphs, Codes, and Their Links*. Cambridge, UK: Cambridge University Press.

[13] Carmichael, R. D. 1956. *Introduction to the Theory of Groups of Finite Orders*. New York, NY: Dover.

[14] Chandrasetty, V. A., Johnson, S. J., and Lechner, G. 2013. Memory efficient decoders using spatially coupled quasi-cyclic LDPC codes. *CoRR*, **arXiv:abs/1305.5625**.

[15] Chang, B. Y., Dolecek, L., and Divsalar, D. 2011 (Nov.). EXIT chart analysis and design of non-binary protograph-based LDPC codes. Pages 566–571 of: *IEEE Military Commun. Conf. (Milcom)*. Baltimore, MD, USA, November 7–10, 2011.

[16] Chang, B. Y., Divsalar, D., and Dolecek, L. 2012 (September 3–7). Non-binary protograph-based LDPC codes for short block-lengths. Pages 282–286 of: *Proc. IEEE Inf. Theory Workshop*. Lausanne, Switzerland, September 3–7, 2012.

[17] Chen, C. L., Peterson, W. W., and Weldon, E. J., Jr. 1969. Some results on quasi-cyclic codes. *Inf. Control*, **15**(5), 407–423.

[18] Chen, J., and Fossorier, M. P. C. 2002. Near optimum universal belief propagation based decoding of low-density parity check codes. *IEEE Trans. Commun.*, **50**(3), 406–414.

[19] Chen, L., Xu, J., Djurdjevic, I., and Lin, S. 2004. Near Shannon-limit quasi-cyclic low-density parity-check codes. *IEEE Trans. Commun.*, **52**(7), 1038–1042.

[20] Chen, Y., and Parhi, K. K. 2004. Overlapped message passing for quasi-cyclic low-density parity check codes. *IEEE Trans. Circuits Syst. I*, **51**(6), 1106–1113.

[21] Colbourn, C. J., and Dintz, J. H. 1996. *The Handbook of Combinatorial Design*. Boca Raton, FL: CRC Press.

[22] Costello, D. J., Jr., Dolecek, L., Fuja, T., Kliewer, J., Mitchell, D. G. M., and Smarandache, R. 2014. Spatially coupled sparse codes on graphs: Theory and practice. *IEEE Commun. Mag.*, **52**(7), 168–176.

[23] Davey, M. C., and MacKay, D. J. C. 1998. Low-density parity check codes over GF(q). *IEEE Commun. Lett.*, **2**(6), 165–167.

[24] Di, C., Proietti, D., Telatar, I. E., Richardson, T. J., and Urbanke, R. L. 2002. Finite-length analysis of low-density parity-check codes on the binary erasure channel. *IEEE Trans. Inf. Theory*, **48**(6), 1570–1579.

[25] Diao, Q., Huang, Q., Lin, S., and Abdel-Ghaffar, K. 2011 (Feb.). A transform approach for analyzing and constructing quasi-cyclic low-density parity-check codes. Pages 1–8 of: *Proc. IEEE Inf. Theory Applic. Workshop*. La Jolla, CA, USA, February 6–11, 2011.

[26] Diao, Q., Huang, Q., Lin, S., and Abdel-Ghaffar, K. 2012a. A matrix-theoretic approach for analyzing quasi-cyclic low-density parity-check codes. *IEEE Trans. Inf. Theory*, **58**(6), 4030–4048.

[27] Diao, Q., Zhou, W., Lin, S., and Abdel-Ghaffar, K. 2012b (Feb.). A transform approach for constructing quasi-cyclic Euclidean geometry LDPC codes. Pages 204–211 of: *Proc. IEEE Inf. Theory Applic. Workshop*. San Diego, CA, USA, February 5–10, 2012.

[28] Diao, Q., Tai, Y. Y., Lin, S., and Abdel-Ghaffar, K. 2013. LDPC codes on partial geometries: Construction, trapping set structure, and puncturing. *IEEE Trans. Inf. Theory*, **59**(12), 7898–7914.

[29] Diao, Q., Li, J., Lin, S., and Blake, I. F. 2016. New classes of paritial geometries and their associated LDPC codes. *IEEE Trans. Inf. Theory*, **62**(6) 2947–2965.

[30] Divsalar, D., Dolinar, S., and Jones, C. 2005a (Sep.). Low-rate LDPC codes with simple protograph structure. Pages 1622–1626 of: *Proc. IEEE Int. Symp. Inf. Theory*. Adelaide, SA, USA, September 4–9, 2005.

[31] Divsalar, D., Jones, C., Dolinar, S., and Thorpe, J. 2005b (Nov.). Protograph based LDPC codes with minimum distance linearly growing with block size. Page 5 of: *Proc. IEEE Glob. Commun. Conf.*, vol. 3. St. Louis, MO, USA, November 28–December 2, 2005.

[32] Divsalar, D., Dolinar, S., and Jones, C. 2006 (Jul.). Construction of protograph LDPC codes with linear minimum distance. Pages 664–668 of: *Proc. IEEE Int. Symp. Inf. Theory*. Seattle, WA, USA, July 9–14, 2006.

[33] Divsalar, D., Dolinar, S., and Jones, C. 2007 (Oct.). Short protograph-based LDPC codes. Pages 1–6 of: *IEEE Military Commun. Conf. (Milcom)*. Orlando, FL, USA, October 29–31, 2007.

[34] Divsalar, D., Dolinar, S., Jones, C. R., and Andrews, K. 2009. Capacity-approaching protograph codes. *IEEE J. Sel. Areas Commun.*, **27**(6), 876–888.

[35] Djurdjevic, I., Xu, J., Abdel-Ghaffar, K., and Lin, S. 2003. A class of low-density parity-check codes constructed based on Reed-Solomon codes with two information symbols. *IEEE Commun. Lett.*, **7**(7), 317–319.

[36] Dolecek, L., Divsalar, D., Sun, Y., and Amiri, B. 2014. Non-binary protograph-based LDPC codes: Enumerators, analysis, and designs. *IEEE Trans. Inf. Theory*, **60**(7), 3913–3941.

[37] El-Khamy, M., and McEliece, R. J. 2006. Iterative algebraic soft-decision list decoding of Reed–Solomon codes. *IEEE J. Sel. Areas Commun.*, **24**(3), 481–490.

[38] Fan, J. L. 2000 (Sep.). Array codes as low-density parity-check codes. Pages 543–546 of: *Proc. 2nd Int. Sym. on Turbo Codes and Related Topics*. Brest, France, September 4–7, 2000.

[39] Fossorier, M. P. C. 2004. Quasi-cyclic low-density parity-check codes from circulant permutation matrices. *IEEE Trans. Inf. Theory*, **50**(8), 1788–1793.

[40] Gallager, R. G. 1962. Low-density parity-check codes. *IRE Trans. Inform. Theory*, **IT-8**(Jan.), 21–28.

[41] Han, Y., and Ryan, W. E. 2009. Low-floor decoders for LDPC codes. *IEEE Trans. Commun.*, **57**(6), 1663–1673.

[42] Horn, R. A., and Johnson, C. R. 1985. *Matrix Analysis*. Cambridge, UK: Cambridge University Press.

[43] Hu, X. Y., Eleftheriou, E., and Arnold, D. M. 2001. Progressive edge-growth Tanner graphs. Pages 995–1001 of: *Proc. IEEE Glob. Commun. Conf.*, vol. 2. San Antonio, TX, USA, November 25–29, 2001.

[44] Hu, X. Y., Eleftheriou, E., and Arnold, D. M. 2005. Regular and irregular progressive edge-growth Tanner graphs. *IEEE Trans. Inf. Theory*, **51**(1), 386–398.

[45] Huang, J., Liu, L., Zhou, W., and Zhou, S. 2010. Large-girth nonbinary QC-LDPC codes of various lengths. *IEEE Trans. Commun.*, **58**(12), 3436–3447.

[46] Huang, Q., Diao, Q., Lin, S., and Abdel-Ghaffar, K. 2012. Cyclic and quasi-cyclic LDPC codes on constrained parity-check matrices and their trapping sets. *IEEE Trans. Inf. Theory*, **58**(5), 2648–2671.

[47] Iyengar, A. R., Papaleo, M., Siegel, P. H., Wolf, J. K., Vanelli-Coralli, A., and Corazza, G. E. 2012. Windowed decoding of protograph-based LDPC convolutional codes over erasure channels. *IEEE Trans. Inf. Theory*, **58**(4), 2303–2320.

[48] Jiang, J., and Narayanan, K. R. 2008. Algebraic soft-decision decoding of Reed–Solomon codes using bit-level soft information. *IEEE Trans. Inf. Theory*, **54**(9), 3907–3928.

[49] Jimenez Felstrom, A., and Zigangirov, K. S. 1999. Time-varying periodic convolutional codes with low-density parity-check matrix. *IEEE Trans. Inf. Theory*, **45**(6), 2181–2191.

[50] Kang, J., Huang, Q., Zhang, L., Zhou, B., and Lin, S. 2010. Quasi-cyclic LDPC codes: An algebraic construction. *IEEE Trans. Commun.*, **58**(5), 1383–1396.

[51] Kang, J., Huang, Q., Lin, S., and Abdel-Ghaffar, K. 2011. An iterative decoding algorithm with backtracking to lower the error-floors of LDPC codes. *IEEE Trans. Commun.*, **59**(1), 64–73.

[52] Karlin, M. 1969. New binary coding results by circulants. *IEEE Trans. Inf. Theory*, **15**(1), 81–92.

[53] Kasami, T. 1974. A Gilbert-Varshamov bound for quasi-cycle codes of rate 1/2. *IEEE Trans. Inf. Theory*, **20**(5), 679.

[54] Köetter, R., and Vardy, A. 2003. Algebraic soft-decision decoding of Reed–Solomon codes. *IEEE Trans. Inf. Theory*, **49**(11), 2809–2825.

[55] Kou, Y., Lin, S., and Fossorier, M. P. C. 2000a (Sep.). Construction of low density parity check codes: A geometric approach. Pages 137–140 of: *Proc. 2nd Int. Sym. on Turbo Codes and Related Topics*. Brest, France, September 4–7, 2000.

[56] Kou, Y., Lin, S., and Fossorier, M. P. C. 2000b. Low density parity check codes based on finite geometries: A rediscovery. Page 200 of: *Proc. IEEE Int. Symp. Inf. Theory*. Sorrento, Italy, June 25–30, 2000.

[57] Kou, Y., Lin, S., and Fossorier, M. P. C. 2000c (Nov./Dec.). Low density parity check codes: Construction based on finite geometries. Pages 825–829 of: *Proc. IEEE Glob. Commun. Conf.*, vol. 2. San Francisco, CA, USA, November 27–December 1, 2000.

[58] Kou, Y., Lin, S., and Fossorier, M. P. C. 2001. Low-density parity-check codes based on finite geometries: A rediscovery and new results. *IEEE Trans. Inf. Theory*, **47**(7), 2711–2736.

[59] Kudekar, S., Richardson, T. J., and Urbanke, R. L. 2011. Threshold saturation via spatial coupling: Why convolutional LDPC ensembles perform so well over the BEC. *IEEE Trans. Inf. Theory*, **57**(2), 803–834.

[60] Kudekar, S., Richardson, T., and Urbanke, R. L. 2013. Spatially coupled ensembles universally achieve capacity under belief propagation. *IEEE Trans. Inf. Theory*, **59**(12), 7761–7813.

[61] Kumar, S., and Pfister, H. D. 2015. Reed–Muller codes achieve capacity on erasure channels. *CoRR*, **arXiv:abs/1505.05123**.

[62] Laendner, S., and Milenkovic, O. 2005. Algorithmic and combinatorial analysis of trapping sets in structured LDPC codes. Pages 630–635 of: *Wireless Networks, Communications and Mobile Computing, 2005 International Conference on*, vol. 1. Maui, HI, USA, June 13–16, 2005.

[63] Laendner, S., and Milenkovic, O. 2007. LDPC codes based on Latin squares: Cycle structure, stopping set, and trapping set analysis. *IEEE Trans. Commun.*, **55**(2), 303–312.

[64] Lan, L., Zeng, L., Tai, Y. Y., Chen, L., Lin, S., and Abdel-Ghaffar, K. 2007. Construction of quasi-cyclic LDPC codes for AWGN and binary erasure channels: A finite field approach. *IEEE Trans. Inf. Theory*, **53**(7), 2429–2458.

[65] Lan, L., Tai, Y. Y., Lin, L., Behshad, M., and Honary, B. 2008. New constructions of quasi-cyclic LDPC codes based on special classes of BIBDs for the AWGN and binary erasure channels. *IEEE Trans. Commun.*, **56**(1), 39–48.

[66] Lentmaier, M., Sridharan, A., Zigangirov, K. S., and Costello, D. J., Jr. 2005 (Sep.). Terminated LDPC convolutional codes with thresholds close to capacity. Pages 1372–1376 of: *Proc. IEEE Int. Symp. Inf. Theory*. Adelaide, SA, USA, September 4–9, 2005.

[67] Lentmaier, M., Sridharan, A., Costello, D. J., Jr., and Zigangirov, K. S. 2010. Iterative decoding threshold analysis for LDPC convolutional codes. *IEEE Trans. Inf. Theory*, **56**(10), 5274–5289.

[68] Li, J., Liu, K., Lin, S., and Abdel-Ghaffar, K. 2014a. Algebraic quasi-cyclic LDPC codes: Construction, low error-floor, large girth and a reduced-complexity decoding scheme. *IEEE Trans. Commun.*, **62**(8), 2626–2637.

[69] Li, J., Liu, K., Lin, S., and Abdel-Ghaffar, K. 2014b (Feb.). Decoding of quasi-cyclic LDPC codes with section-wise cyclic structure. Pages 1–10 of: *Proc. IEEE Inf. Theory Applic. Workshop*. San Diego, CA, USA, February 9–14, 2014.

[70] Li, J., Liu, K., Lin, S., and Abdel-Ghaffar, K. 2014c (Jun.). Quasi-cyclic LDPC codes on two arbitrary sets of a finite field. Pages 2454–2458 of: *Proc. IEEE Int. Symp. Inf. Theory*. Honolulu, HI, USA, June 29–July 4, 2004.

[71] Li, J., Lin, S., and Abdel-Ghaffar, K. 2015 (Jun.). Improved message-passing algorithm for counting short cycles in bipartite graphs. In: *Proc. IEEE Int. Symp. Inf. Theory*. Hong Kong, China, June 14–19, 2015.

[72] Li, Z., Chen, L., Zeng, L., Lin, S., and Fong, W. H. 2006. Efficient encoding of quasi-cyclic low-density parity-check codes. *IEEE Trans. Commun.*, **54**(1), 71–81.

[73] Lidl, R., and Niederreiter, H. 1997. *Finite Fields*. Cambridge, UK: Cambridge University Press.

[74] Lin, S., and Costello, D. J., Jr. 2004. *Error Control Coding: Fundamentals and Applications*, 2nd edition. Upper Saddle River, NJ: Prentice Hall.

[75] Lin, S., Kasami, T., Fujiwara, T., and Fossorier, M. P. C. 1998. *Trellis and Trellis-Based Decoding Algorithm for Linear Block Codes*. New York, NY: Springer-Verlag New York.

[76] Lin, S., Xu, J., Djurdjevic, I., and Tang, H. 2002 (Oct.). Hybrid construction of LDPC codes. Pages 1149–1158 of: *Proc. 40th Annual Allerton Conf. Commun., Control, Computing*. Monticello, IL, USA, October 1–3, 2002.

[77] Lin, S., Diao, Q., and Blake, I. F. 2014a (Aug.). Error floors and finite geometries. Pages 42–46 of: *Proc. 8th Int. Sym. on Turbo Codes and Iterative Inf. Processing*. Bremen, Germany, August 18–22, 2014.

[78] Lin, S., Liu, K., Li, J., and Abdel-Ghaffar, K. 2014b (Nov.). A reduced-complexity iterative scheme for decoding quasi-cyclic low-density parity-check codes. Pages 119–125 of: *Proc. 48th Annual Allerton Conf. Commun., Control, Computing*. Pacific Grove, CA, USA, November 2–5, 2014.

[79] Liu, K., Lin, S., and Abdel-Ghaffar, K. 2013. A Revolving iterative algorithm for decoding algebraic cyclic and quasi-cyclic LDPC codes. *IEEE Trans. Commun.*, **61**(12), 4816–4827.

[80] Liva, G., and Chiani, M. 2007 (Nov.). Protograph LDPC codes design based on EXIT analysis. Pages 3250–3254 of: *Proc. IEEE Glob. Commun. Conf.* Washington, DC, USA, November 26–30, 2007.

[81] MacKay, D. J. C. 1999. Good error-correcting codes based on very sparse matrices. *IEEE Trans. Inf. Theory*, **45**(2), 399–431.

[82] MacKay, D. J. C., and Davey, M. C. 2001. Evaluation of Gallager codes for short block length and high rate applications. *The IMA Volumes in Mathematics and its Applications*, **123**(Jun.), 113–130.

[83] MacKay, D. J. C., and Neal, R. M. 1996. Near Shannon limit performance of low density parity-check codes. *Electro. Lett.*, **32**(18), 1645–1646.

[84] Mann, H. 1949. *Analysis and Design of Experiments*. New York, NY: Dover.

[85] Mitchell, D. G. M., Smarandache, R., and Costello, D. J., Jr. 2014. Quasi-cyclic LDPC codes based on pre-lifted protographs. *IEEE Trans. Inf. Theory*, **60**(10), 5856–5874.

[86] Mitchell, D. G. M., Lentmaier, M., and Costello, D. J., Jr. 2015. Spatially coupled LDPC codes constructed from protographs. *IEEE Trans. Inf. Theory*, **61**(9), 4866–4889.

[87] NASA, Standards. 2008. *GSFC-STD-9100*. https://standards.nasa.gov/documents/viewdoc/3315856/3315856. Accessed May 4, 2015.

[88] Nguyen, T. V., Nosratinia, A., and Divsalar, D. 2012. The design of rate-compatible protograph LDPC codes. *IEEE Trans. Commun.*, **60**(10), 2841–2850.

[89] Peterson, W. W., and Weldon, E. J., Jr. 1972. *Error-Correcting Codes*, 2nd edition. Cambridge, MS, USA: MIT Press.

[90] Pishro-Nik, H., and Fekri, F. 2004. On decoding of low-density parity-check codes over the binary erasure channel. *IEEE Trans. Inf. Theory*, **50**(3), 439–454.

[91] Pishro-Nik, H., and Fekri, F. 2007. Results on punctured low-density parity-check codes and improved iterative decoding techniques. *IEEE Trans. Inf. Theory*, **53**(2), 599–614.

[92] Pusane, A. E., Smarandache, R., Vontobel, P. O., and Costello, D. J., Jr. 2011. Deriving good LDPC convolutional codes from LDPC block codes. *IEEE Trans. Inf. Theory*, **57**(2), 835–857.

[93] Reed, I. S., and Solomon, G. 1960. Polynomial codes over certain finite fields. *J. Soc. Indust. Appl. Math.*, **8**(2), 300–304.

[94] Richardson, T. 2003 (October 1–3). Error-floors of LDPC codes. Pages 1426–1435 of: *Proc. 41st Annual Allerton Conf. Commun. Control, Computing*. Monticello, IL, USA, October 1–3, 2003.

[95] Richardson, T., and Urbanke, R. L. 2008. *Morden Coding Theory*. Cambridge, UK: Cambridge University Press.

[96] Richardson, T. J., Shokrollahi, M. A., and Urbanke, R. L. 2001. Design of capacity-approaching irregular low-density parity-check codes. *IEEE Trans. Inf. Theory*, **47**(2), 619–637.

[97] Ryan, W. E., and Lin, S. 2009. *Channel Codes: Classical and Modern*. New York, NY: Cambridge University Press.

[98] Ryser, H. J. 1996. *Combinatorial Mathematics*. New York, NY: Wiley.

[99] Sassatelli, L., and Declercq, D. 2010. Nonbinary hybrid LDPC codes. *IEEE Trans. Inf. Theory*, **56**(10), 5314–5334.

[100] Song, S., Zhou, B., Lin, S., and Abdel-Ghaffar, K. 2009. A unified approach to the construction of binary and nonbinary quasi-cyclic LDPC codes based on finite fields. *IEEE Trans. Commun.*, **57**(1), 84–93.

[101] Tai, Y. Y., Lan, L., Zeng, L., Lin, S., and Abdel-Ghaffar, K. 2006. Algebraic construction of quasi-cyclic LDPC codes for the AWGN and erasure channels. *IEEE Trans. Commun.*, **54**(10), 1765–1774.

[102] Tang, H., Xu, J., Lin, S., and Abdel-Ghaffar, K. 2005. Codes on finite geometries. *IEEE Trans. Inf. Theory*, **51**(2), 572–596.

[103] Tanner, R. M. 1981. A recursive approach to low complexity codes. *IEEE Trans. Inf. Theory*, **27**(5), 533–547.

[104] Tanner, R. M., Sridhara, D., Sridharan, A., Fuja, T. E., and Costello, D. J., Jr. 2004. LDPC block and convolutional codes based on circulant matrices. *IEEE Trans. Inf. Theory*, **50**(12), 2966–2984.

[105] Thorpe, J. 2003. Low density parity check (LDPC) codes constructed from protographs. *JPL INP Progress Report*, August 15, 42–154.

[106] Townsend, R., and Weldon, E. 1967. Self-orthogonal quasi-cyclic codes. *IEEE Trans. Inf. Theory*, **13**(2), 183–195.

[107] Vasic, B., and Milenkovic, O. 2004. Combinatorial constructions of low-density parity-check codes for iterative decoding. *IEEE Trans. Inf. Theory*, **50**(6), 1156–1176.

[108] Vellambi, H., and Fekri, F. 2007. Results on the improved decoding algorithm for low-density parity-check codes over the binary erasure channel. *IEEE Trans. Inf. Theory*, **53**(4), 1510–1520.

[109] Xu, J., and Lin, S. 2003 (Jun.). A combinatoric superposition method for constructing low density parity check codes. Page 30 of: *Proc. IEEE Int. Symp. Inf. Theory*. Pacifico Yokohama, Yokohama, Japan, June 29–July 4, 2003.

[110] Xu, J., Lin, S., and Blake, I. F. 2003 (Mar.). On products of graphs for LDPC codes. Pages 6–9 of: *Proc. IEEE Inf. Theory Workshop*. La Sorbonne, Paris, France, March 31–April 4, 2003.

[111] Xu, J., Chen, L., Zeng, L., Lan, L., and Lin, S. 2005. Construction of low-density parity-check codes by superposition. *IEEE Trans. Commun.*, **53**(Feb.), 243–251.

[112] Xu, J., Chen, L., Djurdjevic, I., Lin, S., and Abdel-Ghaffar, K. 2007. Construction of regular and irregular LDPC codes: Geometry decomposition and masking. *IEEE Trans. Inf. Theory*, **53**(1), 121–134.

[113] Zhang, L., Huang, Q., Lin, S., Abdel-Ghaffar, K., and Blake, I. F. 2010. Quasi-cyclic LDPC Codes: An algebraic construction, rank analysis, and codes on Latin squares. *IEEE Trans. Commun.*, **58**(11), 3126–3139.

[114] Zhang, L., Lin, S., Abdel-Ghaffar, K., Ding, Z., and Zhou, B. 2011. Quasi-cyclic LDPC codes on cyclic subgroups of finite fields. *IEEE Trans. Commun.*, **59**(9), 2330–2336.

[115] Zhang, Z., Dolecek, L., Nikolic, B., Anantharam, V., and Wainwright, M. J. 2008 (November 30–December 4). Lowering LDPC error floors by postprocessing. Pages 1–6 of: *Proc. IEEE Glob. Commun. Conf.* New Orleans, LO, USA, November 30–December 4, 2008.

[116] Zhou, B., Kang, J., Tai, Y. Y., Lin, S., and Ding, Z. 2009. High performance non-binary quasi-cyclic LDPC codes on Euclidean geometries. *IEEE Trans. Commun.*, **57**(5), 1298–1311.

Index

Printed in the United Kingdom by TJ Clays Ltd.